Academic Genealogy of
Mathematicians

Academic Genealogy of
Mathematicians

Sooyoung Chang

Pohang University of Science & Technology,
South Korea

 **World Scientific**

NEW JERSEY · LONDON · SINGAPORE · BEIJING · SHANGHAI · HONG KONG · TAIPEI · CHENNAI

Published by

World Scientific Publishing Co. Pte. Ltd.

5 Toh Tuck Link, Singapore 596224

USA office: 27 Warren Street, Suite 401-402, Hackensack, NJ 07601

UK office: 57 Shelton Street, Covent Garden, London WC2H 9HE

British Library Cataloguing-in-Publication Data

A catalogue record for this book is available from the British Library.

ACADEMIC GENEALOGY OF MATHEMATICIANS

ISBN-13 978-981-4282-29-1

Printed in Singapore by Mainland Press Pte Ltd.

Dedicated to Dr. Tae Joon Park,
Founder of Pohang University of Science and Technology,
Pohang Iron and Steel Company
and Former Prime Minister of the Republic of Korea

Foreword

There are two groups of mathematicians, namely those who are keenly interested in their academic genealogy, thus tracing their academic lineage, their forefather(s), their ancestry; and those to whom such issues do not matter. The first group is surely connected with those mathematicians who have a firm interest in the history of mathematics.

As the author of "History of German Universities and Science" (Hakmunsa, 2000) and "Academic Genealogy of Physicists" (Seoul National University Press, 2005), Professor Chang already has amassed considerable experience for his latest rich and rewarding genealogical project, namely the genealogy of mathematicians.

He skillfully develops his own approach to an incredibly broad subject, covering a considerable number of the eminent mathematicians (from 18 countries) who played critical roles in the advancement of mathematics and helped shape our discipline as it exists today. The book contains not only the genealogies of these mathematicians but also short, well-balanced accounts of their lives and work, often in historical context.

This unique and complex work provides a wealth of information for mathematicians of every age. In my own experience students generally greatly appreciate hearing some highlights about the inventors of the new mathematical concepts they learn in class; learning the historical background of such concepts helps in understanding them.

Paul L. Butzer, Professor Emeritus
Rheinisch Westfälische Technische Hochschule, Aachen, Germany

Looking at the history of sciences, especially at that of mathematics, two circumstances become evident. First, science develops in centers, in schools based on long traditions and often formed around great scholars. These are the places where people mainly know which problems are interesting to be studied, --- or they even create them, --- and they are also aware of the best techniques: scientific, psychological and sociological ways to approach them. Second, even in old times, being restricted to the old ways of communication (or in modern times when communication was restricted by politics) the best centers and most scholars got to know very well and pretty fast what was happening in other centers. One can or cannot agree with these thoughts but it is for sure that it is an exciting adventure to study the genealogy of mathematicians during which one encounters great surprises and interesting connections. For certain, the reader will enjoy this fascinating book by Professor Chang.

Domokos Szász, Professor of Mathematics
Budapest University of Technology

While being an electrical engineer, Professor Chang has clearly shown deep interest and knowledge in mathematics. This book will not only serve as a nice reference book on the biographics of all these great mathematicians, but also will inspire many young minds. I wish to express my sincere gratitude and appreciation for his love and passion in mathematics and for doing such a great job in writing this book.

Dohan Kim, President, Korean Mathematical Society

Professor Chang's book is very unique of this kind; it contains tremendous amount of information on the genealogy of mathematicians and still the description of each mathematician remains concise enough to allow the reader a good access who may have given up reading a tedious biography. All levels of readers will benefit from reading it in their own ways concerning to their and needs. I myself have learned much on mathematics even in some of the closest areas to mine.

Yuichiro Taguchi
Department of Mathematics, University of Kyushu

Preface

I was fascinated by the Fourier Series when I learned about it during my sophomore year at Seoul National University in 1958. When I learned functional analysis in the United States, I found that some professors did not know the difference between Schwarz and Schwartz. This lack of awareness surprised me and contributed to my interest in the in academic genealogy of mathematicians.

Selecting approximately 750 mathematicians for this book proved extremely difficult. The American Mathematical Society has 33,000 members. The great MacTutor Archives list more than 2,100 mathematicians. And the Mathematics Genealogy Project run by the University of North Dakota has 126,000 records on mathematicians. Consequently, this book omits many, many tremendously accomplished men and women mathematicians. All of the decisions about whom to include were mine.

Since I am interested in academic genealogy, mathematicians are not strictly grouped by their nationality. Instead, they are organized by the nation in which they received their doctoral degrees with a few exceptions. For example, Sofya Kovalevskaya is included in the German School but not in the Russian School; and Alfred Haar is included in the German School but not in the Hungarian School. Heisuke Hironaka is listed in the American School although he is a Japanese.

While writing this book, I had the pleasure of meeting Professors John J. O'Connor and Edmund F. Robertson, developers of the MacTutor Archives, at the University of St. Andrews in March 2007, I also had the privilege of meeting many of the mathematicians listed in this book; Jack Milnor, Stephen Smale, Jacques Louis Lions, Jean Christoph Yoccoz, Efim Zelmanov, John Coates, Paul Butzer, Victor Sadovnichy, Peter Swinnerton-Dyer, Domokos Szasz, Bélla Bollóbas, Benoit Mandelbrot, Rimhak Ree and Heisuke Hironaka.

I would like to express my gratitude to all of the people who helped me with this effort. Professors Dohan Kim and Yuichiro Taguchi provided extremely helpful comments on my draft. My children, Patrick and Katherine, both lawyers in the United States, carefully proofread and improved my English.

My secretary Juae Choi and Young Min Lee of Pohang University of Science and Technology (postech) did the tedious word processing. Mr. Seong-Taek Kim, of the Dream of Network Korea, Messrs. Bae Geun Kim and Kyoo Sam Lee of postech helped providing me the photos in this book. Mr. Young Chul Hong, Chaiman of Kiswire, Ltd.

and Mr. Woon Hyung Lee, Chairman of the SeAH Steel Corporation, provided me with financial support for my sabbatical year at Churchill College, University of Cambridge in 2007.

I want to especially thank my wife, Choonhyung, who spent many weekends alone while I was writing this book.

I would be happy to hear of any criticisms and corrections, in the hope of publishing a better second edition.

Sooyoung Chang

Contents

German School

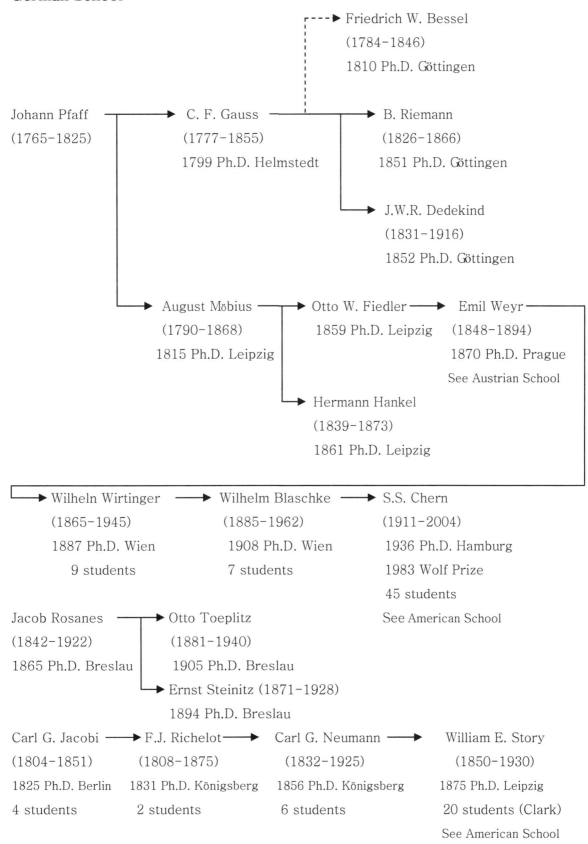

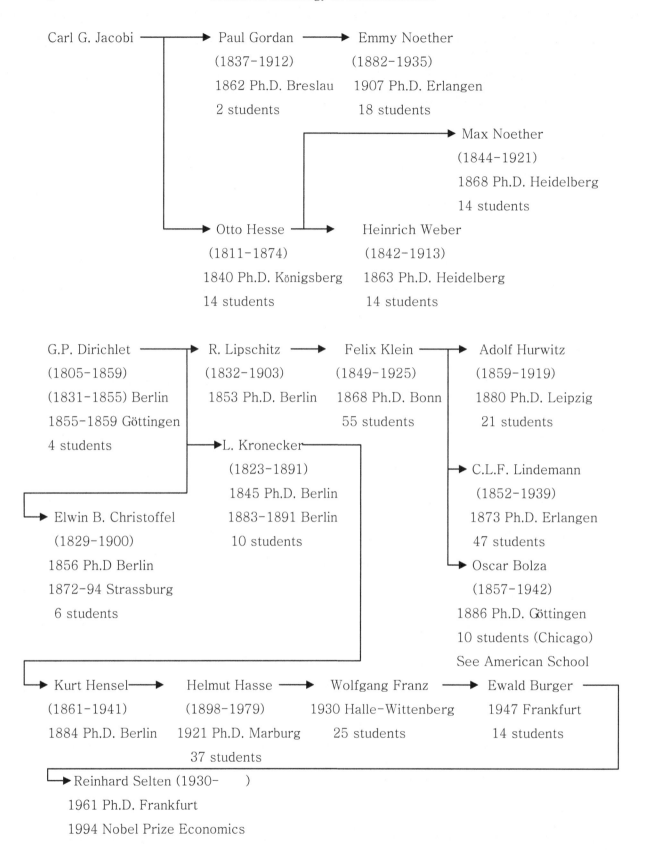

Carl G. Jacobi ⟶ Paul Gordan ⟶ Emmy Noether
(1837-1912)　　　(1882-1935)
1862 Ph.D. Breslau　1907 Ph.D. Erlangen
2 students　　　　　18 students

Max Noether
(1844-1921)
1868 Ph.D. Heidelberg
14 students

Otto Hesse ⟶ Heinrich Weber
(1811-1874)　　　(1842-1913)
1840 Ph.D. Königsberg　1863 Ph.D. Heidelberg
14 students　　　　　14 students

G.P. Dirichlet ⟶ R. Lipschitz ⟶ Felix Klein ⟶ Adolf Hurwitz
(1805-1859)　　　(1832-1903)　　(1849-1925)　　(1859-1919)
(1831-1855) Berlin　1853 Ph.D. Berlin　1868 Ph.D. Bonn　1880 Ph.D. Leipzig
1855-1859 Göttingen　　　　　　　55 students　　21 students
4 students

L. Kronecker
(1823-1891)
1845 Ph.D. Berlin
1883-1891 Berlin
10 students

C.L.F. Lindemann
(1852-1939)
1873 Ph.D. Erlangen
47 students

Elwin B. Christoffel
(1829-1900)
1856 Ph.D Berlin
1872-94 Strassburg
6 students

Oscar Bolza
(1857-1942)
1886 Ph.D. Göttingen
10 students (Chicago)
See American School

Kurt Hensel ⟶ Helmut Hasse ⟶ Wolfgang Franz ⟶ Ewald Burger
(1861-1941)　　　(1898-1979)　　1930 Halle-Wittenberg　1947 Frankfurt
1884 Ph.D. Berlin　1921 Ph.D. Marburg　25 students　　14 students
　　　　　　　37 students

Reinhard Selten (1930-　　)
1961 Ph.D. Frankfurt
1994 Nobel Prize Economics

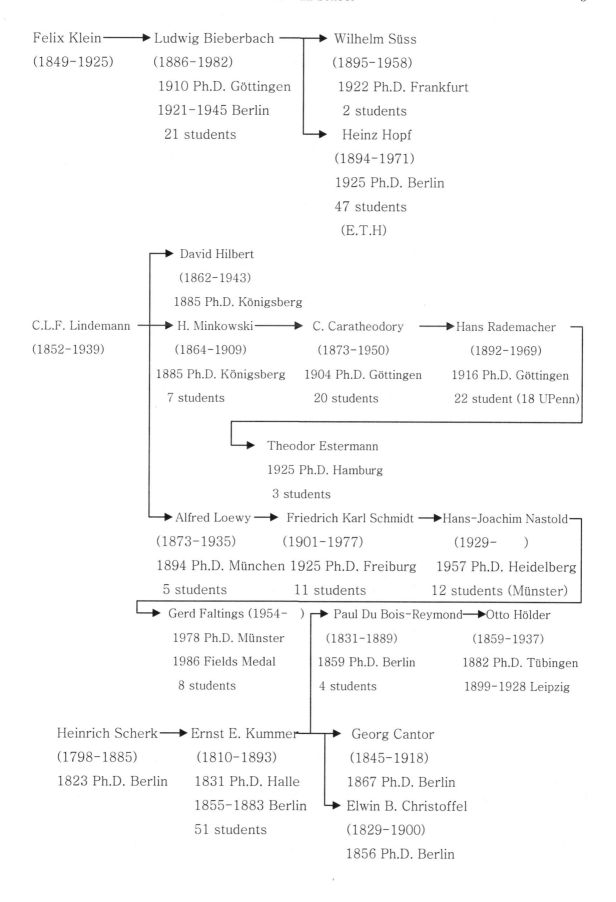

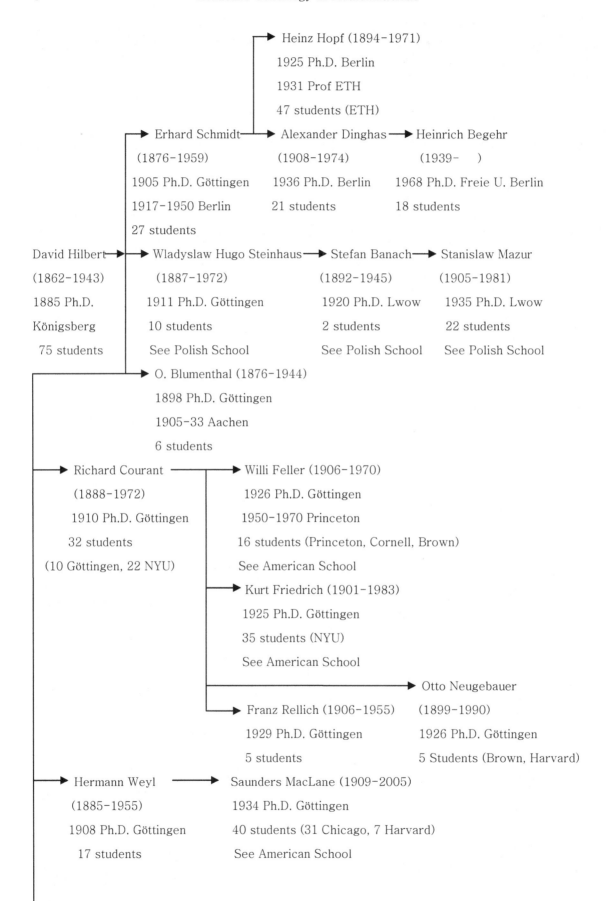

Heinz Hopf (1894-1971)

1925 Ph.D. Berlin

1931 Prof ETH

47 students (ETH)

Erhard Schmidt   Alexander Dinghas   Heinrich Begehr

(1876-1959)   (1908-1974)   (1939-)

1905 Ph.D. Göttingen   1936 Ph.D. Berlin   1968 Ph.D. Freie U. Berlin

1917-1950 Berlin   21 students   18 students

27 students

David Hilbert   Wladyslaw Hugo Steinhaus   Stefan Banach   Stanislaw Mazur

(1862-1943)   (1887-1972)   (1892-1945)   (1905-1981)

1885 Ph.D.   1911 Ph.D. Göttingen   1920 Ph.D. Lwow   1935 Ph.D. Lwow

Königsberg   10 students   2 students   22 students

75 students   See Polish School   See Polish School   See Polish School

O. Blumenthal (1876-1944)

1898 Ph.D. Göttingen

1905-33 Aachen

6 students

Richard Courant   Willi Feller (1906-1970)

(1888-1972)   1926 Ph.D. Göttingen

1910 Ph.D. Göttingen   1950-1970 Princeton

32 students   16 students (Princeton, Cornell, Brown)

(10 Göttingen, 22 NYU)   See American School

Kurt Friedrich (1901-1983)

1925 Ph.D. Göttingen

35 students (NYU)

See American School

Otto Neugebauer

Franz Rellich (1906-1955)   (1899-1990)

1929 Ph.D. Göttingen   1926 Ph.D. Göttingen

5 students   5 Students (Brown, Harvard)

Hermann Weyl   Saunders MacLane (1909-2005)

(1885-1955)   1934 Ph.D. Göttingen

1908 Ph.D. Göttingen   40 students (31 Chicago, 7 Harvard)

17 students   See American School

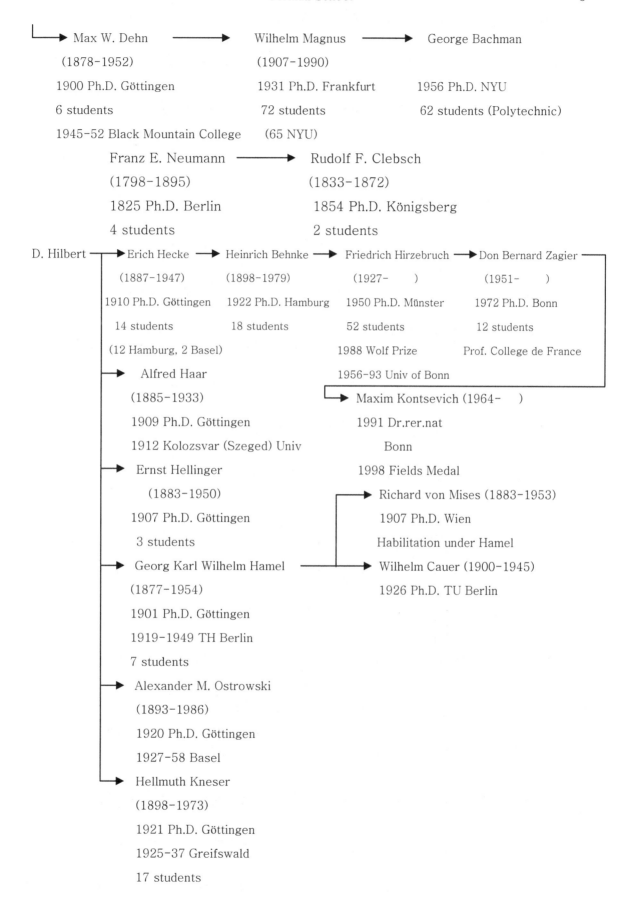

Max W. Dehn ⟶ Wilhelm Magnus ⟶ George Bachman

(1878-1952) (1907-1990)

1900 Ph.D. Göttingen 1931 Ph.D. Frankfurt 1956 Ph.D. NYU

6 students 72 students 62 students (Polytechnic)

1945-52 Black Mountain College (65 NYU)

Franz E. Neumann ⟶ Rudolf F. Clebsch

(1798-1895) (1833-1872)

1825 Ph.D. Berlin 1854 Ph.D. Königsberg

4 students 2 students

D. Hilbert ⟶ Erich Hecke ⟶ Heinrich Behnke ⟶ Friedrich Hirzebruch ⟶ Don Bernard Zagier

(1887-1947) (1898-1979) (1927-) (1951-)

1910 Ph.D. Göttingen 1922 Ph.D. Hamburg 1950 Ph.D. Münster 1972 Ph.D. Bonn

14 students 18 students 52 students 12 students

(12 Hamburg, 2 Basel) 1988 Wolf Prize Prof. College de France

⟶ Alfred Haar 1956-93 Univ of Bonn

(1885-1933)

1909 Ph.D. Göttingen ⟶ Maxim Kontsevich (1964-)

1912 Kolozsvar (Szeged) Univ 1991 Dr.rer.nat

⟶ Ernst Hellinger Bonn

(1883-1950) 1998 Fields Medal

1907 Ph.D. Göttingen

3 students ⟶ Richard von Mises (1883-1953)

⟶ Georg Karl Wilhelm Hamel ⟶ 1907 Ph.D. Wien

(1877-1954) Habilitation under Hamel

1901 Ph.D. Göttingen ⟶ Wilhelm Cauer (1900-1945)

1919-1949 TH Berlin 1926 Ph.D. TU Berlin

7 students

⟶ Alexander M. Ostrowski

(1893-1986)

1920 Ph.D. Göttingen

1927-58 Basel

⟶ Hellmuth Kneser

(1898-1973)

1921 Ph.D. Göttingen

1925-37 Greifswald

17 students

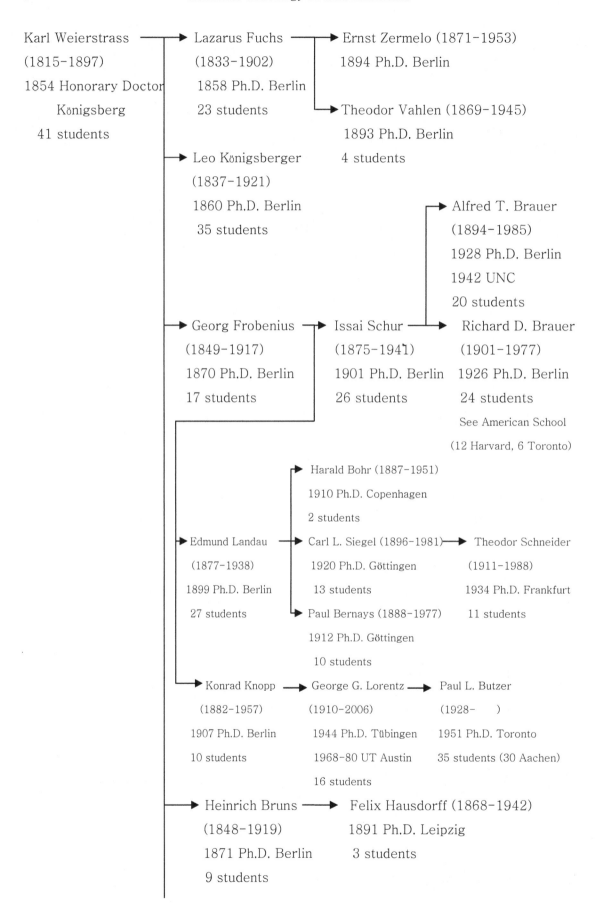

Karl Weierstrass
(1815-1897)
1854 Honorary Doctor
    Königsberg
  41 students

Lazarus Fuchs
(1833-1902)
1858 Ph.D. Berlin
23 students

Ernst Zermelo (1871-1953)
  1894 Ph.D. Berlin

Theodor Vahlen (1869-1945)
  1893 Ph.D. Berlin
  4 students

Leo Königsberger
(1837-1921)
1860 Ph.D. Berlin
35 students

Alfred T. Brauer
(1894-1985)
1928 Ph.D. Berlin
1942 UNC
20 students

Georg Frobenius
(1849-1917)
1870 Ph.D. Berlin
17 students

Issai Schur
(1875-1941)
1901 Ph.D. Berlin
26 students

Richard D. Brauer
(1901-1977)
1926 Ph.D. Berlin
24 students
See American School
(12 Harvard, 6 Toronto)

Harald Bohr (1887-1951)
1910 Ph.D. Copenhagen
2 students

Edmund Landau
(1877-1938)
1899 Ph.D. Berlin
27 students

Carl L. Siegel (1896-1981)
1920 Ph.D. Göttingen
13 students

Theodor Schneider
(1911-1988)
1934 Ph.D. Frankfurt
11 students

Paul Bernays (1888-1977)
1912 Ph.D. Göttingen
10 students

Konrad Knopp
(1882-1957)
1907 Ph.D. Berlin
10 students

George G. Lorentz
(1910-2006)
1944 Ph.D. Tübingen
1968-80 UT Austin
16 students

Paul L. Butzer
(1928-   )
1951 Ph.D. Toronto
35 students (30 Aachen)

Heinrich Bruns
(1848-1919)
1871 Ph.D. Berlin
9 students

Felix Hausdorff (1868-1942)
1891 Ph.D. Leipzig
3 students

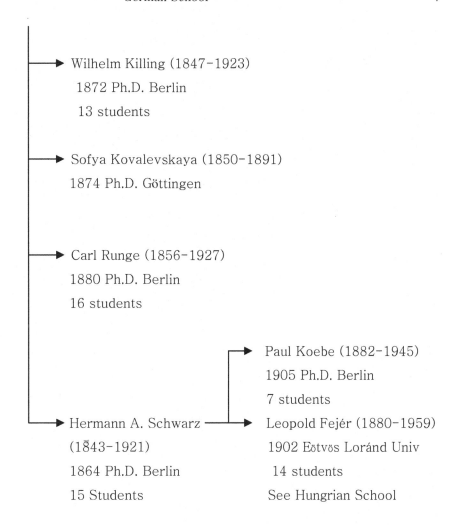

Wilhelm Killing (1847–1923)

1872 Ph.D. Berlin

13 students

Sofya Kovalevskaya (1850–1891)

1874 Ph.D. Göttingen

Carl Runge (1856–1927)

1880 Ph.D. Berlin

16 students

Paul Koebe (1882–1945)

1905 Ph.D. Berlin

7 students

Hermann A. Schwarz

(1843–1921)

1864 Ph.D. Berlin

15 Students

Leopold Fejér (1880–1959)

1902 Eötvös Loránd Univ

14 students

See Hungrian School

Hugo von Seeliger

(1849–1924)

1872 Ph.D. Leipzig

15 students (Munich)

Gustav Herglotz

(1881–1953)

1900 Ph.D. München

27 students

(25 Leipzig)

Emil Artin (1898–1962)

1921 Ph.D. Leipzig

1925 Prof Hamburg

34 students

(Princeton, Hamburg)

See American School

Wilholm Franz Meyer

1899–1924 Königsberg

Theodor Kaluza (1885–1954)

1907 Ph.D. Königsberg

1935–54 Göttingen

German School

Although Germany was unified in 1869, Prussia, Bayern, Hessen, Hannover, Baden, Sachsen, Mecklenburg and Württemberg are all included when we discuss German universities. Until the first half of the 20th Century, the standard of education in German gymnasiums (high school) had been very high. After studying six semesters at any university, students could submit a doctoral dissertation to a university. Students were free to move around universities.

A committee of three or four professors examine the dissertation and the candidate takes an oral examination. At some universities, the candidate was required to defend his dissertation in front of many students. Universities gave no examination until the final doctoral examination. Consequently there were no bachelors or masters degrees. Some universities award Ph.D and magister (master) degrees at the same time.

But students at Technische Hochschules (TH) must receive Diplom degrees first and THs were allowed to award doctoral degrees in 1899.

In 1925, there were 23 universities and 10 Technische Hochschules. The total number of doctorates awarded in 1925 was 4698.

The following table shows Ph.D. degrees and habilitations awarded by German universities in mathematics.

Universities	Ph.D.	Hab	Period	1893-1910
Berlin	145	1	1835-1893	29
Bonn	51	0	1853-1891	7
Breslau	55	10	1853-1890	27
Erlangen	33	3	1855-1891	16
Freiburg	15	3	1862-1890	5
Giessen	25	5	1863-1892	22
Göttingen	160	1	1851-1892	76
Greifswald	15	0	1860-1887	21
Halle	93	9	1851-1892	27
Heidelberg	22	7	1851-1892	9
Jena	109	6	1863-1893	13
Kiel	23	1	1857-1892	20
Königsberg	34	1	1854-1892	13
Leipzig	79	16	1867-1892	37
Marburg	104	5	1851-1893	11
München	37	5	1855-1893	41
Münster	18	0	1850-1893	11
Rostock	70	0	1868-1893	30
Strassburg	22	0	1876-1892	58

Tübingen	34	0	1865–1890	22
Würzburg	7	1	1878–1893	5
Sum	**1151**	**74**		**500**

Source: Mathematik in Berlin Geschichte und Dokumentation, Erster Halbband, herausgegeben von Heinrich Begehr, Shaker Verlag, Aachen 1998, p. 185

In order to teach at universities, a habilitation thesis must be accepted. Then he or she can be appointed as a privatdozent which is a lecturer that is not paid by the university but receives fees from the students.

After spending several years as privatdozent, one can be appointed as extraordinary (associate) professor or ordinary professor. Promotion to ordinary (full) professor is not automatic. One must apply for that position in a nationwide search.

Extraordinary professors now are called C-3 professors and ordinary professors are called C-4 professors.

Only C-4 professors can be appointed Dean (Dekan) or Rector of a university.

Number of Doctorates in Mathematics in Germany from 1907 to 1945.

Universities/TH	Domestic Men	Domestic Women	Foreign Men	Foreign Women
Aahen TH	3	1		
Berlin TH	16	1	6	
Berlin Univ	85	9	13	
Bonn Univ	41	14	3	
Braunschweig TH	1			
Breslau TH	7	1	1	
Breslau Univ	47	9	1	
Danzig TH	1			
Darmstadt TH	17		2	
Dresden TH	58	7		
Erlangen Univ	21	2		
Frankfurt/M	29	7	2	
Freiburg Univ	21	3	1	
Giessen Univ	48	5	5	
Göttingen Univ	119	7	37	1
Greifswald Univ	20			
Halle Univ	56	9	2	
Hamburg Univ	58	5	5	
Hannover TH	5	1		
Heidelberg Univ	32	3	6	
Jena Univ	41	1	2	
Karlsruhe TH	5			

Kiel Univ	33	2	2	
Köln Univ	8	1	1	
Königsberg Univ	37	5		
Leipzig Univ	68	3	10	1
Marburg Univ	31	2	5	1
München TH	45	1	3	
München Univ	53	4	16	1
Münster Univ	60	9		
Rostock Univ	35	1		
Strassburg Univ	30	1	3	
Stuttgart TH	5			
Tübingen Univ	49	1	1	
Würzburg Univ	30	1	4	
	1215	116	131	4

Source: Renate Tobies

www.mathematik.uni-bielefeld/DMV/archiv/dissertation.html

Johann Friedrich Pfaff

Johann Friedrich Pfaff was born on 22 December 1765 in Stuttgart, Germany. He had studied mathematics on his own and began to study the works of Euler. He also spent two years studying mathematics and physics at the University of Göttingen.

Pfaff held the chair of mathematics at the University of Helmstedt from 1788 until 1809. Helmstedt was dissolved under the French occupation in 1809. Napoleon closed 22 universities in Germany including Köln, Ingolstadt, Trier, Mainz, Dillingen, Helmstedt, Rinteln, Altdorf, Fulda, Passau, Lemberg, Bonn and Strassburg.

Pfaff held the chair of mathematics at the University of Halle-Wittenberg from 1810 to 1825. He is best known for his 1813 memoir on the integration of systems of differential equations. This posed the Pfaffian problem of classifying certain differential expressions; this was treated as a problem involving alternating bilinear forms with congruent variables.[2]

Gauss went to Helmstedt in 1798 to attend Pfaff's lectures and even lived in his house.[3] Gauss always retained a friendly memory of Pfaff both as a teacher and as a friend. Gauss received his Ph.D. at Helmstedt in 1799.

Pfaff died on 21 April 1825 in Halle, Saxony.

Carl Friedrich Gauss

Carl Friedrich Gauss was born on 30 April 1777 in Brunswick (Braunschweig), Germany to a very poor, uneducated family. His father was a gardener and brick layer, but his mother was eager to educate her son. When he was 14 years old, his skill at computing impressed the Duke of Brunswick who supported Gauss generously with a grant so that he could study at the Brunswick Collegium Carolinum in Hanover.

Gauss studied at the University of Göttingen from 1795 to 1798. Gauss's teacher there was Abraham G. Kästner (1719–1800), whom Gauss often ridiculed. Gauss obtained his doctorate with a thesis entitled *New Demonstration of the Theorem that Every Rational Integral Algebraic Function in one Variable can be Resolved into Real Factors of First or Second Degree* in 1799 at the University of Helmstedt.[2]

Gauss married his childhood friend Johanna Ostoff on 9 October 1805 but Johanna died after giving birth to their second son in 1809. Gauss married for a second time to Frederica Wilhelmine (Minna) Waldeck, the daughter of a law professor at the University of Göttingen. After bearing three children between 1811 and 1816, Minna became a permanent invalid. She died in 1831.

Gauss's daughter Therese stayed with her father and kept house for him until his death in 1855. Two sons, Eugene and Wilhelm, emigrated to the United States. The eldest son Joseph, from the first marriage, became a railroad engineer.

Gauss' father died in 1808, and his mother died in 1839 at the age of ninety-seven. Gauss was appointed director of the Göttingen Observatory and professor of mathematics in 1807. He did not like to teach. Between 1810 and 1820 much of his time was spent building and outfitting a new observatory. During the 1820's he was charged

Figure 1. Johann F. Pfaff

Figure 2. Carl F. Gauss

with the geographical survey of the Kingdom of Hannover, which opened a new direction in differential geometry and geodesy.

In 1831 Wilhelm Weber was appointed professor of physics at the University of Göttingen. Gauss and Weber worked together for six years building a primitive telegraph device which could send messages over a distance of 1500 meters.

There is a statue of Gauss and Weber in Göttingen. Gauss became a good friend of Hungarian Farkas Bolyai. In 1831 Farkas Bolyai sent his son Janos Bolyai's work on non-Euclidean geometry to Gauss. Gauss replied "to praise it would mean to praise myself." Bolyai resented it very much.

The collected works of Gauss include over 300 papers, many written in Latin. He taught himself to read and write in Russian.

He was elected a foreign member of the Royal Society of London in 1804, mainly based on his calculations of the orbits of the asteroids Ceres and Pallas, and he was awarded the Copley Medal in 1838. He also became a Geheimrat (Privy Councilor). Gauss was featured on the 10 Deutsche Mark note.

He died on 23 February 1855 in Göttingen, Germany.

Friedrich Wilhelm Bessel

Friedrich Wilhelm Bessel was born on 22 July 1784 in Minden, now Nordrhein· Westfalen, Germany. In January 1799, at the age of 14, he left school to become an apprentice to the commercial firm of Kulenkamp in Bremen[3]. The company was involved in the import-export business so Bessel became interested in navigation and finding the position of a ship at sea. This in turn led him to study astronomy and mathematics.

In 1804 Bessel wrote a paper on Halley's comet which was sent to Heinrich Olbers (1758-1840). Olbers was so impressed that he arranged its publication in Monatliche Correspondenz and proposed Bessel as assistant at the Lilienthal Observatory. After only four years at Lilienthal, the Prussian government requested that he construct the first big observatory in Königsberg (now Kalingrad). In 1810 at the age of 26, Bessel was appointed professor of astronomy at Königsberg.

It was not possible for Bessel to be appointed professor without a doctoral degree. A doctorate was awarded by the University of Göttingen in 1810 on the recommendation of Gauss, who had met Bessel in 1807 in Bremen.

When he was at the University of Königsberg, Carl Jacobi and Franz Neumann were also professors there. Many great mathematicians and physicists came from Königsberg including Carl Neumann, Otto Hesse, David Hilbert, Hermann Minkowski, Rudolf Clebsch, Theodor Kaluza, Gustav Kirchhoff and Arnold Sommerfeld.

In 1812 Bessel was elected to the Berlin Academy and in 1825 he received the honor of being elected as a Fellow of the Royal Society. Bessel determined the positions and proper motions of over 5000 stars which led to the discovery in 1838 of the parallax of 61 Cygni as 0.314". The correct value is 0.292".[3]

Bessel died of cancer on 17 March 1846 in Königsberg.

Figure 3. Friedrich W. Bessel   **Figure 4. Julius W. R. Dedekind**

Julius Wilhelm Richard Dedekind

Richard Dedekind was born on 6 October 1831 in Brunswick (Braunschweig) where Gauss had been born in 1777. Dedekind's father was a professor at the Collegium Carolinum where Gauss studied. Dedekind also studied two years at the Collegium Carolinum before entering the University of Göttingen in 1850.

After only four semesters in Göttingen, he received his doctorate supervised by Gauss in 1852 on the theory of Eulerian integrals. However he was not well trained in advanced mathematics and fully realized the deficiencies in his mathematical education.[3] He spent two more years learning the latest mathematical developments and working for his habilitation. In 1854 both Dedekind and Riemann were awarded their habilitation degrees which qualified Dedekind as a University Lecturer.

In 1858 Dedekind was appointed professor of mathematics at Polytechnikum in Zürich (now E.T.H). While in Zürich, Dedekind discovered the idea of a Dedekind cut. The idea was that every real number r divides the rational numbers into two subsets, namely those greater than r and those less than r.

The Collegium Carolinum in Brunswick had been upgraded to the Brunswick Polytechnikum (now TU Braunschweig) by the 1860's, and Dedekind was appointed to the Polytechikum in 1862. He remained there for the rest of his life retiring on 1 April 1894. He was never married and lived with sister Julie who also remained unmarried.

Dedekind edited Dirichlet's Lectures on number theory and published these as *Vorlesungen über Zahlen theorie* in 1863. In 1879 he published *Über die Theorie der ganzen algebraischen Zahlen*.

He died on 12 February 1916 in Braunschweig where he was born.

Georg Friedrich Bernhard Riemann

Bernhard Riemann was born on 17 September 1826 in Breselenz, Hannover. His father was a Lutheran minister. While Riemann was a student in the Johanneum Gymnasium in Lüneburg, he read Legendre's book on the theory of numbers (900 pages) in six days.[3]

In 1846 Riemann enrolled at the University of Göttingen as a theology student then later transferred to the Faculty of Philosophy to study mathematics.

Until the early 20th century, mathematics and physics belonged the Faculty of Philosophy in German Universities. Riemann took courses in mathematics from Moritz Stern and Gauss. Then Riemann moved to the University of Berlin in 1847 to study under Jacob Steiner, Carl Jacobi, Peter Lejeune Dirichlet and Gotthold Eisenstein. But Dirichlet was the most influential professor to Riemann in Berlin.

In 1849 Riemann returned to Göttingen and received his Ph.D. supervised by Gauss in 1851. His thesis *Grundlagen für eine allgemeine Theorie der Funktionen einer veränderlichen Komplexe grösse* (Foundations for a General Theory of Functions of One Complex Variable).[4]

Riemann's thesis studied the theory of complex variables and what we now call Riemann surfaces. It introduced topological methods into complex function theory and examined geometric properties of analytic functions, conformal mappings and the connectivity of surfaces.

After two years of intense work Riemann presented his habilitation thesis in December 1853. It translates as "On the representability of function by a trigonometric series." In the thesis he introduced the type of integral we know as the Riemann integral.

In his inaugural lecture on 10 June 1854 as privatdozent (lecturer), he presented on the hypotheses that lie at the foundations of geometry. Of those in the audience at the lecture, only Gauss would have been able to appreciate the depth of Riemann's thinking. Riemann raised the question of physical space. His results were so significant that Bertrand Russell described him as "logically the immediate predecessor of Einstein."

Without the concepts of the curvature of a Riemann space or manifold, the theory of general relativity could not have been formulated.

Riemann worked for 18 months as an assistant to physicist Wilhelm Weber and he gained a strong background in theoretical physics.

After Gauss died in 1855, Dirichlet succeeded him, but Dirichlet died soon after on 5 May 1859. Riemann was appointed to the chair of mathematics (Erster Lehrstuhl) on 30 July 1859.

In June 1862 Riemann married Elise Koch and they had one daughter Ida.

Riemann sent a report "On the number of primes less than a given magnitude" to the Belin Academy of Sciences. In this paper Riemann examined the zeta function

$$\zeta(s) = \frac{1}{1^s} + \frac{1}{2^s} + \frac{1}{3^s} + \ldots\ldots + \frac{1}{n^s} + \ldots\ldots$$

where $s = \sigma + i\tau$ is a complex variable.

The Riemann conjecture is that all of the imaginary zeros of the zeta function have real part $\sigma = \frac{1}{2}$. Mathematicians have not yet been able to prove or disprove this famous conjecture.

Figure 5.   Georg F. B. Riemann

Figure 6.   August Möbius

Riemann died of tuberculosis on 20 July 1866 in the small village of Selasca, at the northern end of Lake Maggiore in Italy. He was less than forty years old.

August Ferdinand Möbius

August Möbius was born on 17 November 1790 in Schulpforta, near Naumburg, Saxony. In 1809 he enrolled at the University of Leipzig to study mathematics, astronomy and physics. In those days German students freely moved from one university to another, so he studied astronomy under Gauss for two semesters at the University of Göttingen.

Then he went to Halle where he studied under Johann Pfaff. In 1815 Möbius wrote his doctoral thesis on *The occultation of fixed stars* at the University of Leipzig.

He was appointed extraordinary professor (Associate Professor) in 1816 at Leipzig and remained there until his death on 26 September 1868.[1]

The Möbius strip, presented in a paper discovered after his death, was devised by him to illustrate the properties of one-sided surfaces, and consists of a length of paper with connected ends and a half-twist in the middle.[5]

Johann Benedict Listing (1808–1882) also discovered the strip independently at the same time.

Hermann Hankel

Hermann Hankel was born on 14 February 1839 in Halle. He studied at the Universities of Leipzig, Göttingen and Berlin. Hankel received his doctorate at Leipzig for a thesis *Über eine besondere Klasse der symmetrischen Determinanten*[3] (On the Special Class of Symmetrical Determinants) in 1862.

After one semester as an extraordinary professor at Leipzig, he was appointed ordinary (full) professor at Erlangen in 1867. He married Marie Dippe in Erlangen and moved to University of Tübingen in 1869. Hankel was there only four years before he died on 29 August 1873.

The Hankel functions provide a solution to the Bessel differential equation, which was originally produced in connection with astronomical studies.

Wilhelm Wirtinger

Wilhelm Wirtinger was born on 15 July 1865 in Ybbs, a town on the Danube River about half way between Vienna and Linz. He received his doctorate in 1887 and his habilitation in 1890 at the University of Vienna.

After ten years at the University of Innsbruck he returned to a chair at the University of Vienna in 1905.

In 1896 he published a work of major importance on the general theta function which brought him recognition as a leading mathematician.

He was awarded the Sylvester Medal of the Royal Society of London in 1907.

Wirtinger died on 15 January in Ybbs, Austria.

Figure 7.   Hermann Hankel   Figure 8.   Wilhelm Wirtinger

Wilhelm Johann Eugen Blaschke

Wilhelm Blaschke was born on 13 September 1885 in Graz, Austria. He studied architectural engineering at the Technische Hochschule in Graz for two years before going to Vienna. In 1908 he received a doctorate supervised by Wirtinger at the University of Vienna. He went to Bonn where he worked with Eduard Study (1862–1930) and submitted his habilitation thesis in 1910.[3]

After working at TH Prague, Leipzig, and Königsberg, he was appointed in 1919 to a chair at the University of Hamburg from where he retired in 1953.

Blaschke wrote an important book on differential geometry *Vorlesungen Über Differential Geometrie* (1921-1929). He wrote two more major texts in 1938 and 1955.

In 1936 Blaschke joined the Nazi Party and took a leading role in German mathematics until the end of World War II. In September 1945 the allies dismissed Blaschke from his chair at Hamburg but he was reinstated in October 1946.

He died on 17 March 1962 in Hamburg, Germany.

Figure 9. Wilhelm J. E. Blaschke

Figure 10. Otto Toeplitz

Otto Toeplitz

Otto Toeplitz was born on 1 August 1881 in Breslau, Germany. Breslau is now Wroclaw, Poland.

His father and grandfather both taught mathematics in a Gymnasium. Toeplitz was awarded his doctorate in 1905 at the University of Breslau.

Composer Johannes Brahms received an honorary doctorate at the same University in 1879, composing the Academic Festival Overture.

In 1906 Toeplitz went to Göttingen and completed his habilitation in 1907. He was a personal assistant of Felix Klein but was greatly influenced by Hilbert's works. Toeplitz discovered the basic idea of what is now called the 'Toeplitz Operator'.

Toeplitz, Max Born, Ernst Hellinger and Richard Courant all came from Breslau. Born and Hellinger came to Göttingen in 1904, Toelitz in 1906 and Courant in 1907.

Although Courant was younger than the other three, they all became good friends.

After seven years in Göttingen, Toeplitz was appointed extraordinary professor at the University of Kiel in 1913 and then was promoted to ordinary professor at Kiel in 1920.[3] From 1928 to 1935 he was at the University of Bonn. Because he was Jewish, Teoplitz was dismissed from his chair by the Nazis in 1935 and emigrated to Palestine in 1939. He died on 15 Februnary 1940 in Jerusalem.

Carl Gustav Jacob Jacobi

Carl Jacobi was born of Jewish parents at Potsdam on 10 December 1804. His father was a banker and his family was prosperous. Jacobi was a child prodigy and he was awarded his doctorate in 1825 at the University of Berlin. His Ph.D thesis was an analytical discussion on the theory of partial fractions. Jacobi presented his habilitation thesis in 1825 to became a privatdozent (lecturer).

In 1827 he was appointed extraordinary professor at the University of Königsberg (now Immanuel Kant State University in Kaliningrad, Russia) and was then promoted to ordinary professor two years later.

Friedrich Bessel and Franz Neumann were professors at Königsberg at that time and many great mathematicians and physicists had studied there previously including David Hilbert, Hermann Minkowski, Otto Hesse, Alfred Clebsch, Carl Neumann, Gustav Robert Kirchhoff, and Arnold Sommerfeld.

Jacobi was one of the great mathematical teacher of his generation and he stimulated and influenced his pupils. Jacobi's most celebrated investigations were those on elliptic functions, and the modern notations on elliptic functions are due to him. He introduced elliptic functions into the number theory and into the theory of integration, which in turn connected with the theory of differential equations.[5]

Sylvester has given the name "Jacobian" to the functional determinant in order to pay resepect to Jacobi's work on algebra and elimination theory.[6]

In 1831 Jacobi married Marie Schwink. Around 1843 he became ill and was diagnosed with diabetes. He took a year's sick leave in Italy and returned to Germany in 1844. Because of the more severe climate of Königsberg he decided to move to Berlin. Since he had been elected to the Berlin Academy he was entitled to give lectures at the University of Berlin. Jacobi became a good friend of Dirichlet.

Early in 1851 Jacobi caught influenza, followed by smallpox, and this led to his death on 8 February 1851. He left a wife, five sons and three daughters.[4]

His great friend Dirichlet delivered a memorial lecture at the Berlin Academy of Science on 1 July 1852, in which he described Jacobi as "the greatest mathematician among members of the Academy since Joseph Lagrange."

Carl G. Neumann

Carl Neumann was born on 7 May 1832 in Königsberg, the son of Franz Neumann and Bessel's sister-in-law. He received his doctorate in 1856 at Königsberg and his habilitation in 1858 at the University of Halle-Wittenberg. He was appointed extraordinary professor at Halle in 1863 after five years as a lecturer.

After two years in Basel and three years in Tübingen he was appointed a chair at the University of Leipzig in 1868 succeeding A.F. Möbius.

He married Mathilde Elise Kloss in Basel but she died in 1875.

In 1868 he and Alfred Clebsch founded *Mathematische Annalen* and he became an editor.

During his Leipzig years, Felix Klein, Sophus Lie and Felix Hausdorff were in the same department.

Neumann's name is honored in the Neumann boundary condition and the Bessel function of the second kind is called the Neumann function.

He died on 27 March 1925 in Leipzig, Germany.

Figure 11. Carl G. J. Jacobi

Figure 12. Carl G. Neumann

Paul Albert Gordan

Paul Gordan was born on 27 April 1837 in Breslau, Germany (now Wroclaw, Poland). He studied at the Universities of Berlin and Königsberg under Jacobi and he received his doctorate in 1862 at the University of Breslau. His dissertation was on geodesics of spheroids. He received his habilitation in 1863 at the University of Giessen supervised by Alfred Clebsch. Gordan was appointed extraordinary professor at Giessen in 1865.

In 1869 Gordan married Sophie Deuer, the daughter of a professor of law at Giessen.

The Clebsch–Gordan coefficients used in quantum mechanics were introduced by them while working on *"Theorie der Abelschen Funktionen"* in 1866.

The topic for which Gordan is most famous is invariant theory, a topic to which Clebsch introduced him in 1868.[3]

Gordan moved to Erlangen in 1875 and retired there in 1910. His predecessor was Felix Klein. One of Gordan's student was Emmy Noether, the daughter of Max Noether who was Professor of Mathematics at Erlangen from 1888 to 1919.

Gordan died on 21 December 1912 in Erlangen, Germany.

Emmy Amalie Noether

Emmy Noether was born on 23 March 1882 in Erlangen. Her father was a distinguished professor of mathematics at Erlangen. Her mother was Ida Kaufmann and both Emmy's parents were of Jewish origin. She was the eldest of their four children.

Figure 13. Paul A. Gordan   **Figure 14. Emmy A. Noether**

In 1904 she became the first woman to enroll at the University of Erlangen and received her doctorate in 1908 under Paul Gordan. Her dissertation was *On complete systems of invariants for ternary biquadratic forms.*

Hilbert and Klein invited Emmy to Göttingen but many professors objected to granting her habilitation. Hilbert provoked his famous outburst at a faculty meeting: "I do not see that the sex of the candidate is an argument against her admission as Privatdozent. After all, we are a University, not a bathing house."[4]

In 1919 she finally became a privatdozent. In 1922 she was appointed *"nicht beamte ausserordentliche Professor"* (an unofficial associate professor) with no salary. There is a Göttingen saying to the effect that "an extraordinary professor knows nothing ordinary and an ordinary professor knows nothing extraordinary."

Noether's theorem (1918) demonstrated the intimate relationship between the symmetry properties of the dynamical equations (i.e., their invariance under infinitesimal Lie transformations) and the conservation laws recognized to be valid in the given dynamics.[7]

In 1924 B.L. van der Waerden came to Göttingen to spend a year studying with Noether. After returning to Amsterdam he wrote "Modern Algebra" in two volumes. The major part of the second volume consists of Noether's work.[3]

She received, jointly with Emil Artin, the Alfred Ackermann-Teubner Memorial Prize for the Advancement of Mathematical Knowledge in 1932.

In 1933 the Nazis government dismissed her from the University of Göttingen because she was Jewish. She accepted a visiting professorship at Bryn Mawr College in Pennsylvania, USA. After surgery for a uterine tumor, she died suddenly in 14 April 1935 in Bryn Mawr.

Hermann Weyl said at her funeral: "The memory of her work in science and of her personality among her fellows will not soon pass away. She was a great mathematician, the greatest, I firmly believe, her sex has ever produced, and a great woman."[4]

Ludwig Otto Hesse

Otto Hesse was born on 22 April 1811 in Königsberg. He taught physics and chemistry at a trade school before he received his doctorate at Königsberg in 1840. His thesis was *De ecto punctis intersectionis trium superficium secundi ordinis.*[3] In the same year he submitted his habilitation thesis and was appointed as a privatdozent.[1] Hesse married Marie Sophie Emilie Dulk in 1841, and they had one son and five daughters.

In 1845 Hesse was appointed extraordinary professor and stayed in that position until 1856. The years he was at Königsberg were very productive, and published most of his work in Crelle's Journal. Hesse had many outstanding students including Gustav Kirchhoff, Carl Neumann, Alfred Clebsch and Rudelph Lipschitz.

After one year at Halle, Hesse was appointed ordinary professor at Heidelberg in 1856. At Heidelberg he also had many famous students including Adolph Mayer, Heinrich Weber and Max Noether.

He then moved to Technische Hochschule München in 1868 and stayed there until his death.

Hesse introduced the "Hessian Determinant" in 1842 during an investigation of cubic and quadratic curves.[3]

He also wrote two textbooks in mathematics. Some recent research has shown that Hesse not only improved Jacobi's results but also made many original contributions.

Hesse died on 4 August 1874 in Munich, and he was buried in Heidelberg.

Figure 15.   Ludwig O. Hesse

Figure 16.   Heinrich M. Weber

Heinrich Martin Weber

Heinrich Weber was born on 5 May 1842 in Heidelberg. Weber was awarded a doctorate in 1863 supervised by Otto Hesse at the University of Heidelberg three years after enrollment. He went to Königsberg for his habilitation work under Franz Neumann and Friedrich Richelot, but his thesis was submitted to Heidelberg and was granted in 1866.

In 1869 he was appointed as extraordinary professor at Heidelberg. After only one year at Heidelberg, he went to Zurich Polytechnikum (1870-1875), Königsberg (1875-1883), Technische Hochschule Chalottenburg (now TU Berlin) (1883-1884), Marburg (1884-1892), Göttingen (1892-1895) and Strassburg (1895-1912).[1] He is best known for his textbook *"Lehrbuch der Algebra"* published in 1895.

Weber died on 17 May 1913 in Strassburg, Germany (now Strasbourg, France).

Max Noether

Max Noether was born on 24 September 1844 in Mannheim, Baden. He suffered an attack of polio when he was 14 years old which left him handicapped for the rest of his life.[3]

He obtained a doctorate under Otto Hesse in 1868 at Heidelberg. From 1870 to 1874 he was Privatdozent and in 1874 he was appointed extraordinary professor.

Noether moved to Erlangen in 1875 and was then appointed ordinary professor in 1888. He retired in 1919.[1] In 1882, his daughter Emmy Norther was born.

He was one of the leaders of 19th century algebraic geometry. In 1873 he proved an important result on the intersection of two algebraic curves.[3]

He died on 13 December 1921 in Erlangen.

Figure 17.   Max Noether

Figure 18.   J. Peter G. Lejeune Dirichlet

Johann Peter Gustave Lejeune Dirichlet

Lejeune Dirichlet was born on 13 February 1805 in Düren, halfway between Aachen and Köln.

In 1805, Düren was a part of France, but it belongs to Germany now. He attended the Jesuit college in Köln (secondary school) and there one of his teacher was Georg Ohm (1789–1854), discoverer of the Ohm's law.

His family originally came from Belgium town of Richelet. His name means "Le jeune de Richelet" (Young from Richelet).[3]

Dirichlet went to Paris in May 1822 to attend lectures at the Collége de France and the Sorbonne Faculté des Sciences. He was fortunate to contact Biot, Fourier, Laplace, Legendre and Poisson. From the summer of 1823 Dirichlet was employed as a tutor by general Maximilien Sebastien Foy (1775–1825), living in his house in Paris.[3] When the general died in 1825, Dirichlet returned to Germany. In 1825 he presented his first paper to the French Academy of Sciences.

To be a professor he needed a doctorate and habilitation, but he had none. Fortunately an honorary doctorate was given at the University of Bonn in 1827.[10] Several authors claim it was the University of Cologne, but it was closed by Napoleon and reopened in 1919. Dirichlet submitted his habilitation thesis at the University of Breslau in 1827 to become a privatdozent. He was appointed extraordinary professor at Breslau in 1828 when he was 23 years old. He was extraordinary professor at the University of Berlin (1831–1839) and ordinary professor (1839–1855).

He became a full member of the Prussian Academy of Science at the age of 27 and a corresponding member of the Academie des Sciences at the age of 28, as well as of the Academy of Sciences of Russia at 32.[117]

After Gauss died, Dirichlet succeeded him at Göttingen in 1855. Dirichlet considered Gauss as his teacher. When he went to France, Dirichlet carried with him Gauss's *Disquisitiones arithmeticae* (Arithmetic Inquiry).

In 1831 Dirichlet married Rebecca Mendelssohn, one of the composer Felix Mendelssohn's two sisters. In the summer of 1858 he suffered a heart attack in Montreux, on the lake of Genera. Not long afterwards his wife Rebecca died after a stroke and Dirichlet also died on 5 May 1859 in Göttingen at the age of 54.[4]

In 1837 he suggested a very broad definition of function and he gave also the first rigorous proof of the convergance of Fourier series for a function subject to certain restrictions, known as Dirichlet's conditions. His name arises as Dirichlet boundary condition in thermodynamics and electrodynamics.

He was a good friend with Carl Jacobi who said: "Dirichlet alone, not I, nor Cauchy, nor Gauss knows what a completely rigorous proof is. Rather we learn it first from him. When Gauss says he has proved something it is very clear; when Cauchy says it, one can wager as much pro as con; when Dirichlet says it, it is certain."[4]

In his address on the occasion of the Dirichlet centenary in Göttingen in 1905, Minkowski listed Ferdinand Eisenstein, Bernhard Riemann, Leopold Kronecker, Richard Dedekind and Rudolf Lipschitz as students of Dirichlet. At the University of Berlin, Dirichlet was a formal first or second referee for doctoral dissertations of 14 students including Kronecker and Lipschitz. Dirichlet also supervised or influenced five doctoral dissertations at Göttingen.[117]

Elwin Bruno Christoffel

Elwin Christoffel was born on 10 November 1829 in Monschau near Aachen, Germany. After one year of military service he continued his study at the University of Berlin where he recevied his doctorate in 1856. His dissertation was on the motion of electricity in homogeneous bodies (De motu permaneti electricitatis in corporibus homogeneis).[10] Although the referees of his dissertation were Martin Ohm, Ernst Kummer and Gustav Magnus, Christoffel was a student of Dirichlet.[117]

After three years taking care of his mother in Monschau, his habilitation thesis was approved in 1859. He served as privatdozent 1859-1862. From 1862 to 1868 he was at the Polytechnicum in Zurich (now E.T.H) succeeding Dedekind.

Then he moved to the Gewerbeakademie in Berlin (now TU Berlin) in 1869 and stayed there until 1872. He and his colleague Siegfried Aronhold tried to attract high quality students but they could not compete with the prestigions University of Berlin with Weiesstrass, Kummer and Kronecker.

After three years in Berlin he was offered the chair of mathematics at the University of Strassburg, where he retired in 1892.

Christoffel died on 15 March 1900 in Strassburg. He was a lonely man, shy, distrustful, unsociable, irritable and brusque.[3]

He supervised six doctoral candidates at Strassburg. Rikitaro Fujisawa (1861-1933) received his Ph.D. in 1886 to become the Japanese father of mathematics and Paul Epstein (1871-1939) received his doctorate in 1895. Epstein committed suicide after dismissed from his chair at the University of Frankfurt in 1939.

Christoffel generalized Gauss's method of quadrature and expressed the polynomials which are involved, as a determinant. This is now called Christoffel's theorem.[3] His best paper *"Über die transformation der homogen differentialausdrücke zweiten grades"* published in 1869 introduced the symbols that later became known as Christoffel symbols of the first and second order.

Rudolf Otto Sigmund Lipschitz

Rudolf Lipschitz was born on 14 May 1832 in Königsberg. He started his university education at the University of Königsberg then moved to Berlin where he received his doctorate in 1853 supervised by Dirichlet. After teaching at gymnasiums he submitted his habilitation thesis at the University of Bonn to become a privatdozent there in 1857.

In 1862 Lipschitz was appointed extraordinary professor at the University of Breslau and stayed there two years. In 1864 he was appointed ordinary professor at Bonn where he spent the rest of his career.

For an ordinary differential equation of the first order

$$x^{'} = f(t, x)$$

If there exists a constant $k > 0$ such that for every (t, x_1) and (t, x_2) in domain D

$$\mid f(t, x_1) - f(t, x_2) \mid \; \leq \; k \mid x_1 - x_2 \mid$$

This is the Lipschitz condition for uniqueness of solution for the differential equation. k is called the Lipschitz constant.

Lipschitz died on 7 October 1903 in Bonn, Germany.

Figure 19. Elwin B. Christoffel   **Figure 20. Rudolf O. S. Lipschitz**

Felix Christian Klein

Felix Klein was born on 25 April 1849 in Düsseldorf. Klein received his doctorate in 1868 at the University of Bonn when he was only nineteen years old. His dissertation was *Über die Transformation der allgemeinen Gleichung des zweiten Grades zwischen Linien-Koordinaten auf eine kanonische Form* supervised by Julius Plücker (1801–1868) and examined by Rudolf Lipschitz.[3]

Klein submitted his habilitation thesis at Göttingen in 1871 to became a privatdozent. Klein was appointed ordinary professor in 1872 at the age of 23 at the University of Erlangen. He was perhaps the youngest German professor of mathematics.

Klein was the successor of Hermann Hankel who had left for Tübingen in 1869. Klein delivered his famous antrittsrede (inangural speech) *"Vergleichende betrachtungen über neuere geometrische Forschungen"* (Comparative review of recent researches in geometry), which later known as the Erlangen Program.

Klein was unable to build a school because Erlangen was a small university. Even in 1925 there were only 220 students there in mathematics and natural sciences.

After three years at Erlangen, he was at Technische Hochschule München from 1875 to 1880. Then Klein was appointed chair of geometry at the University of Leipzig in 1880 where Carl Neumann was a professor. When Klein left for Göttingen in 1886 his successor was Sophus Lie.

Felix Klein stayed in Göttingen for 27 years and established a mathematical research center which became a model throughout the world. He brought David Hilbert, Hermann Minkowski, Heinrich Weber and Edmund Landau to Göttingen.

Many excellent students came to Göttingen to learn from these mathematicians. Klein was a good friend of Friedrich Althoff (1839–1908), director of higher education in the Prussian Government, Althoff stayed in that position for 27 years and he had control over budget allocation for universities.

Most universities had two or three chairs of mathematics, but Berlin and Göttingen had four. While Klein was in Leipzig, the effort of trying to compete with Poincaré imposed a great strain on his mental health.[4] He had a nervous breakdown and his career as a research mathematician was essentially over around 1884. Mathematicians in Berlin; Weierstrass, Kronecker and Kummer, considered Klein as superficial and sometimes a charlatan.

In the 1890's he worked on mathematical physics and engineering, and wrote a textbook on the theory of gyroscope with Arnold Sommerfeld who was Klein's assistant at Göttingen. Through the effort of Klein, Technische Hochschules were allowed to award doctorate degrees beginning in 1899.

Klein was elected a Fellow of the Royal Society in 1885 and received the Copley Medal in 1912.

He died on 22 June 1925 in Göttingen and was buried in the Stadt Friedhof in Göttingen.

Figure 21. Felix C. Klein

Figure 22. Adolf Hurwitz

Adolf Hurwitz

Adolf Hurwitz was born on 26 March 1859 in Hildesheim, Niedersachsen, Germany into a Jewish family. He entered the Andreanum Gymnasium in Hildesheim in 1868 where he was taught mathematics by Hermann C.H. Schubert (1848-1911). Hurwitz and Schubert wrote a joint paper when Hurwitz was a pupil at Andreanum.

Hurwitz received his Ph.D. in 1881 at the University of Leipzig with his dissertation *Grundlagen einer independenten Theorie der elliptischen Modulfunktionen und Theorie der Multiplikator-Gleichungen 1. Stufe.* His adviser was Felix Klein.

Hurwitz was at Göttingen in 1882-84 as a privatdozent. In 1884 he was appointed extraordinary professor at the University of Königsberg where he stayed for eight years. In Königsberg, he became a lifelong friend with his students David Hilbert, 22 years old, and Hermann Minkowski, 20 years old. Although Hurwitz was their teacher, he was only 25 years old. The three young men met at 5:00 PM each evening for a walk "to the apple tree" in Königsberg discussing mathematics.

After Minkowski left for Bonn in 1885, Hurwitz and Hilbert continued to take the walks. Whenever Minkowski returned to Königsberg, he joined their daily walk. In Königsberg Hurwitz married Ida Samuel, and they had three children.

Königsberg had only one ordinary professor of mathematics until 1899 when another ordinary professor was appointed. But Königsberg produced many excellent mathematicians.

Hurwitz accepted a chair at Eidgenössische Polytechnicum Zürich in 1892 and remained there for the rest of his life.

He played piano very well but continually suffered from ill health.

Hurwitz published in 1895 the criterion for the equation

$$F(s) = a_n s^n + a_{n-1} s^{n-1} + \ldots + a_1 s + a_0 = 0,$$

where s is a complex variable to have all roots to lie in the left-half plane of the s-plane. The necessary and sufficient condition is that Hurwitz determinants D_k, $k = 1, 2, \cdots, n$ must all be positive.

E.J. Routh (1831-1907) later published a similar but easier method, so it is now known as Routh-Hurwitz Criterion in automatic control theory.

Hurwitz invited George Polya to ETH in 1914 where Polya stayed until 1940. Hurwitz played music with his daughter Lisi and Albert Einstein. Hurwitz died on 18 November 1919 in Zürich.

Carl Louis Ferdinand von Lindemann

Ferdinand von Lindemann was born on 12 April 1852 in Hannover. He received his doctorate in 1873 at the University of Erlangen supervised by Felix Klein who was only three years older than Lindemann.

His dissertation was *Über unendlich kleine Bewegungen under über Kraftsysteme bei allgemeiner projektivischer Massbestimmung.*

Lindemann was a privatdozent at the University of Würzburg in 1877. In the same years he was appointed extraordinary professor at the University of Freiburg and he was promoted to ordinary professor in 1880.

In 1883 he became a professor at the University of Königsberg where he stayed for ten years. He was lucky to have David Hilbert, Hermann Minkowski and Arnold Sommerfeld as his students. Lindemann was the first to prove in 1882 that π is transcendental, that is π is not the root of any algebraic equation with rational coefficients.[3]

Figure 23.   Carl L. F. von **Lindemann**   Figure 24. Leopold Kronecker

Lindemann was appointed a chair at the University of München in 1893 and retired there in 1923. While he was in München, Werner Heisenberg asked him to be his adviser, but Lindemann did not accept him because Heisenberg told him he read Weyl's Raum-Zeit-Materie. Heisenberg instead became a student of Sommerfeld who was a pupil of Lindemann.

Some people confuse British physicist Frederick Alexander Lindemann (1886-1957) with the German mathematician Lindemann.

Lindemann died on 6 March 1939 in München.

Leopold Kronecker

Leopold Kronecker was born on 7 December 1823 in Liegnitz, Prussia (now Legnica, Poland) into a wealthy Jewish family.

Kronecker was taught mathematics at the local gymnasium by Ernst Kummer who later became a professor at the University of Berlin.

In 1845 he was awarded a Ph.D at Berlin with a thesis on the complex roots of unity. His adviser Dirichlet said that "Kronecker demonstrated unusual penetration, great assiduity, and an exact knowledge of the present state of mathematics."[4] Kronecker was a good athlete; a gymnast, swimmer and mountain climber. He was also a good pianist and vocalist. However, Kronecker was very small in stature and was very self-conscious of his height.

After receiving his doctorate Kronecker spent many years to manage his uncle's extensive farming enterprises. He married Fanny Prausnitzer, his late uncle's daughter and they had six children.

Even though he did not hold a university chair, he published many papers on the number theory, elliptic functions and algebra. He was nominated to the Berlin Academy by his teacher Kummer in 1860. Carl Borchardt and Karl Weierstrass seconded the nomination, and in 1861 Kronecker was elected to the Berlin Academy. Members of the Berlin Academy were entitled to lecture at the University of Berlin. He therefore lectured there from 1861 to 1883 without a university appointment. When Kummer retired in 1883 Kronecker was appointed ordinary professor at Berlin.

Kronecker criticized Weierstrass' mathematics although they were close friends in their early years. He particularly disliked Georg Cantor's revolutionary ideas on the transfinite set theory and tried to prevent or at least delay publication of Cantor's work.

The Deutsche Mathematiker-Vereinigung was founded in 1890 and Cantor invited Kronecker to address the first meeting.

He was elected to the Paris Academy in 1868 and to the Royal Society of London in 1884.

δ_{mn} is called the Kronecker delta.

His wife died on 23 August 1891 after injury in a climbing accident and Kronecker also died on 29 December 1891 in Berlin.

Helmut Hasse

Helmut Hasse was born on 25 August 1898 in Kassel, Germany. He was in the German Navy during World War I, Hasse received his doctorate in 1921 at the University of Marburg with his dissertation *Über die Darstellbarkeit von Zahlen durch quadratische Formen im Körper der rationalen Zahlen.* He also obtained his habilitation at Marburg in 1922 and he moved to the University of Kiel later that year to become a privatdozent there. In 1925 Hasse was appointed ordinary professor at Halle.

After staying five years at Halle, he succeeded his teacher Kurt Hensel at Marburg where he stayed for four years.

When Hermann Weyl left Göttingen for the United States in 1933, Hasse was appointed to his chair. Initially there was fierce opposition for his appointment from the Nazis in the Mathematics Institute in Göttingen. Hasse tried not to compromise his mathematics for political reasons, but he made no secret of his strong approval of many of Hitler's policies.

From 1939 until 1945 Hasse was on leave from Göttingen to work in Berlin on problems in ballistics. He returned to Göttingen in 1945 but he was dismissed by the British occupation forces. After one years at Humboldt University (former University of Berlin) in East Berlin, he was appointed in 1950 to a chair at the University of Hamburg and he retired in 1966.

He is best known as the editor of Crelle's Journal for 50 years and author of his textbook *"Zahlentheorie".*

Hasse died on 26 December 1979 in Ahtensburg near Hamburg, Germany.

Figure 25.   Helmut Hasse

Figure 26.   Ludwig G. E. M. Bieberbach

Ludwig Georg Elias Moses Bieberbach

Ludwig Bieberbach was born on 4 December 1886 in Goddelau near Darmstadt, Germany. He received a doctorate in 1910 at the University of Göttingen with his thesis *Zur Theorie der automorphen Funktionen* supervised by Felix Klein. Bieberbach submitted his habilitation thesis in 1911 at the University of Königsberg on his solution of the first part of Hilbert's eighteenth problem. After staying two years as a privatdozent in Königsberg, he was appointed full professor at the University of Basel in Switzerland in 1913. In 1915 he moved to Frankfurt and stayed there for six years. When Bieberback left for Berlin in 1921 his successor was Max Dehn.[1]

Bieberbach was at the University of Berlin from 1921 to 1945. His colleagues in Berlin were Issai Schur, Erhard Schmidt and Richard von Mises.

After Hitler came to power on 30 January 1933, Bieberbach was converted to the views of the Nazis and persecuted his Jewish colleagues. Von Mises left Berlin for Istanbul in 1934 and Schur was forced to retire in 1934 and emigrated to Palestine in 1939.

Bieberbach wore the Nazi uniform during doctoral oral examinations after 1933.

He was dismissed from his professorship in 1945 and died on 1 September 1982. Before he died he lived in the house of one of his sons.

Bieberbach is best remembered for the Bieberbach conjecture (1916) which stated a necessary condition on a holomorphic function to map the open disk of the complex plane injectively to the complex plane.[8] He also wrote an important book with Issai Schur titled *Über die Minkowskische Reduktiontheorie der Positiven quadratischen Formen* in 1928.

Biebach and another anti-semite Theodore Vahlen (1869-1945) founded the "Deutsche Mathematik" journal.

Heinz Hopf

Heinz Hopf was born on 19 November 1894 in Gräbschen near Breslau, Germany (now Wroclaw, Poland). His studies in Breslau was interrupted by the outbreak of World War I in 1914. He fought on the Western front as a lieutenant, wounded twice and received the iron cross (first class) in 1918.

Figure 27.   Heinz Hopf

Figure 28.   Wilhelm Süss

Hopf received his Ph.D in 1925 at the University of Berlin with his thesis *Über zusammenhänge zwischen Topologie and Metrik von Mannigfaltigkeiten* (Connections between topology and metric of manifolds). His advisers were Ludwig Bieberbach and Erhard Schmidt. Hopf stayed in Berlin as a privatdozent from 1926 to 1931 when he was appointed full professor at ETH in Zürich succeeding Hermann Weyl. He spent the 1927-28 academic year at Princeton University on a Rockefeller fellowship.

Hopf married Anja von Mickwitz in October 1928.[8] He and Russian mathematician Pavel Aleksandrov (1896-1982) published a book on topology in 1935 originally planned in three volumes.

Wilhelm Süss

Wilhelm Süss was born on 7 March 1895 in Frankfurt am Main. He fought for three years during World War I. Süss received his Ph.D in 1920 at Frankfurt with his thesis *Begründung der Inhaltslehre in Raum ohne Benetzung von Stetigkeitsaxiomen.*[10]

From 1923 to 1928 he taught in Kagoshima, Japan.

He submitted his habilitation thesis at the University of Greifswald to became a privatdozent. Süss became a lifelong friend of Hellmuth Kneser (1898-1973) at Greifswald. Süss was appointed President of the German Mathematical Society in 1937 and joined the Nazi party in 1938.

In August 1944 Süss created the *Mathematisches Forschungsinstitut Oberwolfach* in the tiny village of Oberwolfach in the Black Forest and became its first director.

In 1945 Oberwolfach was occupied by the Moroccan army who tried to use the books as fuel. It was mathematician John Todd who prevented the destruction of the books. Todd was a Lieutenant Commander in the British Navy who arrived Oberwolfach in July 1945.

Süss died on 21 May 1958 in Freiburg im Breisgau, Germany.

Friedrich Ernst Peter Hirzebruch

Friedrich Hirzebruch was born on 17 October 1927 in Hamm between Dortmund and Münster, Germany. He received his Ph.D from the University of Münster in 1950 for his thesis *Über vierdimensionale Riemannsche Flächen Mehrdeutiger analytischer Funktionen von zwei komplexen Veränderlichen.* He also studied the algebraic topology and algebraic geometry with Heinz Hopf at ETH Zürich from 1949 to 1950.

Hirzebruch spent the two years 1952-54 at the Institute for Advanced Study in Princeton with Armand Borel, Kunihiko Kodaira and D.C Spencer.[3] He married Ingeborg Spitzley in 1952 and they had one son and two daughters.

His most famous achievement, the Hirzebruch-Riemann-Roch theorem, appeared in his 1954 paper. After spending one year at Münster and one year at Princeton University he was appointed ordinary professor at the University of Bonn 1956 where he retired in 1993.

He founded the *Max-Planck-Institut für Mathematik* in Bonn in 1980 and served as director of the institute from 1980 to 1995.[3] Hirzebruch was President of the German Mathematical Society in 1961-62 and again in 1990.[3]

Hirchbruch received many prizes including the Wolf Prize, the Lobachevsky Prize, the Lomonosov Gold Medal, the Albert Einstein Medal, the Stefan Banach Medal and Georg Cantor Medal. He also received thirteen honorary doctorates.

Figure 29.   Friedrich E. P. Hirzebruch

Figure 30.   Maxim Kontsevich

Maxim Kontsevich

Maxim Kontsevich was born on 25 August 1964. After attending Moscow State University he received his Ph.D (Dr.rer.nat) at the University of Bonn under Don Bernard Zagier in 1992. His thesis *Intersection Theory on the Moduli Space of Curves and the Matrix Airy Function* proves a conjecture by Edward Witten that two quantum gravitational models are equivalent.

He has been influenced by the work of Richard Feynman and Edward Witten. Kontsevich is an expert in the string theory and quantum field theory.

Another result of Kontsevich relates to the knot theory. He is a professor at the Institute des Hautes Etudes Scientifique (IHES) in France and Visiting Professor at Rutgers University, New Jersey in the United States. Kontsevich was awarded the Fields Medal in 1998 for his work in mathematical physics, algebraic geometry and topology opening up new approaches in research.

David Hilbert

David Hilbert was born on 23 January 1862 in Königsberg, Prussia (now Kaliningrad, Russia). His father Otto was a judge and his mother, Maria Theresa Erdtmann, was interested in mathematics. He received his doctorate in 1885 for a thesis *Über invariante Eigenschaften specieller binärer Formen, insbesondere der Kugelfunktionen.*[10] In Königsberg he became a good friend of Adolf Hurwitz and Hermann Minkoski. Hilbert was a privatdozent at Königsberg (1886-1892), extraordinary professor (1892-1893) and ordinary professor (1893-1895).

In 1895 he was appointed to the chair of mathematics at Göttingen succeeding Heinrich Weber. In 1892 Hilbert married Käthe Jerosh and their only son Franz was born in 1893. Franz suffered from mental illness from the age of twenty-one and never lived a normal life.[4]

Hilbert completed *"Zahlbericht"* in 1897 a brilliant synthesis of the number theory works of Kummer, Kronecker and Dedekind also containing a wealth of Hilbert's own ideas. He published *"Grundlagen der Geometrie"* in 1899. Paul Bernays published the 10th edition of the book in 1999.

Hilbert proposed 23 problems to be solved in 20th century at the Second International Congress of Mathematicians in Paris in 1900. Mathematicians who solved these problems gained great reputation among their colleagues.

His name is best remembered through the concept of Hilbert space. Richard Courant and Hilbert published *"Methoden der Mathematischen Physik"* in 1924 which became very useful in quantum mechanics after 1925.

He retired in 1929, although he continued to lecture occasionally. His house was at 29 Wilhelm Weber Strasse in Göttingen. A street in Göttingen was named after Hilbert. In 1930 Königsberg made him an honorary citizen. In his address at the ceremony *"Naturerkennen und Logik"*, he said *"Wir müssen wissen, wir werden wissen"* (We must know, we shall know). These words are engraved in his tombstone in StadtFriedhof, Göttingen.

He was well known for his forgetfulness. Once he invited guests to his house, but Mrs. Hilbert noticed that he needed to change his shirts. He went upstairs to his bedroom, took off his shirts and went to bed.

After many Jewish professors and liberal non-Jewish professors were expelled from Göttingen he was seated at a banquet next to Bernhard Rust, the Nazis Minister of Education. The minister asked Hilbert whether the Mathematical Institute had suffered from the removal of those professors. Hilbert replied bitterly, 'It hasn't suffered, Herr Minister. It just doesn't exist any more.'

Hilbert died on 14 February 1943 in Göttingen, and was buried in the Stadt Friedhof in Göttingen.

Figure 31.   David Hilbert

Figure 32.   Hermann Minkowski

Hermann Minkowski

Hermann Minkowski was born on 22 June 1864 in Alexotas, Russia (now Kaunas, Lithuania) but his parents were German. When Hermann was eight years old the family moved to Königsberg.

Minkowski received the Grand Prize in Mathematics by the Paris Academy of Sciences in 1883 jointly with Henry Smith (1826–1883) for a paper on the theory of quadratic forms with integral coefficients.

Minkowski received his Ph.D in 1885 at Königsberg with his thesis *Untersuchungen über quadratische Formen Bestimmung der Anzahl Verschiedener Formen, welche ein gegebenes Genus enthält.* He presented his habilitation thesis *Raumliche Anschauung und Minima positiver definiter quadratischer Formen* in 1887 to the University of Bonn.

Minkowski became Heinrich Hertz's assistant and intended to be a physicist. When Hertz died in 1894, Minkowski returned to mathematics. After spending five years as a privatdozent, he was appointed extraordinary professor in 1892 at Bonn. In 1895 Minkowski succeeded Hilbert as ordinary professor at Königsberg.

Minkowski married Auguste Adler in Strassburg in 1897; they had two daughters.[3] After only one year in Königsberg, he accepted a chair at the Eidgenössische Polytechnikum Zurich (now ETH) where he became a colleague of his former teacher and friend Hurwitz who had come to Zürich in 1892. Einstein took several courses from Minkowski at Zürich and the two would become interested in the relativity theory.

Klein and Hilbert created a new chair at Göttingen with Althoff's help, and Minkowski was appointed to the new chair in 1902. Therefore Klein, Hilbert, Minkowski, Carl Runge and Felix Bernstein were all at the Göttingen mathematics department in the first decade of 20th century.

In September 1908 at the annual meeting of the Society of German Scientists and Physicians in Köln, Minkowski lectured in "Space and Time." He said, "Henceforth space by itself, and time itself are doomed to fade away into mere shadows, and only a kind of union of the two will preserve an independent reality." His theory of four-dimensional manifold was necessary for the working out of the general theory of relativity. His most original achievement was his *"Geometric der Zahlen"* (geometry of numbers) first published in 1896. It was reprinted in 1953 and 1968 by Chelsea, New York.

Figure 33. **Constantin Carathéodory**

Figure 34. **Hans Rademacher**

Minkowski died in 12 January in Göttingen from a ruptured appendix. His successor at Göttingen was Edmund Landau.

Constantin Carathéodory

Constantin Carathéodory was born on 13 September 1873 in Berlin. His father, Stephanos Carathéodory, was First Secretary at the Ottoman Embassy in Berlin. Two years later Stephanos was appointed the Ottoman Ambassador in Belgium, so the family went to Brussels where Constantin attended secondary school and trained as an engineer. He worked on the construction of the Assiut dam in Egypt until April 1900.

He entered the University of Berlin in May 1900 where he learned from Frobenius and Schwarz. Caratheodory became a good friend with Hungarian Leopold Fejér, in Berlin.[3] Carathéodory moved to Göttingen where he received his Ph. D. in 1904 with his thesis *Über die diskontinuierlichen Lösungen in der Variationsrechnung*. He also wrote his habilitation thesis *Über die starken Maxima und Minima bei einfachen Integralen* in 1905. He was a privatdozent in Göttingen until 1908. After one year at Bonn, he was appointed ordinary professor in 1909 at the Technische Hochschule Hannover.

On 5 February 1909 Carathéodory married Euphrosyne Carathéodory in Istanbul who was his aunt and eleven years his junior.[3] They had two children.

Carathéodory moved to TH Breslau from 1910 to 1913 where his successor was Max Dehn. Then in 1913 he succeeded Felix Klein at Göttingen when Klein retired. In 1918 Carathéodory moved to Berlin to succeed Georg Frobenius.

Carathéodory was appointed professor of Analytical and Higher Geometry at the University of Athens in 1920 and tried to establish another university in Smyrna. Because the Turks attacked Smyrna, which is now called Izmir in Turkey, the plan completely failed. He taught at Athens until 1924 when he moved to München to succeed Lindemann.

Carathéodory retired in 1938 at München. He made significant contributions to the calculus of variations, the theory of point set measure, and the theory of functions of a real variable.[3] He wrote many good books. Carathéodory died on 2 February 1950 in München.

Hans Rademacher

Hans Radenacher was born on 3 April 1892 in Hamburg, Germany. He served for two years in the German Army during World War I. He was awarded his doctorate in 1916 at Göttingen with his thesis *Eindeutige Abbildungen und Messbarkeit* (Single-valued mappings and measurability) under Caratheodory's supervision. Radermacher was a privatdozent at Berlin from 1919 to 1922 and an extraordinary professor at Hamburg from 1922 to 1925.[1]

In 1925 he was appointed ordinary professor at Breslau where he stayed until 1933. He was not Jewish but he was the Chairman of the local German Peace Society (Deutsche Friedens Gesellschaft) in Breslau which was not acceptable to the Nazis. He left Germany for the United States to be a visiting professor at the University of Pennsylvania in 1934. He had married Suzanne Gaspary during his Berlin years but the marriage ended in divorce in 1929 and he remarried Olga Prey. After divorcing Olga Prey in 1947 he married Irma Wolpe in 1949.

Although he was a full professor for eight years in Germany, the University of Pennsylvania appointed him assistant professor in 1935 which he gladly accepted. He supervised 18 doctoral students in Pennsylvania and four in Germany.

Rademacher was invited to deliver an address to the International Congress of Mathematicians in Cambridge, Massachusetts in 1950.[3] He wrote many fine books.

After retiring from the University of Pennsylvania in 1962, he taught at New York University and the Rockefeller University.[3]

He died on 7 February 1969 in Haverford, Pennsylvania.

Klaus Friedrich Roth

Klaus Roth was born on 29 October 1925 in Breslau, (now Wroclaw, Poland). He came to England when he was young and graduated from St. Paul's School in London in 1943. He received his BA degree in 1945 at Peterhouse, Cambridge where Harold Davenport was his teacher.

Roth was awarded his doctorate in 1950 under Theodor Eastmann's supervision at the University College London. He became a professor in 1961.

He solved the major open problem of approximating algebraic numbers by rationals in 1955. This work made Roth be awarded a Fields Medal at the International Congress of Mathematicians in Edinburgh in 1958.[3]

In 1952 Roth also proved a conjecture made by Erdös and Turán, in 1935.[3]

Roth moved to the chair of Pure Mathematics at Imperial College, London in 1966 and stayed there until 1988.

He was elected a Fellow of the Royal Society of London in 1960 and received the DeMorgan Medal in 1983 and the Sylvester Medal in 1991.

Figure 35.   Klaus Friedrich Roth   **Figure 36.   Gerd Faltings**

Gerd Faltings

Gerd Faltings was born on 28 July 1954 in Gelsenkirchen-Buer between Essen and Dortmund, Germany. He received his doctorate (Dr. rer. nat) in 1978 at the University of Münster where he also received his habilitation in 1981.

After two years as a professor at the University of Wuppertal (1982–1984), he was appointed professor at Princeton University in 1985.

In 1986 he was awarded a Fields Medal at the International Congress of Mathematicians at Berkeley for his proof of the Mordell Conjecture.

Louis Mordell (1888–1972) in 1922 stated that a given set of algebraic equations with rational coefficients defining an algebraic curve of genus greater than or equal to 2 must have only finite number of rational solutions.

In 1995 he moved to the University of Bonn and received the Gottfried Wilhelm Leibniz Prize in 1996.

Ernst Eduard Kummer

Eduard Kummer was born on 29 January 1868 in Sorau, Brandenburg, Prussia. He was awarded a doctorate in 1831 at the University of Halle-Wittenberg with his thesis *De cosniuum et sinuum potestatibus secundum cosinus et sinus arcuum multiplicium evolvendis.*[10]

He taught for ten years at the Gymnasium in Liegnitz, now Legnica in Poland where Kronecker was a student.

Kummer's paper on hypergeometric series published in Crelle's Journal in 1836 impressed Jacobi and Dirichlet. Kummer was elected to the Berlin Academy of Science in 1839 on Dirichlet's recommendation. In 1840 Kummer married a cousin of Dirichlet's wife, but she died in 1848.

In 1842 he was appointed ordinary professor at Breslau with strong support from Jacobi and Dirichlet.

When Dirichlet left Berlin to succeed Gauss at Göttingen in 1855, Kummer filled Dirichlet's chair in Berlin.

Kummer was a popular professor because of his charm and sense of humor. He was concerned with the well-being of his students and willingly aided them when financial difficulties arose.[3]

While he was in Berlin, Weierstrass and Kronecker were also teaching there. Although Kronecker was appointed to a chair in 1883 when Kummer retired, Kronecker taught from 1863 as a member of the Berlin Academy.

Kummer's successor was Lazarus Fuchs. The Paris Academy of Sciences awarded Kummer the Grand Prize in 1857 for his work relating to Fermat's Last Theorem. He supervised 51 doctoral students in Berlin.

He died on 14 May 1893 in Berlin.

Figure 37. Ernst E. Kummer

Figure 38. Georg F. L. P. Cantor

Georg Ferdinand Ludwig Philipp Cantor

Georg Cantor was born on 3 March 1845 in St. Petersburg, Russia. When Cantor was eleven years old the family moved to Germany. He became a good friend with Hermann Schwarz in Berlin while attending lectures by Kummer, Weierstrass and Kronecker. He was awarded a doctorate in 1867 at the University of Berlin with his thesis *De aequationibus secondi gradus indeterminatis.*[10]

In 1869 in Halle he received his habilitation with his thesis on number theory. He was a privatdozent for three years, extraordinary professor for seven years and ordinary professor for 34 years (1879-1913) at Halle.

In 1872 he began a friendship with Dedekind whom he had met in Switzerland. Cantor married Vally Guttmann in 1874. They had six children.

At the end of May 1884 Cantor had his first episode of depression.[3] The bitter antagonism between Cantor and Kronecker began in 1887 when Kronecker opposed Cantor's paper on dimension submitted to Crelle's Journal. The mathematical correspondence between Cantor and Dedekind was ended in 1882 as a result.

Cantor co-founded the Deutsche Mathematiker-Vereinigung in 1890. He was the first President then stepped down in 1893 and was succeeded by Paul Gordan in 1894.

His last major papers on the set theory were published in 1895 and 1897 in *Mathematische Annalen.*

Hilbert said, about Cantor's work "It is, I think, the finest product of mathematical genius and one of the supreme achievements of purely intellectual human activity."[9]

Cantor's mother died in 1876, and his younger brother and youngest son died in 1899. He developed depression and suffered from it for the rest of his life.

Cantor retired in 1913 and died of heart attack on 6 June 1918 in Halle, Germany.

Paul David Gustav du Bois-Reymond

Paul du Bois-Reymond was born on 2 December 1831 in Berlin. His parents had five children and they spoke French at home. He received a doctorate in 1853 at the University of Berlin for his thesis *De aequilibrio fluidorum.*

After teaching mathematics and physics at a secondary school for 12 years, he became a privatdozent at Heidelberg in 1865.[1] Three years later he was appointed extraordinary professor at Heidelberg.

In 1870 he was appointed ordinary professor at Freiburg where he stayed for three years.[1] He then moved to Tübingen in 1874 where he was until 1884. At Tübingen, he supervised a few doctoral students including Otto Hölder (1859-1937) who became professor at Leipzig (1899-1928). Paul du Bois-Reymond moved to TH Berlin in 1884 and stayed until 1889.

He published *Die allgemeine Funktiontheorie* in 1882 and many other important papers. He died on 7 April 1889 in Freiburg, Germany.

Figure 39. Paul D. G. du Bois-Reymond

Otto Ludwig Hölder

Otto Hölder was born on 22 December 1859 in Stuttgart, Germany. At the University of Berlin, he was a fellow student of Carl Runge and Hölder attended lectures by Weierstrass, Kummer and Kronecker. Hölder received a Ph.D. at the University of Tübingen in 1882 for his thesis *Beiträge zur Potentialtheorie* (Contributions to potential theory) supervised by Paul DuBois-Reymond.

Hölder received his habilitation in 1884 at the University of Göttingen and became a privatdozent there. He became an extraordinary professor at the University of Tübingen in 1889. In 1896, he succeeded Hermann Minkowski as an ordinary professor at the University of Königsberg, where he stayed for three years.[1] In 1899 he was appointed to a chair (Zweiter Lehrstuhl für Geometrie) at the University of Leipzig succeeding Sophus Lie.[1] He retired from Leipzig in 1928 where he had supervised 41 doctoral students.[10]

Hölder is famous for Hölder inequality (1889), the Jordan-Hölder theorem and he also proved that the Gamma function satisfies no algebraic differential equation.

The Hölder's inequality was first found by Leonard James Rogers (1862-1933), a graduate of Oxford University and professor at the Yorkshire College, later the University of Leeds. Rogers discovered the inequality in 1888, one year before Hölder.[105]

In the case of Euclidean space, for all x and y in R^n (or in C^n)[106]

$$\sum_{k=1}^{n} |x_k y_k| \leq \left(\sum_{k=1}^{n} |x_k|^p\right)^{1/p} \left(\sum_{k=1}^{n} |y_k|^q\right)^{1/q}$$

For $p = q = 2$, Hölder's inequality results in the Cauchy-Schwarz inequality.

Hölder died on 29 August 1937 in Leipzig, Germany.

Ludwig Otto Blumenthal

Otto Blumenthal was born in 20 July 1876 in Frankfurt am Main, Germany. When he was 18 years old, he converted from Judaism and became a Protestant.

He became the first doctoral student of Hilbert at Göttingen where he received a doctorate in 1898. He was much influenced by Arnold Sommerfeld (1868-1951) who was Klein's assistant at that time. His thesis was on the Stieltjes continued fraction expansions.[3]

Blumenthal received his habilitation at Göttingen in 1901 with his thesis *Über Modalfunktionen von mehreren Veränderlichen.* He stayed in Göttingen for three more years as a privatdozent.

Bludementhal was appointed ordinary professor at TH Aachen in 1905 and he was forced by the Nazis to retire in 1933.

In 1908 he married Mali Ebstein and they had two children, one daughter and one son. He became a member of the German League for Human Rights and the Society of Friends of the New Russia which would later be considered a crime by Nazis.[3]

He was the executive editor of the famous journal *Mathematischen Annalen* from 1905 to 1938. He was also editor of the *Jahresbericht der Deutschen Mathematiker-Vereinigung* (DMV). Blumenthal was the President of DMV in 1924.

The Blumenthals were forced to move to Utrecht, Netherlands in July 1939. With Dutch mathematician Julius Wolff Blumenthal conducted a weekly mathematical colloquium for two years in their private rooms until October 1942.[3]

The Blumenthals were sent to the concentration camp at Westerbork in the Netherlands where Mali died. Otto also died on 12 November 1944 in Theresienstadt (now in Czech Republic) after suffering from pneumonia, dysentery and tuberculosis.

Figure 40.   Ludwig O. Blumenthal

Figure 41.   Erhard Schmidt

Erhard Schmidt

Erhard Schmidt was born on 13 January 1876 in Dorpat, Germany (now Tartu, Estonia). He received his doctorate in 1905 at Göttingen with his thesis *Entwickelung willkürlicher Funktionen nach Systemen vorgeschriebener* which was on integral equations. He was awarded his habilitation in 1906 at Bonn where he stayed for two years as a privatdozent.

After two years as full professor in Zürich he was appointed ordinary professor in 1910 at Erlangen. From 1912 to 1917 he was at Breslau. And in 1917 he moved to the University of Berlin where he retired in 1950.

His colleagues in Berlin in the 1930's were Issai Schur, Richard von Mises and Ludwig Bieberwach.

After the end of World War II, the University of Berlin which was located in East Berlin was renamed Humboldt University. Schmidt remained in East Berlin to become the first editor of *Mathematische Nachrichten.*

His main interest was in integral equations and the Hilbert space. He is best remembered for the Gram-Schmidt orthonormalisation process for constructing an orthonormal set of functions from a linearly independent set. However, it should be noted that Laplace presented this idea before either Gram or Schmidt.[3]

Schmidt died on 6 December 1959 in East Berlin, East Germany.

Alexander Dinghas

Alexander Dinghas was born on 9 February 1908 in Smyrna (now Izmir), Turkey. He graduated from the Athens Technical University in 1930 majoring in electrical and mechanical engineering. He married Fanny Grafiadou in 1931.

Dinghas received his Ph.D. in 1936 at Berlin with his thesis *Beiträge zur Theorie der meromorphen Funktionen.*[10]

In 1939 he submitted his habilitation thesis to obtain *venia legendi* (right to lecture) at the same university.

Being a non-German, Dinghas found it difficult to obtain a permanent university position during World War II. He became professor at the reopened University of Berlin 1947 which was renamed Humboldt University in 1949.

Many professors and students opposing Communist rule in East Germany (Deutsche Demokratische Republik) moved to West Berlin where a new university Freie Universität Berlin was founded in December 1948.

Dinghas was appointed professor and director of the Mathematical Institute in the new University.

He wrote three books and 121 papers.[3] Dinghas commanded the respect owed to some of his position, but he was also extremely hospitable and generous. He felt profound sympathy for less fortunate.[3]

He died on 19 April 1974 in West Berlin.

Richard Courant

Richard Courant was born to Jewish parents on 8 January 1888 in Lublinitz, Germany (now Lubliniec, Poland). When he was nine his family moved to Breslau where he attended the university for two years.

He arrived at Göttingen in November 1907 and obtained his Ph.D. in 1910 with his thesis *Über die Anwendung des Dirichletschen Prinzipes auf die Probleme der konformen Abbildung* under Hilbert's supervision.[3]

Courant submitted his habilitation thesis to Göttingen and he gave his inaugural lecture as a privatdozent "On Existence Proof in Mathematics" on 23 February 1912.[3] He married Nelly Neumann in 1912 and they divorced in 1915.

During World War I he was wounded on a battle field in September 1915. After the war Courant married Nerina (Nina) Runge on 22 January 1919. She was a daughter of Professor Carl Runge and granddaughter of Emil du Bois-Reymond, eminent physiologist and former Rector of the University of Berlin.

Courant was appointed ordinary professor at Göttingen in 1920 succeeding Erich Hecke who left for Hamburg. Courant was elected to the Göttingen City Council.

In the 1920's Göttingen did not have a separate building for mathematics. In fact mathematics, physics, philosophy, history, classics, and philology all belong to the Philosophical Faculty. Professors did not even have offices.

Courant decided to construct a new building for the Mathematics Institute and contacted the International Education Board in the United States. The Board provided $350,000 for the construction and equipment of a building for the Mathematical Institute and an addition to the building for the Physics Institute at Göttingen, with the understanding that the German government would provide $25,000 annually for the maintenance of the buildings.[3]

The formal dedication of the Mathematics Institute, which was the first such building in Germany, took place on 2 December 1929. It was a three-level T-shaped building. The spacious lobby containing a bust of Hilbert today is known as "the Hilbert space." Some students would ask, "How come Hilbert is not in the infinite dimensional space?"

Figure 42. Alexander Dinghas

Figure 43. Richard Courant

In 1922 Courant published a book on the function theory co-written with Hurwitz although Hurwitz died in 1919. In 1924 he also published, jointly with Hilbert, *Methoden der mathematischen Physik* which was very useful in quantum mechanics.

After Hitler came to power in January 1933, all Jewish and liberal non-Jewish professors and lecturers were dismissed. Courant's family moved to New York on August 21, 1934.

Courant slowly built a graduate mathematics program at New York University attracting many German refugees including Friedrichs, Lewy, Busemann, Fritz John, and Willi Feller as well as American mathematicians.

From 1953 to 1958 he was the director of the new Institute of Mathematical Science at New York University, which in 1964 was re-named the Courant Institute.[3] His creation of the Courant Institute at NYU, starting with nothing was a great contribution to mathematics in the United States.

After suffering a stroke in November 1971, he died on 27 January in New Rochelle, New York.

Franz Rellich

Franz Rellich was born on 14 September 1906 in Tramin, South Tyrol, Austro-Hungarian Empire. He received a Ph.D. at the University of Göttingen in 1929 for his thesis *Verallgemeinerung der Riemannschen Integrationmethode auf Differentialgleichungen n-ter Ordnung in Zwei Veränderlichen* (Generalisation of Riemann's integration method on differential equations of *n*-th order in two variables) under Courant's supervision. He was habilitated in 1933 at Göttingen.[1] In 1934 he became a privatdozent in Marburg and in 1940 a professor at the Technische Hochschule Dresden, and in 1946 director of the Mathematical Institute in Göttingen. He was instrumental in rebuilding the mathematical tradition in Göttingen after many Jewish and liberal professors left.

Among his mathematical contributions, his works in perturbation theory of linear operators in Hilbert Space are important.

He also showed that the Monge-Ampère differential equation in the elliptic care, it can have at most two solutions.

In 1940, he proved that a differential equation $W' = f(z,w)$ has at most countably many entire solutions $W(z)$, if $f(z,w)$ is a linear entire function in z.

This is now known as Rellich's theorem.[92]

He died on 25 September 1955.

Hermann Klaus Hugo Weyl

Hermann Weyl was born on 9 November 1885 in Elmshorn near Hamburg, Germany. He received his doctorate in 1908 at Göttingen with his thesis *Singluläre Integral-gleichungen mit besonder Berücksichtigung des Fourierschen Integraltheorems* under Hilbert's supervision. He submitted his habilitation thesis *Über gewöhnliche Differentialgleichungen mit Singularitäten und die zugehörigen Entwicklungen willkürlicher Funktionen* in Göttingen in 1910 and stayed there until 1913 as a privatdozent.[3]

Figure 44. Hermann K. H. Weyl

Weyl was an ordinary professor at Eidgenössische Technische Hochschule Zürich from 1913 to 1930 supervising 15 doctoral candidates. In 1918 he published *"Raum-Zeit-Materie"* (Space-Time-Matter) which was based in his lectures at ETH in the summer term of 1917.

In 1916 he published one of his best papers on number theory on the definition of the uniform distribution modulo 1, which was to be of great significance in the later work of G.H. Hardy and John Littlewood.[5]

He also published *Gruppentheorie und Quantemmechanik* in 1928. Weyl wrote 15 important books on mathematics and mathematical physics.

Weyl succeeded Hilbert at Göttingen 1930. In 1933 he was forced to leave Germany for the United States because his wife Helene (Hella) Joseph was Jewish. They had married in 1913. In 1921 Erwin Schrodinger was appointed professor of physics at the University of Zürich and became a close friend of Weyl.

Hella Weyl was a beautiful philosopher and literatus. But Schrödinger's wife Annemarie (Anny) was almost an exact opposite of Hella Weyl. Anny Schrödinger was madly in love with Weyl, while Hella Weyl was infatuated with physicist Paul Scherrer (1890-1969).[3] Schrödinger invited "an old girl-friend in Vienna" to join him skiing in Arosa while his wife Annemarie remained home in Zürich.[11]

Weyl was appointed professor at the Institute for Advanced Study at Princeton from 1933 to 1952. His wife Hella died in 1948, and he married the sculptor Ellen Lohnstein Bär in 1950.[3] After his retirement he and his wife Ellen split their time between Princeton and Zürich. He was walking home in Zürich after mailing letters to those who had sent greeting for his 70th birthday when he collapsed and died on 9 December 1955.

Max Wilhelm Dehn

Max Dehn was born on 13 November 1878 in Hamburg. He obtained his doctorate in 1900 with his thesis *Die Legendreschen Sätze über die Winkelsumme in Dreieck* at Göttingen under Hilbert's supervision. He wrote his habilitation thesis at Münster in 1901 and stayed there for ten years. After two years (1911-1913) at Kiel and eight years (1913-1921) at TH Breslau, he was appointed ordinary professor at Frankfurt in 1921.

Figure 45. Max W. Dehn

Figure 46. Alexander M. Ostrowski

His colleagues at Frankfurt were Carl L. Siegel, Ernst Hellinger, Otto Szász and Paul Epstein. Together they created an important research center in mathematics.

Dehn was 22 years old when he solved Hibert's third problem on the congruence of polyhedra. He served in the German Army from 1915 to 1918.

After the Nazis came to power, the Frankfurt Mathematics Institute was completely destroyed. Siegel was not Jewish but because of the Nazis policies, he went to the United States followed by Dehn, Hellinger and Szász. Epstein committed suicide when he received a summons from the Gespato.

Dehn arrived in the United States in 1941 after traveling through Norway, Siberia, and Japan. He was an assistant professor at the University of Idaho, Illinois Institute of Technology and St. John's College in Annapolis, Maryland.

In 1945 he joined the faculty at Black Mountain College in North Carolina. It never had more than 90 students, but there was a large faculty, all of whom lived at the College.[12] Dehn taught Greek and philosophy in addition to mathematics.

He was the only mathematician at the college. Most of the mathematicians fleeing Europe were able to obtain prestigious academic position in the United States, but Dehn was unable to do so.

He was paid only $40 a month at Black Mountain College but he loved the dogwood forest nearby. One summer day in June 1952 he saw some loggers cutting down trees and he ran up a steep path to try to stop them. That effort probably caused the 'embolism' that killed Dehn the next day. He woke the next morning with severe chest pain and asked for aspirin. By the time his wife came back with the aspirin. He was dead on 27 June 1952 in Black Mountain.[12]

The Black Mountain College closed in 1956.

Alexander Markowich Ostrowski

Alexander Ostrowski was born into a Jewish family on 25 September 1893 in Kiev, Ukraine. While he was a teenager he participated B.N. Grave's seminar at the University of Kiev and wrote his first paper with Grave's assistance. Since he was a graduate of Kiev College of Commerce not a high school, he could not be admitted to Russian universities.

Kurt Hensel (1861–1941) invited Ostrowski to the University of Marburg in 1912. After the outbreak of World War I in 1914, Ostrowski was interned as a hostile foreigner. Fortunately he was allowed to use the library at the University of Marburg during that time so he could read through mathematical journals.

When the war ended in 1918, he went to Göttingen and he was awarded his doctorate in 1920 under Hilbert and Landau. His doctoral dissertation, his fifth paper, was published in *Mathematische Zeitschrift.*[3] The results of this work allowed Ostrowski to solve Hilbert's 18th Problem in part.

Ostrowski succeeded in proving that the Dirichlet zeta series

$$\zeta(x,s) = 1^{-s}x + 2^{-s}x^2 + 3^{-s}x^3 + \cdots\cdots$$

does not satisfy an algebraic partial differential equation.[30]

Ostrowski left Göttingen for Hamburg, where as an assistant of Erich Hecke he worked on his habilitation thesis. He was awarded habilitation in 1922, then he became a privatdozent at Göttingen from 1923 to 1927.

During the academic year 1925-26 he visited Oxford, Cambridge, and Edinburgh as a Rockefeller Research Fellow. He accepted the chair of mathematics at the University of Basel in 1927 where he remained until he retired in 1958. He married Margaret Sachs, a psychoanalyst from the school of Carl Gustav Jung in 1949.

Ostrowski wrote around 275 papers and books. His works were published in six volumes in 1983-85 by Birkhäuser.

He died on 20 November 1986 in Montagnola, Lugano, Switzerland.

Wilhelm Magnus

Wilhelm Magnus was born in Berlin on 5 February 1907. He received his doctorate in 1931 at Frankfurt with his thesis *Über Unendlich diskontinuierliche Gruppen von einerdefinirenden Relation (der Freiheitssatz).*

Magnus was a privatdozent at Frankfurt from 1933 to 1938. He refused to join the Nazi Party, and as a consequence, he was not allowed to hold an academic post during World War II.[3]

From 1948 to 1950 he collaborated with Arthur Erdélyi on the project of organizing and publishing Harry Bateman's (1882-1946) manuscripts at the California Institute of Technology.[3]

He then spent 23 years at the Courant Institute of Mathematical Sciences at New York University and five years at the Polytechnic Institute of New York.

He supervised 72 doctoral candidates mostly at NYU where he received the Great Teacher Award. Magnus died on 15 October 1990.

He was a real expert on Rainer Maria Rilke.

Figure 47. Wilhelm Magnus

Rudolf Friedrich Alfred Clebsch

Alfred Clebsch was born on 19 January 1833 in Königsberg, Germany (now Kaliningrad, Russia).

He studied at the University of Königsberg where Otto Hesse, Friedrich Richelot and Franz Neumann taught. After receiving his Ph.D. in 1854 at Königsberg, he became a privatdozent at Berlin in 1857.

Clebsch was appointed to Chair of Mechanics and Synthetic Geometry in 1858 at TH Karlsruhe. He moved to Giessen in 1863 where he stayed for five years.

He succeeded Bernhard Riemann at Göttingen in 1868, and died of diphtheria there on 7 November 1872.

Although his Ph.D. thesis was on hydrodynamics, he changed his interests to study the calculus of variations and patial differential equations.[3]

Clebsch and Carl Neumann, the son of his teacher at Königsberg, founded *Mathematische Annalen*. The Clebsch-Gordan coefficients used in quantum mechanics were introduced by them while working on *"Theorie der Abelschen Funktionen"* in 1866.

Figure 48. Rudolf F. A. Clebsch

Figure 49. Alfréd Haar

Erich Hecke

Erich Hecke was born on 20 September 1887 in Buk, Posen, Germany (now Poznan, Poland). He was awarded his Ph.D. at Göttingen for a dissertation *Zur Theorie der Modulfunktionen von Zwei Variablen und ihrer Anwendung auf die Zahleatheorie* under Hilbert's supervision. He stayed three years at Göttingen as a privatdozent (1912–1915). Hecke was appointed to a chair at the University of Basel, Switzerland, in 1915. Then he succeeded Carathéodury at Göttingen in 1918.

He moved to the University of Hamburg in 1919, the year it was founded so he thought he would have had the opportunity to influence the new institution.[3]

When Hecke was Hilbert's assistant he received 50 Marks ($12.50 at that time) a month. One day Hilbert decided that sum was inadequate. He went to meet with Wilhelm Althoff, who was in charge of universities at the Ministry of Culture. At the end of the conversation, he remembered there was another matter he had intended to discuss with Althoff. Hilbert stuck his head out of the window of the Ministry and shouted down to Mrs. Hilbert who was waiting for him outside. "Käthe, Käthe! What was that other matter I wanted to talk about?" "Hecke, David, Hecke!" Hilbert turned back to Althoff, and demanded that Hecke's salary be doubled.[9]

This episode demonstrates how the Ministry of Culture in Germany micromanaged university affairs at that time.

Hecke's most important work was in 1936 with his discovery of the properties of the algebra of Hecke operators and of the Euler products associated with them.[3]

Hecke was President of the Deutsche Mathematiker-Vereinigung (German Mathematical Society) in 1923.

Hecke died of cancer on 13 February 1947 in Copenhagen, Denmark.

Alfréd Haar

Alfréd Haar was born on 11 October 1885 in Budapest, Hungary to an owner of great vineyards.[9] After graduating from high school in Budapest he went to Göttingen in 1904. He received his doctorate in 1909 at Göttingen with his thesis *Zur Theorie der orthogonalen Funktionensysteme* under Hilbert's supervision. Haar was small, delicately built and had the charming quality of feeling at home anywhere in the world. He possessed the kind of extremely quick and precise talent.[9]

Haar was a privatdozent at Göttingen from 1910 to 1912. He was appointed as an extraordinary professor at the Franz Josef Royal Hungarian University in Kolozsvár (now Cluj in Romania) in 1912. Five years later he was appointed ordinary professor there. Since Hungary was defeated in the World War I, Kolozsvár became a part of Romania and the university was moved to Szeged 270 km west from Cluj in 1920. In 1922 Haar and Frigyes Riesz (1880-1956) set up the János Bolyai Mathematical Institute at the University of Szeged because Bolyai had been born in Kolozsvár.

The University of Szeged is now a prestigious institution in Hungary and Szeged is a sister city of Cambridge, England.

Haar and Riesz also founded the famous journal *Acta Scientiarum Mathematicarum* in 1930. Haar introduced an invariant measure on locally compact groups, now called the Haar measure.[3]

John von Neumann had tried to discourage Haar from seeking such a measure since he was certain that no such measure could exist.[3]

Haar died on 16 March 1933 in Szeged, Hungary.

Ernst David Hellinger

Ernst Hellinger was born on 30 September 1883 in Striegau, Silesia, Germany (now Strizegom, Poland). Hellinger grew up nearby in Breslau where he graduated from the Gymnasium in 1902. Otto Toeplitz, Max Born and Richard Courant all grew up in Breslau.

Hellinger received his Ph.D at Göttingen in 1907 for his thesis *Die Orthogonalinvarianten quadratischer Formen von unendlich vielen Variablen* under Hilbert's supervision. He introduced a new type of integral, the Hellinger integral in his doctoral thesis. He was a privatdozent at Marburg from 1909 to 1914.

In 1914 he was appointed extraordinary professor at the new University of Frankfurt. In 1921 Max Dehn, Paul Epstein and Otto Szász joined the faculty. One year later Carl Siegel also came to Frankfurt which became an important mathematical research center with these five mathematicians.

Hitler's regime forced Hellinger to retire in 1936 despite the fact that he served in the German Army during World War I. He was sent to the Dachau concentration camp in 1938. Fortunately, his friends were able to arrange a temporary job for him at Northwestern University in the United States. He was released and emigrated to the United States in late February 1939.

He became a U.S. citizen in 1944 and worked at Northwestern University until 1949 when he retired. He was one of the best-liked professors in the Northwestern Mathematics Department.[3]

With Toeplitz he wrote *"Integralgleichungen und Gleichungen mit unendlich vielen Unbekannten"* in the *Enzyklopädie der Mathematischen Wissenschaften* in 1927.[3]

Hellinger died of cancer on 28 March 1950 in Chicago, Illinois.

Figure 50. Ernst D. Hellinger   **Figure 51. Georg K. W. Hamel**

Georg Karl Wilhelm Hamel

Georg Hamel was born on 12 September 1877 in Düren near Aachen, Germany. He studied at TH Aachen, Berlin and Göttingen where he received his doctorate in 1901 for his thesis *Über die Geometrien, in denen die Geraden die Kürzesten sind* under Hilbert's supervision.

He was a privatdozent at TH Karlsruhe from 1903 to 1905. Hamel was appointed professor of mechanics at the German Technical University of Brünn (now Brno, Czech Republic) in October 1905. In August 1909 he married Agnes Frangenheim in Köln. They had three daughters.[3]

After seven years at Brünn he moved to TH Aachen in October 1912. Then he was appointed in 1919 to a chair at TH Berlin from where he retired in 1949.

During the Third Reich Hamel was clearly associated with the Nazis. In 1935 he was appointed President of *Deutsche Mathematiker-Vereinigung*. He was made editor of the *Neue Deutsche Forschungen* in 1935. He is best known for the Hamel basis, published in 1905. He tried to construct a basis for the real numbers as a vector space over the rational numbers.

He died on 4 October 1954 in Landshut, Germany.

Figure 52. Richard von Mises

Richard von Mises

Richard von Mises was born on 19 April 1883 in Lemburg, Austria (Lwow, Poland and now Lvov, Ukraine). After graduating TH Vienna studying mathematics, physics and engineering, he became an assistant to Georg Hamel at the University of Brünn. Von Mises received his Ph.D. at Vienna and habilitation at Brünn under Hamel. Von Mises was appointed extraordinary professor at Strassburg in 1909.

He became a pilot and taught the first university course on powered flight in 1913 at Strassburg. During World War I he joined the Austro-Hungarian Army as a pilot.[3] He led a team that constructed a 600-horsepower plane for the Austrians in 1915.

In 1919 he moved to TH Dresden but he moved to Berlin one year later. He was appointed to the new Chair for Applied Mathematics (Lehrstuhl für Angewandte Mathematik) in 1920. It was the first Chair on applied mathematics in Germany.

His new Institute of Applied Mathematics became a center for research in probability, statistics, numerical solutions of differential equations, elasticity and aerodynamics. In 1921 he founded the journal *Zeitschrift für Angewandte Mathematik und Mechanick* and became the editor of the journal. He was an excellent lecturer.

After the Nazis came to power he had to give up his Berlin professorship because he fell under the definition of non-Aryan even though he was a Roman Catholic. His successor was the Nazis Theodor Vahlen who promised that von Mises would not lose his pension rights.

In 1934 he was appointed to a chair in Istanbul, Turkey. The mathematician Hilda Geiringer (1893-1973), von Mises lover, went with him to Istanbul in 1934. In 1939 von Mises left Turkey for the United States. He became a professor at Harvard University and was appointed Gordon-Mckay Professor of Aerodynamics and Applied Mathematics in 1944.

Geiringer also moved to the United States and they married in 1943. Before they were married there was a rumor that von Mises was hiding a mistress. He insisted that Geiringer addresse him as "Herr Professor."

Von Mises wrote a book on philosophy "A Study in Human Understanding" in 1951. He was also an international authority on the Austrian poet Rainer Maria Rilke.[3]

In 1950 von Mises was offered honorary membership of the East German Academy of Scince. He declined because it was the McCarthy era in the United States.

He died on 14 July 1953 in Boston, Massachusetts.

Wilhelm Cauer

Wilhelm Cauer was born in Berlin on 24 June 1900. He received a doctorate at TH Berlin under Hamel's supervision in 1926. His dissertation was *Die Verwirklichung von Wechstomwiderständen vorgeschriebener Frequenzabhängigkeit* (The Realisation of Impedance for Prescribed Frequency Dependence) which took the first step toward network synthesis.

The driving-point impedance for any passive electrical network is a rational function of complex variable. Complex variable *s* is called the complex frequency.

Cauer is best known for Cauer Form in network synthesis utilizing the continued fraction expansion which results in a ladder network. For a given driving-point impedance function, a pole at infinity is represented as a series inductor. This process is "remove a pole, invert and remove a pole," to produce a ladder network.

One of his doctoral students at Louvain, Vitold Belevitch, also made important contributions to the circuit theory. Cauer published *Theorie der linearen Wechselstromschaltangen* (Theory of Linear Alternating Current Circuits) Volume I in Leipzig in 1941.

Although he was a civilian and not a member of the Nazi Party, he was executed on 22 April 1945 by the Red Army soldiers of the Soviet Union after their entry into Berlin.[14]

Figure 53. Karl T. W. Weierstrass

Karl Theodor Wilhelm Weierstrass

Karl Weierstrass was born on 31 October 1815 in Ostenfelde, Bavaria. His father wanted Karl to study finance, law and economics despite Karl's love of mathematics.

At the University of Bonn he spent four years of intensive fencing and drinking,[3] and he studied mathematics on his own. In 1839 he enrolled at the Theological and Philosophical Academy of Münster (now University of Münster) to attend lectures by Christoph Gudermann (1798–1851).

In 1841 he passed the examination to teach at secondary schools. Weierstrass published *Zur Theorie der Abelschen Funktionen* in 1854 in Crelle's Journal. F.J. Richelot (1808–1875) at the University of Königsberg was impressed by the article and the university conferred an honorary doctoral degree on Weierstrass in March 1854. The Gewerbeakademie (Industry Institute in Berlin, later TH Berlin) offered him a chair in 1856 so that he could leave his secondary school after teaching there for 15 years.

In 1856 he was elected to the Berlin Academy which gave him the right to teach at the University of Berlin. Until 1864 he taught at the two universities in Berlin. He was finally offered a newly created chair (Dritter Lehrstuhl) at the University of Berlin in 1864, from where he retired in 1892.[1]

In 1861 his emphasis on rigour led him to discover a function that, although continuous, had no derivative at any point.[3]

Most mathematics students know the following two Weierstrass theorems[13]:

Let T be a closed bounded set of real numbers. Then every function $f(x)$, continuous there, is the limit of a sequence of polynomials in x with real coefficients which is uniformly convergent on that set.

Let $f(x)$ be a continuous periodic function of period 2π. For every $\varepsilon \rangle 0$, there exists a trigonometric polynomial $T(x)$ such that $\mid f(x) - T(x) \mid < \varepsilon$ for all x.

Weierstrass never married and lived with his unmarried sisters. A lecture room at the University of Berlin (now Humboldt University) is named after him. He also served as Rector of the University of Berlin in 1873–74.

Weierstrass died on 19 February 1897 in Berlin.

Lazarus Immanuel Fuchs

Lazarus Fuchs was born on 5 May 1833 in Moschin, Germany (now Poznan, Poland). He received his doctorate at Berlin in 1858 with his thesis *"De Superficierum lineis Curveturae"* under Weierstrass' supervision. After teaching for seven years at a secondary school, he submitted his habilitation thesis in 1865 and he was appointed extraordinary professor at Berlin in 1866. In 1869 he moved to Greifswald, then to Göttingen in 1874 where he stayed only one year.

From 1875 to 1884 he held a chair at Heidelberg succeeding Leo Königsberger who was also a pupil of Weierstrass.

Figure 54.   Lazarus I. Fuchs

Figure 55.   Ernst F. F. Zermelo

In 1884 Fuchs succeeded Ernst Kummer at Berlin where he died in 1902.

Fuchs introduced an important class of linear differential equations in the complex domain with analytic coefficients, a class which is called the Fuchsian type.[3]

He died on 26 April 1902 in Berlin.

Ernst Friedrich Ferdinand Zermelo

Ernst Zermelo was born on 27 July 1871 in Berlin. After graduating from gymnasium he studied at Berlin, Halle and Freiburg. He was awarded a doctorate in 1894 at Berlin for his thesis *Untersuchungen zur Variationsrechnung*. He then became Max Planck's assistant. In 1897 he went to Göttingen where he completed his habilitation with his dissertation *Hydrodynamische Untersuchungen über die Wirbelbewegungen in einer Kugelfläche* in 1899.

He stayed in Göttingen for eleven years as a privatdozent performing significant work on the axiomatic set theory. In 1911 Göttingen awarded him 5000 Deutsche Marks for his major contributions to the set theory which enabled him rest to regain his health.

In 1910 he had been appointed to a chair at the University of Zürich. However, due to poor health he resigned his chair in Zürich in 1916 and moved to the Black Forest in Germany.[3]

Zermelo was appointed to an honorary chair at the University of Freiburg in 1921 but he renounced the chair in 1935.

He died on 21 May 1953 in Freiburg, Germany.

Leo Königsberger

Leo Königsberger was born on 15 October 1837 in Posen, Prussia (now Poznan, Poland). He became friendly with Lazarus Fuchs, who came from the same town. Königsberger was awarded his Ph.D. in 1860 at Berlin with his dissertation *De motu puncti versus duo fixa centra attracti.* His main research interest was on elliptic function influenced by his adviser Weierstrass.

After teaching three years at the Berlin Cadet Corps, he was appointed extraordinary professor at Greifswald in 1864. He was promoted to ordinary professor in 1866. In 1869 he moved to Heidelberg where he stayed six years. When he left Heidelberg for Dresden in 1875 his successor was Lazarus Fuchs.

After two years at TH Dresden he moved to Vienna in 1877, then returned to Heidelberg in 1884.

Despite many moves, he spent 36 years at Heidelberg. His colleagues at Heidelberg were Bunsen, Kirchhoff and Helmholtz.

When Königsberger was first appointed to a chair at Heidelberg, Sofia Kovalevskaya attended his lecture even though she was not permitted to be a formal student.

Königsberger died on 15 December 1921 in Heidelberg, Germany.

Figure 56. Leo Königsberger

Figure 57. Ferdinand G. Frobenius

Ferdinand Georg Frobenius

Georg Frobenius was born on 26 October 1849 in Berlin-Charlottenburg. He received his Ph.D. in 1870 at Berlin with his dissertation *De functionum analyticarum unius variab per series infin repraesentatione* supervised by Weierstrass.

After teaching for four years at a secondary school, he was appointed extraordinary professor at Berlin in 1874.

After only one year he was appointed ordinary professor at the Eidgenösische Polytechnikum Zürich in 1875 where he stayed for seventeen years.

Kronecker died on 29 December 1891 and Frobenius was appointed to his chair at Berlin in 1892, and he retired from there in 1916. He considered himself to be a scholar called to contribute to the knowledge of pure mathematics and he believed that applied mathematics belonged to the technical colleges. At Göttingen pure mathematics and applied mathematics were treated equally. Consequently Frobenius disliked Göttingen mathmaticians, Felix Klein in particular.

In 1892 Frobenius was elected to the Prussian Academy of Science. By studying the different representation of groups and their elements, he provided a firm basis for the solution of general problems in the theory of finite groups.[5]

The rank of a matrix was first defined in 1878 by Frobenius. He used matrices in his work on group representations. The power series approach in ordinary differential equations is called the Method of Frobenius, and the series solutions obtained are Frobenius series.

Issai Schur and Edmund Landau were the most important students of Frobenius.

Frobenius died on 3 August 1917 in Berlin.

Issai Schur

Issai Schur was born on 10 January 1875 in Mogilyov 180km east of Minsk, Belarus. After attending the gymnasium in Latvia, he entered the University of Berlin. He was awarded his Ph.D. in 1901 with his thesis *Über eine Klasse von Matrizen die sich einer gegeberen Matrix zuordnen lassen* supervised by Frobenius.

Schur was a privatdozent at Berlin from 1903 to 1909. From 1913 to 1916 he was an extraordinary professor at Bonn. In 1916 he moved back to Berlin and was promoted to ordinary professor. He succeeded Friedrich Schottky in 1921, and held this chair until the Nazis forced him to retire in 1935.

He was elected to the Prussian Academy of Science on Max Planck's recommendation but he was forced to resign from the Academy in 1938.

Schur is best known for his work on the representation theory of groups and Schur's Lemma.[3]

Students of Linear Algebra know that Schur decomposition of matrix $A = PTP^H$ where P is unitary and T is an upper-triangular matrix. And P^H is the Hermitian transpose of P.

His colleagues at Berlin were Ludwig Bieberbach, Erhard Schmidt and Richard von Mises. Bieberbach became a notorious Nazi and Schmidt collaborated with the Nazis. Schur and von Mises were dismissed because they were Jewish.

Schur was an excellent lecturer and supervised many outstanding doctoral candidates in Berlin. He left Germany for Palestine in 1939 suffering the final humiliation of having to find a sponsor to pay the Reichs flight tax that allowed him to leave Germany. He was forced to sell his beloved books to the Institute for Advanced Study in Princeton.[3]

Schur died on 10 January 1941 in Jerusalem, Palestine.

Figure 58. Edmund Landau

Figure 59. Issai Schur

Edmund Landau

Edmund Landau was born into a wealthy Jewish family on 14 February 1877 in Berlin. He received his Ph.D. in 1899 at the University of Berlin for a thesis on number theory supervised by Frobenius. He submitted his habilitation thesis on Dirichlet series, a topic in analytic number theory.[3] He taught at Berlin until 1909.

Minkowski died in January 1909 and Göttingen considered Oskar Perron and Edmund Landau as his successor. Klein preferred Landau because "it is better if we have a man who is not easy."[9] Despite his outstanding talents as both a teacher and researcher, Landau annoyed many of his colleagues with his arrogant manner. For example, if someone asked for his address in Göttingen, he replied "You'll find it easily; it is the most splendid house in the city."[3]

Landau's colleagues from 1909 to 1934 were Hillbert, Carathéodory, Hecke, Courant and Herglotz, Klein and Runge.

After the Nazis came to power in January 1933, Landau's lectures were boycotted by student members of the Sturmabteilung (storm troopers SA) led by Oswald Teichmüller (1913-1943). After receiving his Ph.D at Göttingen, Teichmüler was killed in action on the Russian front in 1943.

Landau published over 250 papers on number theory and his masterpiece of 1909 was the treatise *Handbuch der Lehre von der Verteilung der Primzahlen* a two volume book giving the first systematic presentation of the analytic number theory.

Landau's wife was Marianne, daughter of Paul Ehrlich, who was the 1908 Nobel laureate in medicine and physiology.

Landau officially retired on 7 February 1934. He received full pay until 1 July 1934, then a pension until he died of a heart attack in Berlin on 19 February 1938. Landau's successor in Göttingen was a Nazis party member Erhard Tornier (1894-1982).

Konrad Hermann Theodor Knopp

Konrad Knopp was born on 22 July 1882 in Berlin, Germany. He studied at the University of Berlin and received his doctorate in 1907 for his thesis *Grenzwerte von Reihen bei der Annäherung an die Konvergenzgrenz* supervised by Frobenius and Schottky.

Knopp went to Japan and taught at the Nagasaki College of Commerce during 1908-09. In 1910 he returned to Germany and married the painter Gertrud Kressner. They had one son and one daughter. The couple moved to Tsingtao (Qingdao) and taught at the German-Chinese Academy during 1910-11.

In 1911 they returned to Germany and he taught at the Military Technical Academy and Military Academy and also received his habilitation at the University of Berlin. Knopp was an officer in the German Army during World War I. He was wounded at the beginning of the war, resulting in his discharge from the German Army.

He was appointed extraordinary professor at the University of Königsberg in 1915 and became an ordinary professor there in 1920.

In 1926 he accepted a chair (Erster Lehrstuhl) at the University of Tübingen succeeding Ludwig Maurer, Knopp retired from there in 1950.

Knopp was the cofounder of *Mathematische Zeitschrift* in 1918, and was the editor from 1934 to 1952. He wrote excellent books on complex functions. *Theorie und Anwendung der unendlichen Reihen* was published in 1922 and *Elemente der Funktionentheorie* was published in 1936.[3]

He also produced the sixth edition of Hans von Mangoldt's (1854–1925) famous three volume book *Höhere Mathematik: eine Einführung für studierende und zum Selbststudium*. The book has since then appeared as jointly authored by von Mangoldt and Knopp and has been popular with generations of German students in mathematics, physics and engineering.

Knopp died on 20 April 1957 in Annecy, France.

Figure 60.   Konrad H. T. Knopp

George G. Lorentz

George Lorentz was born on 25 February 1910 in St. Petersburg, Russia. Lorentz attended a Russian secondary school and later a German school, thereby becoming fluent in German. In 1926 he entered Tiflis Institute of Technology in Gruzia (Georgia). Two years later he transferred to Leningrad State University, where he received his diplom in mathematics in 1931. In 1936 Lorentz became a dozent at Leningrad State University. The Leningrad Mathematical Society was disbanded because the communist government believed that mathematics was too theoretical. His father was arrested on false charges in 1937 and died in a concentration camp in 1938. This tragedy interrupted his studies for about five years.

During World War II Lorentz and his newly wedded wife Tanny escaped Leningrad and arrived at a refugee camp at Torun, Poland. The Lorentzes were registered as ethnic Germans.

In 1943 Lorentz submitted two mathematical papers to Professor Knopp for possible publication in *Mathematische Zeitschrift*. With the help of Wilhelm Süss, Knopp was able to relocate the Lorentz family, including their new-born son Rudolph to Tübingen.

Lorentz received his doctorate at Tübingen in 1944 with his thesis *Eine Fragen der Limitierungs theorie* under supervision of Knopp.

He then moved to the University of Toronto in 1949 where he was able to supervise doctoral students even though his title was instructor. In 1953 he moved to Wayne State University and in 1958 to Syracuse University.

In 1968 he was appointed professor at the University of Texas Austin from where he retired in 1980.

He became a United States citizen in 1959. His son Rudolph became a professor of mathematics at the University of Duisburg, Germany.

Lorentz wrote 130 papers and five books. His hobbies include chess and stamp collecting.

He died on 1 January 2006 in Chico, California.

Figure 61. Paul L. Butzer

Figure 62. Paul I. Bernays

Paul L. Butzer

Paul Butzer was born on 15 April 1928 in Muelheim, Ruhr, Germany. Because of his father's anti-Nazi stand, his family left Germany in 1937, settling first in England. From 1941 to 1955, Paul Butzer lived in Montreal, Canada. He studied at Loyola College (now Concordia University) and the University of Toronto. He was naturalized as a Canadian citizen in 1948. Butzer received his Ph.D. in 1951 at the University of Toronto for his thesis *On Bernstein Polynomials* under supervision of George G. Lorentz.

After teaching at McGill University, he returned to Germany in 1955 where he worked at the Universities of Mainz, Freiburg and Würzburg. He was appointed at the Rheinisch-Westfälische Technische Hochschcele (Aachen University of Technology) in 1958, becoming full professor in 1962 and Emeritus Professor in 1993. He is the author of some 270 research papers and six books or monographs, the editor of fifteen conference proceedings, as well as associate editor of a dozen mathematical journals located in six countries. He is a Corresponding/Associate Fellow/Honorary Member of the Societe Royale des Sciences de Liege, the Academie Royale de Belgique, the Mathematische Gesellschaft in Hamburg, and Honorary Professor at the University of Nanjing, China. He has received honorary doctorates from the Universities of Liege, York (United Kingdom) and Timisoara (Romania). He has had 35 doctoral students and 32 of his former students became professors at universities in Germany, Austria and Greece.

Paul Isaac Bernays

Paul Bernays was born on 17 October 1888 in London, England to Swiss parents. He studied at TH Charlottenburg (now TU Berlin), University of Berlin and Göttingen where he received his doctorate in 1912. His dissertation was *Über die Darstellung von positiven, ganzen Zahlen durch die primitiven, binären quadratischen Formen einer nicht-quadratischen Diskriminante* supervised by Landau.

Bernays was at the University of Zürich as Zermelo's assistant until 1917 when Hilbert offered him a position as his assistant. Bernays received two habilitations at Zürich and Göttingen with his thesis on the completeness of propositional logic in 1919. He was then appointed *Nicht-Beamte Ausserordentlichen Professor* (non-civil service associate professor) in 1922 on Hilbert's recommendation.

After Hitler came to power in 1933 Bernays lost his position and moved to Switzerland because he was a Swiss citizen. He is best known for his two volume work *Grundlagen der Mathematik* (1934–1939) with Hilbert. He also revised Hilbert's *Grundlagen der Geometrie* many times. The tenth edition was published in Zürich. Bernays also published Axiomatic Set Theory in 1958.

Bernays died on 18 September 1977 in Zürich, Switzerland.

Harald August Bohr

Harald Bohr was a younger brother of Niels Bohr (1885–1962). Harald was born on 22 April 1887 in Copenhagen, Denmark. Niels and Harald's mother came from a wealthy Jewish family. Harald had been a member of Denmark's soccer team which placed second in the 1908 Olympic games in London.

He obtained his Ph.D. in 1910 at Kobenhavns Universitet with his thesis *Bidrag til de Dirichletske Rækkers Theori* supervised by Edmund Landau. In 1914 they proved the Bohr–Landau theorem on the distribution of zeros of the Zeta function. They proved that all but an infinitesmal proportion of the zeros of the Zeta function lie in a small neighborhood of the line $s = \frac{1}{2}$.

Bohr became professor of mathematics at the Polytechnic Institute in Copenhagen in 1915 and in 1930 he moved to the University of Copenhagen.

Figure 63. Harald A. Bohr   **Figure 64. Carl L. Siegel**

For most of his life Bohr was a sick man who suffered from terrible headaches. He was a man of refined intellect and a very humane person, always eager to help his pupils, colleagues and friends.

He died on 22 January 1951 in Copenhagen, Denmark.

Carl Ludwig Siegel

Carl Siegel was born on 31 December 1896 in Berlin. He studied at Berlin attending lectures by Frobenius and Planck. During World War I he was called to military service. After the war he studied at Göttingen where he received his Ph.D. in 1920 with his thesis *Approximation algebraischer Zahlen* supervised by Landau.

Siegel was a privatdozent at Göttingen from 1921 to 1922. When Arthur Schonflies (1853-1928) retired from Frankfurt in 1922 Siegel succeeded him at the age of 26. Siegel's colleagues at Frankfurt were Max Dehn, Ernst Hellinger, Otto Szasz and Paul Epstein. So Frankfurt became one of the most productive mathematics research centers in Germany together with Göttingen and Berlin. A mathematics seminar initiated by Max Dehn in 1922 continued until 1935. They made a rule that they would study all the mathematical works in their original languages.

The number of Ph.Ds awarded in the Faculty of Natural Sciences and Mathematics at Frankfurt are as follows[1]:

	Natural Sciences	Mathematics
1915-1920	60	4
1921-1925	303	7
1926-1930	267	9
1931-1935	240	8
1936-1940	248	6
1941-1945	115	1
1946-1950	80	3
1951-1955	259	9
1956-1960	342	4

In the autumn of 1935, Epstein, Hellinger and Dehn were dismissed from their chairs and Szasz had been removed earlier. Although Siegel was Aryan, he despised the Nazis regime.

Siegel moved to Göttingen in 1938 succeeding Rolf Nevanlinna. In 1940 Siegel emigrated to the United States. He was at the Institute for Advanced Study at Princeton from 1940 until 1951, being appointed to a permanent professorship there in 1946.[3]

In 1951 he returned to Göttingen for the rest of his career. He never married, and devoted his life to research. Siegel criticized Serge Lang's (1927-2005) book *Diophantine geometry* published in 1962, because Siegel's own contribution to the subject had altered beyond recognition. Louis Mordell (1888-1972) also wrote a critical review of the book in 1964. Siegel received the Wolf Prize in 1978 for his contributions to the theory of number, the theory of several complex variables, and celestial mechanics.

Siegel died on 4 April 1981 in Göttingen and was buried in the Göttingen Stadtfriedhof where Hilbert, Klein, Schwarzschild were also buried.

Felix Hausdorff

Felix Hausdorff was born into a wealthy Jewish family on 8 November 1868 in Breslau, Germany (now Wroclaw, Poland). The family moved to Leipzig when Felix was still young. Hausdorff received his Ph.D. in 1891 at the University of Leipzig with his thesis *Zur Theorie der astronomischen Strahlenbrechung* (On The Theory of Astronomical Refraction) supervised by Ernest Burns (1848-1919).

He submitted his habilitation thesis *On The Absorption Of Light In The Atmosphere* to Leipzig in 1895 and served as a privatdozent until 1901. He married Charlotte Sara Goldschmidt in Leipzig in 1899.[3] In 1901 he was appointed extraordinary professor at Leipzig. He left for Bonn in 1910 to be an extraordinary professor, working with Eduard Study (1862-1930). From 1913 to 1921 he was ordinary professor at Greifswald. He then moved back to Bonn in 1921 where he stayed until he was forced to give up his chair in 1935.

Hausdorff was also interested in literature and philosophy. He published a play *"The Doctor's Honour"* and a collection of poems entitled *Ecstasies* as well as a book of aphorisms.[4] He was an excellent pianist and occasionally composed songs. In fact he had once considered becoming a composer but his father discouraged him.

In 1914 he published his famous text *Grundzuge der Mengenlehre* (Basic Features Of Set Theory) which won him world wide recognition.

Hausdorff is given credit for laying the foundations of set-theoretic topology and the Hausdorff separation axiom. He introduced and investigated a class of measures and a type of dimension that may assume fractional values. These important concepts are known as Hausdorff measure and Hausdorff dimension.[4]

One day after his seventieth birthday on 9 November 1938 a mob came to his house shouting "We are going to send you to Madagascar where you can teach mathematics to the apes." He was deeply shocked.[4]

In January 1942 Hausdorff's family was ordered to move to a monastery in the suburb of Endenich. Felix, his wife Charlotte and his wife's sister, Edith all took barbiturates on 25 January 1942. Felix and Charlotte died on 26 January and Edith, died after a few days. The family had lived on Hindenburgstrasse which is now called Hausdorffstrasse in Bonn.[4]

Wilhelm Karl Joseph Killing

Wilhelm Killing was born on 10 May 1847 in Burbach (near Siegen), Westphalia. He studied at Münster and Berlin where he received his Ph.D. in 1872 with his dissertation *Der Flächenbüschel zweiter Ordnung* (Bundles of surfaces of the second degree) supervised by Weierstrass.[3]

In 1875 he married Anna Conmer. He taught at several gymnasiums in Berlin and Brilon. In 1882 he moved to the Catholic Academy in Braunsberg which is now called Braniewo, Poland located between Danzig and Königsberg. He taught there for ten years. During this period he produced some of the most original mathematics ever produced. He published a book on non-euclidean geometry in Leipzig in 1885. Killing introduced *the Lie algebra in Programmschrift* in 1884 in Braunsberg without knowing Lie's work.

Figure 65. Felix Hausdorff   **Figure 66. Wilhelm K. J. Killing**

Killing eventually met Sophus Lie (1842–1899) and Friedrich Engel (1861–1941) in Leipzig in the summer of 1886 to discuss the simple Lie algebra. Killing also introduced the term "characteristic equation" of a matrix.[3]

In 1892 he was appointed to a chair at the University of Münster from where he retired in 1919. He was Rector of the University of Münster from 1897 until 1898. Killing was honored with the Lobachevsky Prize in 1900.

He is best remembered by the Killing vector and the Killing equation in tensor analysis.

He died on 11 February 1923 in Münster, Germany.

Sofia (Sonja) Vasilyevna Kovalevskaya

Sofia Kovalevskaya was born on 15 January 1850 in Moscow, Russia. Her parents were Vasily Korvin-Krukovsky, an artillery general, and Velizaveta Schubert.[3] Russian middle names show father's first name. She married Vladimir Onufrievich Kovalevski when she was 18 years old because she wanted to go abroad but her father would not allow it without a husband.

When Sofia was eleven years old, her room was covered with pages of Ostrogradski's calculus lecture notes. In 1869 Sofia traveled to Heidelberg where two professors of mathematics taught. Leo Königsberger was ordinary professor just appointed in 1869 and Paul DuBois-Reymond was extraordinary professor appointed one year earlier.

Physicist Kirchhoff, chemist Bunsen and physiologist Helmholtz were also at Heidelberg at that time.

In 1871 she moved to Berlin where Weierstrass, Kummer and Kronecker lectured. She was not allowed to attend lectures, because she was a woman so Weierstrass tutored her privately.

By the spring of 1874, she had completed three papers on partial differential equations, Abelian integrals and Saturn's Rings. Weierstrass considered each of these papers worthy of a doctorate. In 1874 she was awarded her Ph.D. in absentia at the University of Göttingen. This demonstrates that Göttingen was more liberal than Berlin with regard to female academics.

Despite a strong recommendation from Weierstrass, Kovalevskaya was unable to obtain an academic position. She moved back to Russia with her husband and had a daughter Sophia Vladimirovna, nicknamed "Foufie" in 1878.[5]

In the spring of 1883, Sofia's husband, from whom she had been separated for two years, committed suicide.[3] His business partners were swindlers who had put Vladimir deeply into debt.

Gösta Mittag-Leffler (1846-1927) was able to appoint Kovalevskaya privatdozent at the new Hogskola in Stockholm[4] in early 1884. She was appointed extraordinary professor in June 1884. She and Mittag-Leffler's sister Anne-Charlotte wrote a play together, *The Struggle for Happiness*, which was favorably received when it was performed in Moscow.[4]

She became an editor of the new journal *Acta Mathematica*. In 1888 she won the prestigious Bordin prize awarded by the French Academy of Sciences for her paper on the motion of a rigid body. This brought her international fame and as a result she was promoted to full professor in 1889.

In the same year, on the initiative of Pafnuty Chebyshev (1821-1894), she was elected a corresponding member of the Russian Imperial Academy of Sciences. At the height of her career, Kovalevskaya died of influenza complicated by pneumonia on 10 February 1891 in Stockholm, Sweden.

The existence and uniqueness of quasi-linear partial differential equations with appropriately specified boundary conditions is known as the Cauchy-Kovalevski theorem.

Figure 67.   Carl D. T. Runge

Figure 68.   Sofia V. Kovalevskaya

Carl David Tolmé Runge

Carl Runge was born on 30 August 1856 in Bremen, Germany. He was awarded a doctorate in 1880 from Berlin with his thesis *Über die Krümmung, Torsion und geodätische Krümmung der auf einer Fläche gezogenen Curven* supervised by Weierstrass.[10] Runge was a privatdozent at Berlin from 1883 to 1886.

In 1886 he was appointed ordinary professor at TH Hannover where he stayed eighteen years. In 1904 he was appointed to a newly created chair (Vierter Lehrstuhl) at Göttingen from where he retired in 1924. His colleagues at Göttingen were Klein, Minkowski, Landau and Hilbert.

Runge also studied the wavelengths of the spectral lines of elements other than hydrogen. After he moved to Göttingen one of his students were Max Born.

Runge is well known to students due to the Runge-Kutta method for solving ordinary differential equation by numerical analysis. Runge and Kutta never worked together. Wilhelm Kutta (1867-1944) introduced the method in his doctoral dissertation *Beiträge Zur näherungsweisen Integration totaler Differentialgleichungen* in 1900 at the University of Munich (München). His advisers were Ferdinand Lindemann and Gustav Bauer. He taught at TH München, Jena, Aachen and Stuttgart. Runge's daughter Nerina (Nina) married Richard Courant on 22 January 1919. Runge died of a heart attack on 3 January 1927 in Göttingen. His house was on Wilhelm Weber Strasse in Göttingen between houses of Felix Klein and David Hilbert.

Karl Hermann Amandus Schwarz

Hermann Schwarz was born on 25 January 1843 in Riesengebirge, Germany. After studying at a gymnasium in Dortmund he enrolled at the University of Berlin. He received his Ph.D. there in 1864 for his thesis *De superficiebus in planum explicabilibus primorum septem Ordinum*. After completing his habilitation thesis in 1867 he was appointed extraordinary professor at Halle-Wittenberg where he stayed for two years. From 1869 to 1875 he was at the Eidgenosische Technische Hochschule in Zürich. Then in 1875 he succeeded Lazarus Fuchs at Göttingen. Schwarz was appointed to the chair of Weierstrass in 1892. In the same year he was elected to the Berlin Academy of Sciences.

Schwarz's colleagues at Berlin were Frobenius, Fuchs and Schottky. Schwarz failed to produce significant mathematical research after his move to Berlin, and Göttingen became the most eminent German university for mathematics. Schwarz retired in 1917 and his successor was Erhard Schmidt.

Schwarz married Kummer's daughter. He was the captain of the local voluntary fire brigade and he assisted at the railroad station by closing train doors.[3]

Schwarz made an important contribution in 1865 when he discovered what is now known as the Schwarz minimal surface.[3] He also worked on conformal mapping. The Cauchy–Schwarz inequality is well known to students in mathematics. The integral form has the form

$$\left| \int_a^b f^*(x)\, g(x)dx \right|^2 \leq \int_a^b f^*(x) f(x)dx \int_a^b g^*(x) g(x)dx$$

In linear vector space

$$\|(a \cdot b)\| \leq \|a\| \|b\|$$

Schwarz died on 30 November 1921 in Berlin, Germany.

Gustav Herglotz

Gustav Herglotz was born on 2 February 1881 in Wallern, Bohemia (now Volary, Czech Republic), but spent most of his childhood in Vienna. There he formed a close friendship with three other students Heinrich Tietze (1880–1964), Hans Hahn (1879–1934) and Paul Ehrenfest (1880–1933). They became known as the "inseparable four". Herglotz studied astronomy at München where he received his Ph.D. in 1902 with his thesis *Über die Scheinbaren Helligkeitsverhältinisse eines planetarischen Körpers mit drei ungleichen Hauptträgheitsachsen.*[3] He was advised by Hugo von Seeliger (1849–1924) and Ludwig Boltzmann (1844–1906).

Herglotz submitted his habilitation thesis at Göttingen in 1904 to become a privatdozent until 1907. After working one year as an extraordinary professor at TH Wien he was appointed ordinary professor at Leipzig in 1909 where he stayed for sixteen years. At Leipzig, Herglotz supervised at least 25 doctoral candidates including Emil Artin.

In 1925 Herglotz succeeded Carl Runge at Gottingen from where he retired in 1948. Saunders Mac Lane described his experience with Herglotz.[86]

He had a vast knowledge of classical applied mathematics and classical analysis. He delivered stunningly beautiful lectures on mechanics, mathematical optics, functions with positive real parts, and Lie groups.

Physicist Max Born came to Göttingen as professor of physics in 1921 with his wife Hedwig (Hedi), a poet. Around 1925 Max Born became too involved in quantum mechanics and neglected his wife. She had an affair with Herglotz or Gusti as Hedi Born referred to him. Gossip about Hedi and Herglotz was swirling around Göttingen. Hedi wanted to leave Max Born and marry Herglotz.

She complained that Max reduced her to a formula and did not understand her. The affair lasted eight years until the Borns were forced by the Nazis to leave Göttingen for England.[15]

Herglotz suffered a stroke in 1946 and retired in 1948. One week before he died Hedi Born went to see him but he did not recognize her.

Herglotz died on 22 March 1953 in Göttingen, Germany.

Figure 69. Hermann A. Schwarz

Figure 70. Gustav Herglotz

Emil Artin

Emil Artin was born on 3 March 1898 in Vienna, Austria. His father was an art dealer and his mother was an opera singer. Artin was an accomplished musician playing the flute and harpsichord.[3]

His study at the University of Vienna was interrupted by the World War I and he served in the Austrian Army until the end of the War. He entered the University of Leipzig in 1919 and received his doctorate in 1921 under the supervision of Gustav Herglotz. His dissertation was *Quadratische Körper im Gebiete der höheren Kongruenzen*.[10] He had his habilitation in 1923 at the University of Hamburg where he became an ordinary professor in 1926 at the age of 28.

Artin married Natalie Jasny, one of his students, in 1929. He had to leave Germany for the United States in 1937 because his wife was half-Jewish.

Artin made contributions to solve the Hilbert problems No. 9, 11, and 12 and he solved the seventeenth problem.[12] He was most productive in the ten year period 1921-1931. Artin taught eight years at Indiana University at Bloomington from 1938 to 1946, twelve years at Princeton University from 1946 to 1958 and then moved back to the University of Hamburg in 1958.

His main fields of contribution were class field theory, the theory of braids and Artin rings. Artin supervised 34 Ph.D. candidates (13 in Hamburg, 2 in Indiana, 1 in Columbia and 18 in Princeton). David Gilbarg, John Tate, Jr., Max Zorn, Serge Lang and Hans Zassenhaus are some of his students.

Artin wore an black leather jacket or sometimes a long winter coat belted at the waist. Some students emulated him in dress and manner at Princeton.[12]

His son, Michael Artin became a distinguished mathematician serving as President of the American Mathematical Society in 1991-1992.

Artin died on 20 December 1962 in Hamburg, Germany.

Figure 71. Emil Artin

Figure 72. Theodor F. E. Kaluza

Theodor Franz Eduard Kaluza

Theodor Kaluza was born on 9 November 1885 in Opeln, Germany (now Opole, Poland). He received his Ph.D. at Königsberg in 1907 supervised by Wilhelm Franz

Meyer (1856–1934). Two years later he submitted his habilitation thesis on Tschirnhaus transformations and worked as a privatdozent until 1922.

Normally after one or two years as privatdozent one is appointed extraordinary or ordinary professor. In Kaluza's case, however, it took 13 years for him to be promoted to extraordinary professor at Königsberg. In April 1921, encouraged by Einstein he published an original paper to unify Einstein's theory of gravity and Maxwell's electromagnetic theory. Kaluza's ideas involved the introduction of a fifth dimension.[3] This theory is now known as Kaluza–Klein field theory named after Oskar Klein (1894–1977).

Kaluza was appointed ordinary professor at Kiel succeeding Ernst Steinitz in 1929. In 1935 he moved to Göttingen succeeding Richard Courant and he retired from there in 1954. Kaluza and G. Joos published the textbook *Höhere Mathematik für die Praktiker* in 1938.

He died on 19 January 1954 in Göttingen, Germany.

French School

Gaspard Monge (1746–1818) ⟶ Sylvestre Lacroix (1765–1843)
⟶ Victor Poncelet (1788–1867)
⟶ Jean-Baptiste Biot (1774–1862)

André Marie Ampère (1775–1836) ⟶ Jacques-Charles-Francois Sturm (1803–1855)

Adrien-Marie Legendre (1752–1833)

Marc-Antoine Parseval des Chênes (1755–1836)

Pierre-Simon Laplace (1749–1827)

Josef Maria Hoëne-Wronski (1776–1853)

Augustin Louis Cauchy (1789–1857)

Evariste Galois (1811–1832)

Joseph-Alfred Serret (1819–1885)

Jean Frederic Frenet (1816–1900)

Charles Hermite (1822–1901) ⟶ Henri Poincaré (1854–1912)
1879 Ph.D. Sorbonne
⟶ C.Emile Picard (1856–1941)

Camille Jordan (1838–1922)   1877 Ph.D. Sorbonne
1861 Ph.D. Sorbonne

Felix-Edouard-Justin-Emile Borel (1871–1956)
1893 Ph.D. Sorbonne

Henri Lebesgue (1875–1941)
1902 Ph.D. Sorbonne

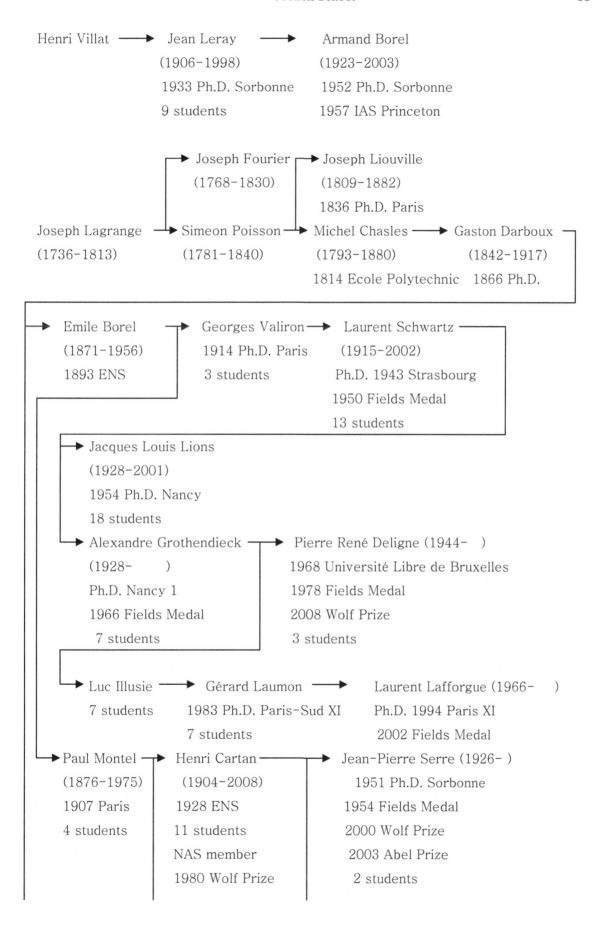

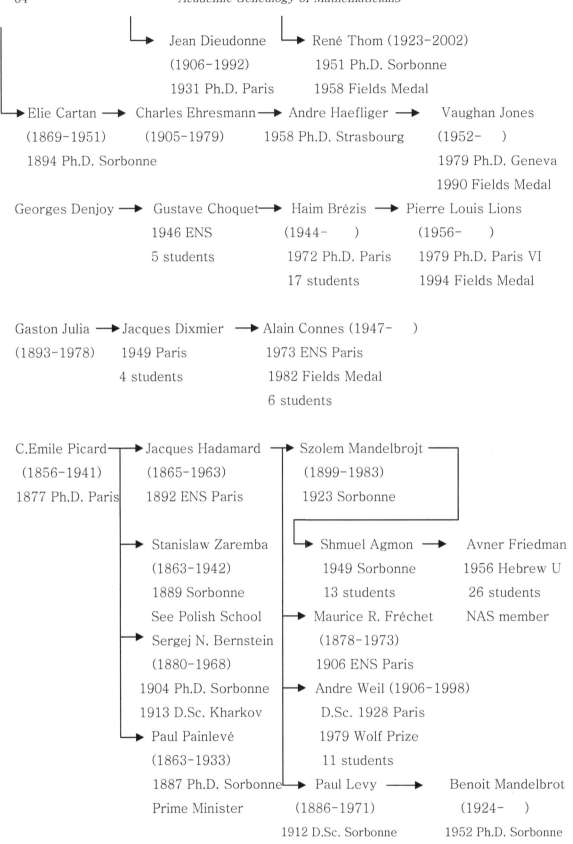

Jean Dieudonne
(1906-1992)
1931 Ph.D. Paris

René Thom (1923-2002)
1951 Ph.D. Sorbonne
1958 Fields Medal

Elie Cartan   Charles Ehresmann   Andre Haefliger   Vaughan Jones
(1869-1951)   (1905-1979)   1958 Ph.D. Strasbourg   (1952-   )
1894 Ph.D. Sorbonne   1979 Ph.D. Geneva
                                                                1990 Fields Medal

Georges Denjoy   Gustave Choquet   Haim Brézis   Pierre Louis Lions
                     1946 ENS   (1944-   )   (1956-   )
                     5 students   1972 Ph.D. Paris   1979 Ph.D. Paris VI
                                        17 students   1994 Fields Medal

Gaston Julia   Jacques Dixmier   Alain Connes (1947-   )
(1893-1978)   1949 Paris   1973 ENS Paris
                     4 students   1982 Fields Medal
                                        6 students

C.Emile Picard   Jacques Hadamard   Szolem Mandelbrojt
(1856-1941)   (1865-1963)   (1899-1983)
1877 Ph.D. Paris   1892 ENS Paris   1923 Sorbonne

                     Stanislaw Zaremba   Shmuel Agmon   Avner Friedman
                     (1863-1942)   1949 Sorbonne   1956 Hebrew U
                     1889 Sorbonne   13 students   26 students
                     See Polish School   Maurice R. Fréchet   NAS member
                     Sergej N. Bernstein   (1878-1973)
                     (1880-1968)   1906 ENS Paris
                     1904 Ph.D. Sorbonne   Andre Weil (1906-1998)
                     1913 D.Sc. Kharkov   D.Sc. 1928 Paris
                     Paul Painlevé   1979 Wolf Prize
                     (1863-1933)   11 students
                     1887 Ph.D. Sorbonne   Paul Levy   Benoit Mandelbrot
                     Prime Minister   (1886-1971)   (1924-   )
                                        1912 D.Sc. Sorbonne   1952 Ph.D. Sorbonne
                                1920-59 Prof. Ecole Polytechnique   IBM Fellow

Michael Herman   ⟶   Jean Christoph Yoccoz (1957–)

1976 Ph.D. Paris XI Orsay   1985 Ph.D. Paris XI Orsay

3 students   1994 Fields Medal

Marc Yor ⟶   Jean-Francois Le Gall ⟶   Wendelin Werner (1968–)

     1982 Ph.D.   1993 Ph.D. Univ Paris VI

     Prof Paris–Sud Orsay   2006 Fields Medal

French School

Mathematicians in France dominated the world of mathematics in the first half of the nineteenth century: Lagrange, Laplace, Legendre, Monge, Ampere, Fourier, Poisson, Galois, Cauchy and Liouville.

Laplace, author of *Mécanique Céleste*, was called the Newton of France. Lagrange was with the Berlin Academy of Sciences for 21 years. Even Riemann studied with Legendre's *"Theory of Numbers"*.

Unlike German mathematicians, most of these French mathematicians were not professors at universities and France did not confer doctorates in mathematics until the middle of the 19th century.

It is therefore very difficult to research teacher-students relationship among French mathematicians, and many of them studied on their own.

Dirichlet, although German, studied in Paris rather than in Germany. He did not receive any degrees in Paris, but he attended lectures by Laplace, Fourier, Legendre, Poisson and others at the *Collége de France*. It was and remains a unique institution, where anyone could attend lectures by renowned scholars without being enrolled in a degree program. Professors offer different courses every year, never repeating the same courses.

Most *grandes écoles* such as *Ecole Normale Supérieure* and *Ecole des Mines* do not grant doctorates. Students at *grandes ecoles* receive degrees at universities. However, *Ecole Polytechnique* which is under jurisdiction of the Ministry of Defense, recently began to award doctorates.

Joseph-Louis Lagrange

Joseph-Louis Lagrange was born on 25 January 1736 in Turin, Sardinian-Piedmont (now Torino, Italy). He was baptized Giuseppe Lodovico Lagrangia.[3]

Lagrange studied mathematics on his own and communicated his results with Leonhard Euler (1707-1783). Lagrange was appointed professor of mathematics at the Royal Artillery School in Torino on 28 September 1755.[3] He was elected to the Prussian Academy of Sciences in Berlin in 1756 on Euler's recommendation.

Lagrange succeeded Euler as Director of Mathematical Physics at the Prussian Academy of Sciences in Berlin in November 1766. In 1767 he married a cousin, Vittoria Conti, and she died just sixteen years later.

He published *Mécanique Analitique* (Analytical Mechanics) in 1788, the application of calculus to the motion of rigid bodies.

After moving to Paris in May 1787 to be a member of the *Academie des Sciences*, he married Renée-Francoise-Adelaide Le Monnier in 1792.

During the French Revolution, his colleague Antoine Laurent Lavoisier was guillotined on 8 May 1794 but Lagrange was saved.

He was appointed professor of analysis at the new *École Polytechnique* in March 1794 but he was not a good lecturer.

When Laplace presented his book *"Mécanique Celeste"* to Napoleon, he was asked by Napoleon why the creator of the universe was not mentioned in it. Laplace replied that he did not need that hypothesis. Napoleon told Lagrange of Laplace's response, and Lagrange replied that it was a beautiful hypothesis that explained many things.

Napoleon made Lagrange a Count of the Empire in 1808 and named him to the Legion of Honour. He died on 10 April 1813 in Paris and buried in the Panthéon.

Figure 73. Joseph-Louis Lagrange

Figure 74. Pierre-Simon Laplace

Pierre-Simon Laplace

Pierre-Simon Laplace was born on 23 March 1749 in Beaumont-en-Auge, Normandy, France. Jean le Rond d'Alembert (1717-1783) was impressed by Laplace's mathematical capability. At just 19 years of age he was appointed professor of mathematics at the *École Militaire* in Paris, teaching elementary mathematics to the cadets.

After two unsuccessful attempts, Laplace was elected an adjoint in the *Acádemie des Sciences* in Paris in 1773. In May 1788, Laplace married Marie-Charlotte de Courty de Romanges who was 20 years younger than him.[3]

In 1785 he examined and passed the 16 year old Napoleon Bonaparte at the Royal Artillery Corps. In 1799 Napoleon, then the First Consul of France, appointed Laplace minister of interior.

But Napoleon dismissed him after only six weeks because he brought "the spirit of the infinitely small into the government"[3]. Apparently he was not a good administrator.

As head of the *Bureau de Longitudes* and the Paris Observatory, he would neglect all the observation except those needed for his formulas.[3]

His great work was the *Traité du Mécanique Céleste* published in five volumes. He presented the first two to Napoleon in 1799. After glancing at it Napoleon asked him why there was no mention of the creator of the universe. Laplace replied that he did not need that hypothesis.

In 1814, Laplace, who had been appointed to the Senate by Napoleon, voted to overthrow Napoleon in favor of a restored Bourbon monarchy. Consequently he became unpopular among political circles.

However, after the restoration of the Bourbons, he was made a marquis. In 1826 he became very unpopular with the liberals for refusing to sign the declaration of the *Academie Française* which supported freedom of the press.

Soon after he was appointed to the *École Militaire*, he discovered the Laplace theorem which states that the sum of the products of the elements of any row (or column) by their respective cofactors is equal to the determinant.

In 1812 he published *Théorie Analytique des Probabilities* which made him one of the founders of probability theory.

He is also credited with the discovery of the orthogonalization procedure although it is known as Gram-Schmidt method. The Laplace transform introduced in his papers on probability became a fundamental tools of analysis.

He considered himself the best mathematician in France and after he became successful he ignored his parents and elder sister.[4]

He died on 5 March 1827 in Paris, France.

Adrien-Marie Legendre

Adrien-Marie Legendre was born on 18 September 1752 in Paris, France. After graduating from the *Collège des Quatre Nations* (a secondary school) he taught at the

École Militaire in Paris until 1780. While studying the attraction of ellipsoids he introduced what we call the Legendre functions today. The general solution of Laplace's equation in spherical coordinates results in a form of Legendre's equation, whose solutions are called Legendre functions.

In 1783 he was appointed an adjoint in the *Académie des Sciences* and two years later to a position as an associate. In 1787 he was elected a Fellow of the Royal Society of London.

In 1794 Legendre published *Eléments de géométrie* which became the leading text on geometry for the next 100 years.

He published *Théorie des nombres* in 1798, then a much improved second edition in 1808, and a third edition in 1830. His other major work was on elliptic functions.

Although Legendre spent 40 years working on elliptic functions, but Carl Jacobi and Niels Abel's works made Legendre obsolete.[3]

Legendre published the method of least squares in 1806. But Gauss had referred to his early discovery of the method in the *Theoria motus*. With great indignation Legendre wrote to Gauss accusing him of dishonest and complaining that Gauss might have had the decency not to appropriate the method of least squares.

Unlike Legrange, Laplace and Monge, Legendre was not ennobled. When Neils Abel was in Paris in 1826, Abel visited Legendre.

After a long and painful illness, Legendre died on 10 January 1833 in Paris, leaving no descendants.

Figure 75. Adrien-Marie Legendre

Figure 76. Gaspard Monge

Gaspard Monge

Gaspard Monge was born on 9 May 1746 in Beaune, Bourgogne, France.

After graduating the *College de la Trinité* in Lyon (a secondary school) in 1764, he was appointed as a draftsman at the *École Royale du Génie Militaire* (Royal School of Military Engineering) in Mézières.

When Charles Bossut left his position as a professor of mathematics at the *École Royal du Génie* at *Mézieres* in 1769, Monge was appointed to succeed him in January 1769 at the age 22.[3]

Monge presented memoirs to the *Académie des Sciences* on a generalization of the calculus of variations, infinitesimal geometry, the theory of partial differential equations and combinatories.[3]

In 1777 Monge married Catherine Huart who brought with her a generous dowry.

Three years later he was elected as an *adjoint géomètre* at the *Académie des Sciences* in Paris. In December 1784 he resigned his professorship at Mézieres after serving for almost 20 years. He became examiner of naval cadets replacing Étienne Bézout (1730–1783).

On 21 September 1792 the French monarchy was abolished and a republic was declared. Monge was a strong republican and supporter of the Revolution, and he was offered the position of Minister of the Navy by the new government. Monge also served on the Commission on Weights and Measures to devise the metric system.

He was appointed as an instructor in descriptive geometry at the new *École Polytechnique* in November 1794.

In October 1799 Monge became director (president) of the *École Polytechnique* which became a model for establishment of higher education because of its "brilliant professors, rigid discipline, demanding examinations and excellent textbooks".[4]

École Polytechnique is one of the most prestigious *grandes ecoles* in France. Students with baccalaureate degree take a two year CPGE (*Classe Préparatoire aux Grandes Écoles*) program, to compete for entrance examination at *École Polytechnique* where students study three years, emphasizing mathematics and physics.

The motto *École Polytechnique,* proclaimed by Napoleon, is *"Pour la patrie, les sciences et la gloire"* (for the nation, science and glory).

École Polytechnique inspired *Eidgenössische Technische Hochschule Zürich* and the Massachusetts Institute of Technology. *École Polytechnique* belongs to the Ministry of Defense and students can be commissioned as second lieutenant upon graduation.

Monge became friendly with Napoleon who appointed him Grand Officer of the Legion of Honour in 1804, president of the Senate in 1806, and Count of Pelure in 1808. After Napoleon was defeated at Waterloo in 1815, Monge remained faithful to him. In March 1816 he was expelled from the *Institut de France* (the new name of the *Academie des Sciences*). He died on 28 July 1818 in Paris.

Monge largely created the field of descriptive geometry and he published his textbook *Géométrie descriptive* in 1799. He is also widely regarded as the father of differential geometry.

Jean Baptiste Joseph Fourier

Joseph Fourier was born on 21 March 1768 in Auxerre Bourgogne, France. By the age of 14 he had completed a study of the six volumes of Bézout's *Cours de mathematique*. He studied at the *École Royale Militaire* of Auxerre and the *Collége Montagu* in Paris.

In 1790 he became a teacher at the *École Royale Militaire* where he had studied. In 1795, Fourier continued his studies at new *École Normale* in Paris where he was taught by Lagrange, Laplace and Monge. In 1797 Fourier succeeded Lagrange when he took the Chair of Analysis and Mechanics at the *École Polytecnique*.

Fourier joined Napoleon's army in its invasion of Egypt in 1798 as scientific adviser.

He conducted archaeological exploration and helped with the founding of the Cairo Institute and was elected its secretary.

Figure 77. Siméon-Denis Poisson

Figure 78. J. B. Joseph Fourier

Napoleon appointed Fourier as Prefect (Governor) of the *Départment of Isère* based at Grenoble in 1801. He worked with 37 different communities and accomplished the drainage of a huge area of marshland, which was then converted to valuable agricultural land. He also constructed the French section of the highway between Grenoble and Torino, Italy.[4] Napoleon conferred on him the titles of Baron in 1808 and Count in 1814.

In Grenoble between 1804 and 1807, he wrote his important monograph on the Propagation of Heat in Solid Bodies In that work Fourier made expansion of functions as trigonometric series, what we now call Fourier series. Lagrange, Laplace, Biot and Poisson initially had objections to Fourier's method.

In 1817 and 1818 Fourier exhibited integral-transform solutions to differential equations. The Fourier series and Fourier transform are basic analysis tools for thermodynamics, vibration study, circuit theory and communication engineering.

Fourier was elected to the *Académie des Sciences* in 1817 and in 1822 he won the position of secretary of the *Académie de Sciences*. While working as director of the Statistical Bureau of the Seine, he investigated the probability theory and the theory of errors.

Toward the end of his life he had peculiar symptoms which are thought to have been due to a disease of thyroid gland called myxoedema, most likely contracted in Egypt.[4]

Fourier died on 16 May 1830 in Paris, France.

Siméon-Denis Poisson

Siméon-Denis Poisson was born on 21 June 1781 in Pithiviers, France. In 1798, Poisson was admitted to the *École Polytechnique* where he was taught by Lagrange, Laplace and Fourier.

After graduation in 1800 he became an assistant tutor and two years later a deputy professor.[4]

In 1806 he became a titular professor at the *École Polytechnique*, taking the place of Fourier who became the Governor of Grenoble.

In 1808 Poisson was appointed to the position of astronomer at the *Bureau des Longitudes,* where later he succeeded Laplace as head of the Bureau.

In 1809 he was also appointed to the Chair of Mechanics in the newly opened *Faculté des Sciences* at Sorbonne. In 1815 he became an examiner at the *École Militaire* and in 1816 he became an examiner for final examination at the *École Polytechnique*.[3]

He married Nancy de Bardi in 1817. They had two sons and two daughters.

Poisson was completely dedicated to mathematics writing about 300 papers and he frequently said "Life is good for only two things, discovering mathematics and teaching mathematics".[4]

In the autumn of 1838, Poisson found that he was suffering from tuberculosis. He died on 25 April 1840 in Sceanz near Paris, France.

The Poisson Bracket was used by Paul Dirac to derive the noncommutative property of canonical coordinates q and momenta p in quantum mechanics.

Poisson's integral formula, Poisson equation and Poisson distribution are frequently used today.

Jean-Victor Poncelet

Jean-Victor Poncelet was born on 1 July 1788 in Matz, Lorraine, France. After graduating from *Metz Lycée* in 1807 he entered the *École Polytechnique* where his teachers included Monge, Lacroix, Hachette and Poisson. After graduation he was admitted to the corps of military engineers. Poncelet took part in Napoleon's disastrous 1812 Russian invasion as an engineer, and he was taken prisoner at the battle of Krasnov in November 1812. He was imprisoned for the eighteen months in Saratov where he wrote a notebook on pure and analytic geometry developing some original ideas concerning the projective properties of conics and systems of conics.[4] This notebook was eventually published as *Application d'analyse et de géométrie* in 1862.

In 1824 he was appointed Professor of Applied Mechanics at the *École du Genie* where Monge had begun his career. In 1835 he married Louise Palmyre Gaudin.[4]

Poncelet was promoted to brigadier-general and was appointed commandant of the *École Polytechnique,* from where he retired in 1850. He was Professor of Mechanics at *College de France* in 1848. He applied mechanics to improve turbines, and more than doubling the efficiency of waterwheels.[3]

Poncelet died on 22 December 1867 in Paris, France.

Jean-Baptiste Biot

Jean-Baptiste Biot was born on 21 April 1774 in Paris, France. At the *École Polytechnique* he was a student of Monge. After graduation in 1797 he became a professor of mathematics at the *École Centrale at Beauvais.*

In 1800 he became a professor of mathematical physics at the *Collège de France* on the recommendation of Laplace. Nine years later he was appointed professor of physical astronomy at the Faculty of Sciences at Sorbonne.

Biot, together with Felix Savart (1791-1841), discovered that the intensity of the magnetic field generated by a current flowing through a wire varies inversely with the square of distance from the wire.

$$dB = \frac{\mu_0 I}{4\pi} \frac{dl \times a_r}{R^2}$$

This is known as Biot-Savart's law.

For his work on the polarization of light passing through chemical solutions. Biot was awarded the Rumford Medal by the Royal Society of London in 1840. He was made a member of the Académie des Sciences in 1856.

Biot died on 3 February 1862 in Paris, France.

Figure 79. Jean-Victor Poncelet

Figure 80. Jean-Baptiste Biot

André-Marie Ampère

André-Marie Ampère was born on 20 January 1775 in Lyon, France. In 1793 Lyon was captured by the republican army and his father Jean-Jacques Ampére – a very wealthy man who was Justice of the Peace in Lyon – was guillotined.

Ampère studied the works of Euler, Bernoulli and Lagrange on his own. In 1797 Ampère married Julie Carron, and their son Jean-Jacques-Antoine later became an eminent historian and philologist. But his wife died in 1803 just five years after they were married.

Ampère taught at *École Centrale in Bourg* and at the Lycée in Lyon.

He was appointed répétiteur (assistant lecturer) in analysis at the *École Polytechnique* in 1804. Five years later he was appointed professor of mathermatics at the *École Polytechnique* where he held posts until 1828.

Ampère married Jenny in August 1806 but they separated two years later.

In 1808 Napoleon appointed him inspector general of the newly formed university system. He was elected to the *Institut National des Sciences* in November 1814. He was appointed to the chair of experimental physics at the *Collége de France* in 1823.

The magnetic force on a complete (closed) circuit of contour C that carries a current I in a magnetic field B is given by

$$F = I \oint_c dl \times B$$

where B is given by the Biot–Savart law.

This is called Ampère's law of force.

His name is honored as the unit of electric current.

Ampère died on 10 June 1836 in Marseilles, France.

Figure 81. André-Marie Ampère

Jacques Charles François Sturm

Charles François Sturm was born on 29 September 1803 in Geneva, Switzerland.

He became a French citizen in 1833 and was elected to the *Académie des Sciences* in 1836.[3] Sturm and Liouville worked together on the theory of second-order ordinary differential equations with boundary conditions. It is known today as the Sturm-Liouville problem, an eigenvalue problem and expansions of functions in series of eigenfunctions.

Sturm was appointed professor of analysis and mechanics in 1840 and in the same year he succeeded Poisson's chair of mechanics at the *Sorbonne Faculté des Sciences*.

He attended lectures by Ampère and also worked in his laboratory.

Sturm died on 18 December 1855 in Paris, France.

Augustin-Louis Cauchy

Augustin-Louis Cauchy was born on 21 August 1789 in Paris, France. After studying classical languages and mathematics he entered the *École Polytechniques* in 1805.

In 1807 he graduated from the *École Polytechnique* and went to the *École des Ponts et Chaussées* to study civil engineering.[3] After graduation he worked at many engineering projects. He won the Grand Prix of the Académie des Sciences for work on waves modulation. He was appointed professor at the *École Polytechnique* in 1816.

That same year, the restoration of the Bourbons to the French throne resulted in the expulsion of Gaspard Monge and Lazare Carnot from the Académie des Sciences. In 1830 King Charles X was overthrown by the July Revolution and when Cauchy refused to take an oath of allegiance to the new regime he was forced to resign his chair. In 1831 he accepted the chair of theoretical physics at the University of Torino. From 1833 to 1838 he served as tutor to Henri, the son of Charles X, then in exile in Vienna.

Cauchy was not thought of as a good lecturer. His lectures tended to be very confusing, skipping from one idea to another without any transition between them.[3]

He returned in 1838 to Paris to resume his professorship at the *École Polytechnique*.

From 1848 to 1852 he was a professor at the Sorbonne.

Cauchy was a devout Catholic and a royalist which caused problems with many of his colleagues. Niels Abel (1802-1829) visited Paris in 1826 and gave his papers to Cauchy. Part of the reason Cauchy delayed the publication of Niels Abel's work was because Abel called him a "bigoted Catholic". Cauchy also repressed the mathematical work of Evariste Galois (1811-1832) because Galois was a radical republican.

Cauchy is famous for two main reasons in the field of mathematics; his works with convergent series and rigor in analysis and all of mathematics. He defined continuity and derivative in terms of the limit. He gave the first good definition of the limit.

Cauchy published seven books and 789 full length papers.

Numerous terms bear Cauchy's name: the Cauchy integral theorem in the theory of complex functions, the Cauchy-Kovalevskaya existence theorem for the solution of partial differential equations, the Cauchy-Riemann equations and Cauchy sequences.[3]

He was elected a Fellow of the Royal Society of London on 1832. He died on 23 May 1857 in Sceaux near Paris, France.

Figure 82. Augustin-Louis Cauchy

Figure 83. Joseph Liouville

Joseph Liouville

Joseph Liouville was born on 24 March 1809 in Saint-Omer near Calais, France.

After studying at the *Collége St Louis* (a secondary school) in Paris he entered the *École Polytechnique* in 1825 and attended courses given by Ampère and François Arago (1786-1853). Upon graduating from the *École Polytechnique* in 1827 he entered the *École des Ponts et Chaussées* as Cauchy had twenty years ealier.

By the time he graduated in 1830 he had concluded that his health was not good enough to endure the life of an engineer in the field and decided instead to become a mathematician.

That same year he married a maternal cousin Marie-Louise Balland and they had three daughters and one son.

In 1831 Liouville was appointed as répétiteur at the *École Polytechnique* and taught at several schools to support his growing family.

In 1836 he founded the *Journal de Mathématiques Pures et Appliquées* to compete with Crelle's Journal in Germany.

He also received his doctorate for a thesis on Fourier series that year.

Two years later he was appointed professor of analysis and mechanics at the *École Polytechnique*. In 1840 he was elected to succeed Poisson at the *Bureau des Longitudes*.

In 1837 Louiville was appointed suppléant (deputy) professor at the *College de France*, which paid only one third of the salary of a full professor. He became a full professor at the *College de France* in 1850.

He was elected to the *Académie des Science* in 1839.

Liouville became good friends with Dirichlet, and the two of them corresponded regularly.

He regretted that he failed to recognize Abel's greatness even though he met Abel in 1827 in Paris. Liouville was the first mathematician to study the works of Évariste Galois in depth. By 1846 he had become convinced of their importance and arranged to have them published in his *Journal de Mathématique Pures et Appliquées.*[4]

Liouville wrote over 400 papers but he is best known as Sturm-Liouville theory. He and Sturm examined properties of eigenvalue and eigenfunction of general linear second-order differential equations.

From 1857 to 1874 he held the post of professor of rational mechanics at the Sorbonne.

He was elected to the Constituent Assembly as a moderate republican in April 1848 but he was defeated in the election to the new Legislative Assembly in 1849.[5]

He died on 8 September 1882 in Paris, France.

Joseph Hoëné de Wronski

Joseph Hoëné was born on 23 August 1778 in Qolsztyn near Poznan, Poland. He became a French citizen in 1800 and married Victoire Henriette Sarrazin de Mountferrier in 1810 in Paris.[3] Then he changed his name to Hoëné Wronski.

He published *Introduction to a Course in Mathematics* in London in 1821. For many years his work was dismissed as rubbish. He exaggerated the importance of his research and reacted aggressively to the slightest criticism.

He is best known for the Wronskian W defined by

$$W(y_1, y_2, \ldots, y_n) = \begin{vmatrix} y_1 & y_2 & \cdots & y_n \\ \dot{y}_1 & \dot{y}_2 & \cdots & \dot{y}_n \\ & & \cdots & \\ y_1^{(n-1)} & y_2^{(n-2)} & \cdots & y_n^{(n-1)} \end{vmatrix}$$

where $y^{(n-1)}$ is the $(n-1)^{th}$ derivative of y.

Then y_1, y_2, ..., y_n are linearly independent on some open interval I if $W \neq 0$.

He died on 8 August 1853 in Neuilly-sur-Seine near Paris, France.

Figure 84.   Joseph Hoëné de Wronski

Figure 85.   Evariste Galois

Evariste Galois

Evariste Galois was born on 25 October 1811 in Bourg -la- Reine near Paris, France.

His father was elected Mayor of Bourg-la-Reine in 1815 and until the age of twelve he was educated solely by his mother. She taught him Greek, Latin and religion.

At the *Lycée Louis-le-Grand* he was taught by Louis Paul Emile Richard who recognized Galois exceptional mathematical abilities. He passed his Baccalaureate examination on 29 December 1829. In July 1829 his father committed suicide.

The priest of Bourg-la-Reine forged Mayor Galois' name on malicious forged epigrams directed at Galois' own relatives.

Galois' father was a good natured man and the scandal that ensued was more than he could stand. He hanged himself in his Paris apartment.[3]

Galois failed the entrance examination at the *École Polytechnique* twice, so he entered *École Préparatoire* (later named *École Normale Supérieure*) instead in 1830.

Galois submitted an article *On the Condition that an Equation be Soluble by Radicals* in February 1830 to the *Académie des Sciences*. The paper was sent to Fourier, the secretary of the academy who died in April 1830, and Galois' paper was never subsequently found.[3]

Galois was expelled from the *École Normale* because he attacked the director of *École Normale* in the *Gazette des École* for locking the students during the insurrection against King Charles X in the summer of 1830.

Galois joined the artillery of the National Guard, a republican branch of the militia, which was then abolished by the new King Louis-Phillips on 31 December 1830.[3]

He was arrested but acquitted on the grounds that he was young and foolish. On 14 July 1831 he was arrested again because he was carrying a loaded rifle, several pistols and a dagger. He was released on 29 April 1832.

Galois fell in love with Stephanie-Felice du Motel and exchanged letters with her. Galois fought a duel with Pescheux d'Herbinville on 30 May 1832 with pistols. The reason for the duel is not clear but it is believed to have been linked with Stephanie. He was wounded in the duel and died in Cochin hospital in Paris on 31 May 1832. Galois's friend Chevalier and Galois's brother Alfred copied his mathematical papers and sent them to Gauss and Jacobi hoping that they could give their opinions on his work.[3]

In September 1843 Liouville found Galois's papers and published them in his Journal in 1846.

Charles Hermite successfully completed one of Galois's investigations in solving the quintic equation by means of elliptic moduler function. Carmille Jordan brought out the group theory governing the behavior of such functions.[4]

Sylvestre François Lacroix

Sylvestre François Lacroix was born on 28 April 1765 in Paris, France. Lacroix attended free courses given by Monge at the *Académie des Sciences*. Monge recommended that Lacroix be appointed as Professor of Mathematics at the *École Gardes de Marine* at Rochefort in 1782.[3]

He returned to Paris, taught at the Lyceé (a high school) and he married.

When Laplace gave up the post of examiner at the Royal Artillery Corps during the French Revolution, Lacroix was appointed to succeed him in October 1793.

Lacroix was appointed as an assistant to Monge at the new *École Polytechnique* in November 1794 for Monges's course in descriptive geometry.

Five years later Lacroeix was appointed to the chair of analysis at the *École Polytechnique*. In May 1799 he was elected to the *Institut National des Sciences* (the former *Académie des Sciences*). In 1809 he was appointed to fill Lagrange's chair at the *École Polytechnique* and the following year as professor of differential calculus at the *Sorbonne Faculté des Sciecnes*. In 1815 he left the *École Polytechnique* for Sorbonne and he was appointed to the chair of mathematics at the *Collége de France* that same year.

His textbooks were translated into English to be used at British universities and the United States Military Academy at West Point.

Lacroix died on 24 May 1843 in Paris, France.

Marc-Antoine Parseval des Chênes

Antoine Parseval was born on 27 April 1755 in Rosières-aux-Saline, France. He was a royalist who was imprisoned during the French Revolution in 1792. He was also against Napoleon and had to flee from France.

He had five publications presented to the *Académie des Sciences*. His paper, dated 5 April 1799, contains the result known today as Parseval's theorem.[3]

Let *f(x)* be a periodic function of period 2π that can be represented by a Fourier series

$$f(x) = a_0 + \sum_{n=1}^{\infty} (a_n \cos nx + b_n \sin nx)$$

The Parseval's equality is given by[17]

$$2a_0^2 + \sum_{n=1}^{\infty} (a_n^2 + b_n^2) = \frac{1}{\pi} \int_{-\pi}^{\pi} f(x)^2 \, dx$$

This form is written in modern notation and differs from the original Parseval paper.

He never received the honor of being elected to the *Académie des Sciences*. He died on 16 August 1836 in Paris, France.

Figure 86.   Joseph A. Serret

Joseph Alfred Serret

Joseph Serret was born on 30 August 1819 in Paris, France. He graduated from the *École Polytechnique* in 1840. In 1861 he was appointed professor of celestrial mechanics at *Collège de France* which was and is a very prestigious institution in France.

He was appointed to the chair of differential and intergral calculus at the Sorbonne in 1863 and he joined the *Bureau des Longitudes* in 1873.

Serret was elected to the *Académie des Sciences* in 1860. He edited the works of Lagrange and Monge.[3]

He is best known for the Frenet-Serret theorem in differential geometry. It was found independently by both Jean Frédéric Frenet (1816-1900) and Serret.

The theorem in matrix form is given by[18]

$$\begin{bmatrix} T' \\ N' \\ B' \end{bmatrix} = \begin{bmatrix} 0 & \kappa & 0 \\ -\kappa & 0 & \tau \\ 0 & -\tau & 0 \end{bmatrix} \begin{bmatrix} T \\ N \\ B \end{bmatrix}$$

T is the tangent vector, N is the normal vector and $B = T \times N$ is the binormal vector field along the given curve. T´, N´, B´ are the derivatives.

$\kappa(s) = |T'(s)|$ is the curvature and $B' = -\tau N$ where τ is the torsion.

Serret died on 2 March 1885 in Versailles, France.

Jean Frédéric Frenet

Jean Frenet was born on 7 February 1816 in Périgueux near Bordeaux, France.

Frenet entered the *École Normale Supérieure* in 1840 and submitted his doctoral thesis *Sur les fonctions qui servent à déterminer l'attraction des sphéroides quelconques. Programme d'une thése sur quelque proprieties des courbes à double courbure* in 1847 at the University of Toulouse.

In that thesis he provided the formulas now known as the Serret-Frenet formulas.[3]

Frenet gave only six formulas while Serret gave all nine. It is believed that Belgian mathematicial G. M. Pagané also independently produced the same results at about the same time in 1847.[3]

In 1848 Frenet was appointed professor of mathematics and director of the astronomical observatory at the University of Lyon. He never married and he lived with his sister as Dedekind and Weierstrass did. Frenet died on 12 June 1900 in Périgueux, France.

Figure 87.   Charles Hermite

Figure 88.   Jules Henri Poincaré

Charles Hermite

Charles Hermite was born on 24 December 1822 in Dieuze, Lorraine, France, but he grew up and was educated in Nancy. He attended secondary school at the *College Louis-le-Grand* where Galois had studied 15 years earlier. His mathematics teacher there was Louis Richard who had also taught Galois.[3] In 1842 Hermite was admitted to the *École Polytechnique* but he was not allowed to graduate on time because of a congenital deformity of his right leg, which resulted in lameness.

He continued his mathematical research on his own and he was eventually able to graduate in 1847. He was appointed répétiteur and admissions examiner at the *École Polytechnique* in 1848.

In 1856 he was elected to the *Académie des Sciecnes,* but he contracted smallpox that same year. Under Cauchy's influence Hermite turned to the Roman Catholic and a royalist.

He showed in 1858 that an algebraic equation of the fifth degree could be solved using elliptic functions.[3]

In 1862 he was appointed *maître de conférenece* (assistant professor) at the *École Polytechnique* and seven years later he was appointed to the chair of analysis at the *École Polytechnique* succeeding Duhamel and at the Sorbonne. He resigned his chair at the *École Polytechnique* in 1876 and retired from the Sorbonne in 1897.

He had married Louise Bertrand and one of their two daughters married Emile Picard.

In 1873 Hermite proved that **e** is a transcendental number. Nine years later Lindemann proved that π was also transcendental.

Many leading French mathematicians in the latter part of the nineteenth century were taught by Hermite, including Paul Appel, Emile Picard, Gaston Darboux, Emil Borel, Paul Painlevé, Jacques Hadamard and Henri Poincaré.

Hermite is best known for Hermite polynomials, Hermite differential equation and Hermitian matrices. He died on 14 January 1901 in Paris, France.

Jules Henri Poincaré

Henri Poincaré was born in Nancy, France on 29 April 1854. His father was a physician and professor of medicine at the University of Nancy. One of Henri's cousins, Raymond Poincaré (1860-1934) was Prime Minister in 1912 and President of France from 1913 until 1920.

Like Euler and Gauss, he was gifted with an exceptional memory, but he was extremely myopic and had very poor motor coordination. He was clumsy in drawing and poor in experimental work. After graduating from the *École Polytechnique* in 1875 he went on to the *École des Mines* for the following four years. It is difficult to understand why he wanted to be a mining engineer with his clumsy motor coordination.

His doctoral dissertation submitted to the University of Paris in 1879 was *Sur les proprieties des fonctions définies par les équations differences.* His adviser was Hermite.

While working at the University of Caen in Normandy, he discovered the occurrence of non-euclidean geometry in the theory of automorphic functions.[4]

The discovery of the underlying geometry put Funchsian function (by Lazarus Fuchs) in a completely new light.

In 1881 he returned to Paris and married Jeanne Louuse Marie Poulain d'Andecy. They had one son and three daughters.

In 1886 he became a professor of mathematical physics and the calculus of probabilities at the Sorbonne where he remained until the end of his life.[4] In 1887 he was elected to the *Académie des Sciences* and six years later to the *Bureau des Longitudes.*

He was elected a Fellow of the Royal Society of London in 1894 and received the Sylvester Medal in 1901. In 1908 he was elected to the literary section of the *Académie française*, for his work on the philosophy of science.

In 1899 he suggested that absolute motion does not exist and in the following year he proposed that nothing could travel faster than light.

The Poincaré conjecture states that a compact, connected 3-manifold S is topologically equivalent to a 3-sphere if it is simply connected. This conjecture remains unresolved in algebraic topology for a long time, but finally solved by Perelman in Russia recently.

He was elected to each the five sections of the Académie française, namely the geometry, mechanics, physics, geography and navigation sections.

At the Sorbonne he lectured in a variety of subjects: optics, electricity, the equilibrium of fluid masses, astronomy, thermodynamics, light, and probability.

He wrote more than 30 books and 500 papers.

A street in Paris is named Rue Henri Poincaré.

He died on 17 July 1912 in Paris due to a post-operative embolism.

Marie Ennemond Camille Jordan

Camille Jordan was born on 5 January 1838 in Lyon France.

After studying at the *Lycée de Lyon* and the *College d'Oullins*, he entered the *École Polytechnique* in 1855. His doctoreal dissertation submitted to the *Sorbonne Faculté des Sciences* in 1861 was *Sur le nombre des valeurs des fonctions* (Part 1), *Sur des periods des fonctions inverses des integrales des differentielles algebriques* (Part 2).[3]

In 1862 Jordan married Marie-Isabelle Munet and they had two daughters and six sons.

Jordan was appointed professor of analysis at the *École Polytechnique* in 1876 and professor of mathematics at the *Collège de France* in 1882. He was elected to the *Académie des Sciences* in 1881 and in 1890 he became an officer of the Legion d'Honneur.

Three of his six sons were killed during World War I, but his remaining three sons became a government minister, professor of history at the Sorbonne and an engineer.[3] He was elected a Fellow of the Royal Society of London in 1919.

G.H. Hardy was greatly impressed by Jordan's *Cours d'analyse* when he read it at Cambridge. He learnt for the first time what mathematics really meant. He was on his way to becoming a real mathematician with a genuine passion for mathematics.[48]

Jordan is best known for the Jordan-Hölder theorem, the Jordan curve theorem and the Jordan canonical form in matrix theory. But the Jordan in the Gauss-Jordan pivoting elimination method for solving the linear first-order simultaneous equations is Wilhelm Jordan (1842-1899), a German mathematician. Camille Jordan died on 22 January 1922 in Paris, France.

Figure 89.   M. E. Camille Jordan

Figure 90.   C. Émile Picard

Charles Émile Picard

Émile Picard was born on 24 July 1856 in Paris, France. His father was killed during the Franch-German war in 1870, so his family faced serious difficulties.

He took the entrance examination for the *École Polytechnique* and *École Normale Supérieure*, he was placed second and first respectively at the two institutions. He chose to attend *École Normale, entering* in 1874.

He received his doctorate in 1877 with his thesis *Applications des complexes lineaires a létude des surfaces et des courbes gauches.*[10] Picard was appointed *maître de conference* (assistant professor) in mechanics and astronomy at the Sorbonne and *École Normale.*

In 1881 he married Hermite's daughter. They had one daughter and two sons, all of whom were killed during World War I.

In 1885 Picard was appointed to the chair of differential calculus at the Sorbonne but he had to wait one year to occupy the chair formally because the university regulation required that professors be at least 30 years old. In 1897 he exchanged his chair for that of analysis and higher algebra so that he could train research students.[3]

Between 1891 and 1896 he published the three volumes masterpiece *Traité d'analyse*. He was elected to the *Académie des Sciences* in 1889, and to the *Académie française* in 1924. Picard received the *Grand Croix de la Légion d'Honneur* in 1932 and the Mittag-Leffler Gold Medal from the Swedish Academy of Sciences in 1937.

He is best known for the Picard-Lindelöf theorem in ordinary differential equation, the Picard iteration for initial value problems and the Picard theorem which states that if $f(z)$ is analytic and has an isolated essential singularity at a point z_0, it takes on every value with at most one exceptional value in an arbitrarily small neighborhood of z_0.[17]

Picard died on 11 December 1941 in Paris, France.

Felix Edouard Justin Émile Borel

Émile Borel was born on 7 January 1871 in Saint Affrique, Aveyron, Midi-Pyrénées, France. After studying at the College of Saint-Barbe and *Lycée Louis-le-Grand*, he took the entrance examination for the *École Normale* and *École Polytechnique*. He placed first in both exams, and he chose to attend the *École Normale* in 1889. In the same year he won first prize in the *National Concours Général* for mathematics.

He graduated in 1892 and in the same year placed first in the examination for agrégation. Agrégation is a unique French system of examination leading to the position of *professeur agrégé* who teach at lycées. Some professeurs agrégé teach in the preparatory classes to the grandes écoles (CPGE) which are approximately equivalent to the first two years at American universities.

But most graduates from *École Normale* today become university professors and researchers.

Borel's doctoral thesis in 1893 was *Sur quelque points de la théorie des fonctions* (On some points in the theory of functions), which contains theory of measure, theory of divergent series, theory of non-analytic continuation and theory of quasi-analytic functions.

That same year, he was appointed lecturer at the University of Lille where he stayed for four years and produced 22 research papers.[4]

In 1897 he returned to the *École Normale,* and in 1909 he was appointed to the chair of theory of function at the Sorbonne where he stayed until 1941. In 1910 he became Deputy Director for research at the *École Normale* and stayed in that position for ten years.

Figure 91. F. E. J. Émile Borel

Figure 92. Henri L. Lebesgue

In 1921 he was elected to the *Académie des Sciences*. He married Marguerite, the 17 years old daughter of his colleague Paul Appell in 1901. She wrote more than 30 novels under the pseudonym Camille Marbo. They had no children but they adopted his nephew Fernand Lebeau.

Borel served in the French Chamber of Deputies from 1924 until 1936 and as Minister of the Navy from 1925 until 1940. When the Nazis invaded France in 1941 he was imprisoned briefly and he later joined the resistance group. He was awarded the Resistance Medal in 1950.

In 1896 he provided a proof of Picard's theorem which had eluded mathematicians for nearly 20 years. His most important works were created in the 1890s on probability, infinitesimal calculus, divergent series and most influential of all, the theory of measures.

He died on 3 February 1950 in Paris, France at the age of eighty-five.

Henri Lion Lebesgue

Henri Lebesgue was born on 28 June 1875 in Beauvais, Oise, Picardie, France. He studied at the *Lycée Saint Louis* and the *Lycée Louris-le-Grand*. Lebesgue graduated from the *École Normale Supérieure* in 1897 and taught at the Lycée Central in Nancy from 1899 to 1902.

He submitted his doctoral dissertation *Integrale, longueur, aire* to the Sorbonne Faculty of Science in 1902. In the same year he was appointed *maître de conferences* at the University of Rennes.

In 1910 he moved to the Sorbonne as *maître de conferences* in mathematical analysis and in 1918 he was promoted to Professor of the Application of Geometry to Analysis. In 1921 he was named the prestigious position of professor of mathematics at the *Collége de France* where he stayed until his death in 1941.

He was elected to the *Académie des Sciences* in May 1922 and to the Royal Society of London on 1934.

Figure 93.   Michel Chasles

With Emile Borel he laid the foundation of modern theory of functions of a real variable. His chief work, however, was his creation of a new approach to the theory of integration. Lebesgue died on 26 July 1941 in Paris, France.

Michel Chasles

Michel Chasles was born on 15 November 1793 in Épernon, 50 km southwest of Paris.

He entered the *École Polytechnique* in 1812 but his studies were interrupted when he joined Napoleon's army in early 1814.

In 1837 he published *Aperçu historique sur l'origine et le développement des méthodes en géometrie* (Historical view of the origin and development of methods in geometry). On the strength of this work he was appointed professor at the *École Polytechnique* in 1841 and he taught geodesy, mechanics and astronomy there.

In 1846 he was appointed to the chair of higher geometry at the Sorbonne where he remained until his death.

He published important texts on geometry in 1852 and in 1865. He never married.

He was elected to the *Académie des Sciences* in 1851 and a Fellow of the Royal Society of London in 1854. Chasles also received the Copley Medal in 1865.

He had great intelligence and a deep interest in history, but he was the victim of a well-known fraud by Denis Vrain-Lucas between 1861 and 1869. He purchased many documents from Vrain-Lucas that purported to be correspondence among Newton, Pascal, Boyle and Galileo. Vrain-Lucao was found guilty of fraud in 1870.[3]

Chasles died on 18 December 1880 in Paris, France.

Figure 94. Élie Joseph Cartan

Figure 95. J. Gaston Darboux

Jean Gaston Darboux

Gaston Darboux was born on 14 August 1842 in Nimes, 50 km northeast of Montpellier, France. After graduating from the *École Normale Supérieure* he submitted his doctoral thesis *Sur les surfaces orthogonales* in 1866 to the *Sorbonne Faculté des Sciences*. He taught at the *Lycée Louis-le-Grand* between 1867 and 1872. From 1872 to 1881 he taught at the *École Normale*. Then in 1878 he became suppléant (deputy) to Chasles in the chair of higher geometry at the Sorbonne. Two years later Chasles died and Darboux assumed the position. He held that chair until his death. Darboux was also the dean of the Sorbonne Faculté des Sciences from 1889 to 1903.[3]

He was elected to the *Académie des Sciences* in 1884, and the Royal Society of London in 1902, winning the Sylvester Medal in 1916.

Darboux managed to complete a proof of the existence of integrals of continuous functions that had eluded Cauchy a generation before. In 1879 he succeeded in defining the "Darboux sums" and used the "Darboux integrals".

He died on 23 February 1917 in Paris, France.

Élie Joseph Cartan

Joseph Cartan was born on 9 April 1869 in Dolomieu, Savoie, Rhône-Alpes, France in a very poor family. Antonin Dubost, inspector of primary schools, obtained state funds for him to attend the Lycée in Lyon and later the *École Normale Supérieure* in Paris in 1888. Cartan received his doctorate with his thesis *Sur la structure des groupes de transformations fins et continus* in 1894 at the Sorbonne.

The lecturers at the *École Normale Supérieure* and the Sorbonne who influenced Cartan the most were Hermite, Darboux, Tannery, Appell, Picard and Goursat. Poincaré introduced Catan to the geometric application of group theory.

From 1894 to 1909 he taught in Montpellier, Lyon and Nancy. In 1912 he was appointed to the Chair of Differential and Integral Calculus at the Sorbonne, then as Professor of Rational Mechanics in 1920 and as Professor of Higher Geometry in 1924.[3]

He married Marie-Louise Bianconi in 1903 and they had three sons and one daughter. Their son Henri Cartan became an outstanding mathematician. Another son Louis was a member of the resistance against the Nazis German forces who beheaded him in

December 1943. Cartan's work represents an extraordinary combination of Lie group theory, classical geometry, differential geometry and topology.[3]

He also initiated the theory of Spinors in 1913. These are complex vectors that are used to transform three-dimentional rotation into two-dimentional representations and Spinors played an important role in quantum mechanics.

Cartan was one of the most influential mathematicians in the first half of the twentieth century along with Poincaré and Hilbert.

He was a modest and kind person. He was elected to the Académie des Sciences in 1931 and became President of the Académie in 1946. He was also elected a Fellow of the Royal Society of London in May 1947.

Cartan received six honorary doctorates including one from Harvard University in 1936.

He died on 6 May 1951 in Paris, France after a long illness.

Vaughan Frederick Randal Jones

Vaughan Jones was born on 31 December 1952 in Gisborne, New Zealand and graduated from Auckland Grammar School. He obtained a B.Sc. in 1972 and an M.Sc. in 1973 at the University of Auckland.

He was awarded Docteur es Sciences (Mathématiques) at the University of Geneva in 1979 for his thesis *Actions of finite groups on the hyperfinite III factor* supervised by André Haefliger.

In 1980, he moved to the United States, where he was a faculty member at the University of California, Los Angeles (1980-1981) and the University of Pennsylvania (1981-1985). Since 1985 he has been Professor of Mathematics at the University of California, Berkeley.[3]

He was awarded a Fields Medal at the 1990 International Congress for Mathematicians in Kyoto for his proving the Index Theorem and discovering a relationship between Von Neumann algebras and geometric topology. He wore a New Zealand rugby jersey when he accepted the prize.

He was elected a Fellow of the Royal Society of London in 1990 and awarded the New Zealand Government Science Medal in 1991. Jones also received Distinguished Companionship of the New Zealand Order of Merit in 2002.

Figure 96.   Vaughan F. R. Jones   Figure 97.   Laurent-Moïse Schwartz

Laurent-Moïse Schwartz

Laurent Schwartz was born on 5 March 1915 in Paris, France. He graduated from the *École Normale Supérieure* in Paris in 1937 with the *Agrégation de Mathématique*.

He received a doctorate with his thesis *Sommes de Fonctions Exponentielles Reeles* at the University of Strasbourg on 1943.[10]

Schwartz's outstanding contribution to mathematics was his work in the theory of distributions in 1948. The theory of distributions is a considerable generalization of the differential and integral calculus. Oliver Heaviside and Paul Dirac introduced the delta function (impulse function) which is the derivative of the unit step function.

These were not built on a rigorous and abstract mathematical foundation. Schwartz's development of the theory of distributions put methods of this type onto a sound basis providing powerful tools for applications in numerous areas in mathematical physics and electrical engineering.

Schwartz received a Fields Medal at the International Congress of Mathematicians on 30 August 1950 for his work on the theory of distributions. He was elected to the Académie des Sciences in 1972.

His hobby was collection of butterflies and he accumulated over 20,000 specimens[3].

He was very active politically believing that nobody can move forward without being subversive.

Schwartz was a Trotskyite for ten years until 1947. He was one of the leaders of the protesting against the Soviet invasion of Hungary in 1956 and he opposed United States involvement in Vietnam, the Soviet invasion of Afghanistan and the Russian war against Chechnya.

In June 1957, when one of his students Augin was killed by French paratoopers in Algiers, Schwartz signed the famous "Manifeste des 121" in favor of military insubordination. Schwartz was dismissed from his professorship at the *École Polytechnique* by the French Minister of Defense as a result, but he was reinstated two years later. Schwartz taught there until 1980 then spent three years at the University of Paris VII until his retirement in 1983. He died on 4 July 2002 in Paris, France.

Jacques-Louis Lions

Jacques-Louis Lions was born on 2 May 1928 in Grasse Alpes-Maritines, France. His father Honoré Lions was the Mayor of Grasse for nearly 30 years.

At the end of 1943, although he was only 15 years old, he joined the French Resistance fighting the German Army. He met Andrée Olivier, also a member of the Resistance and they later married in August 1950. They had one child, Pierre-Louis Lions who became a famous mathematician receiving a Fields Medal.

Lions graduated from the *École Normale Supérieure* in 1950 and received his *Docteur és Sciences* in 1954 at the University of Nancy supervised by Laurent Schwartz. He was appointed *Maître de Conférences* at Nancy and later became a professor there.

In 1963 he became a Professor in the Faculté des Sciences at the University of Paris where he began a weekly numerical analysis seminars and set up a numerical analysis laboratory.

The University of Paris was split into thirteen separate universities in 1970. Lions went to Paris VI. Among the thirteen universities VI, VII and XI are primarily science oriented. In 1973 he was elected to the *Académie des Sciences* and was appointed to the Chair of *Analyse Mathematique des Systems et de leur contrôle* at the *Collège de France*.

He was President of the *Académie des Sciences* in 1996-98 and received many prizes and medals including *Commandeur de la Légion d'Honneurs* in 1993.

His most outstanding contribution was *Mathematical Analysis and numerical methods for science and technology* written in 1984 and 1985 with Robert Dautray and containing approximately 4000 pages in nine volumes.

His published a total of 529 papers and books.

Lions died on 17 May 2001 in Paris, France.

Figure 98.   Jacques–Louis Lions

Figure 99.   Alexander Grothendieck

Alexander Grothendieck

Alexander Grothendieck was born on 28 March 1928 in Berlin, Germany. His father was Russian Jew and he was murdered by the Nazis. He moved to France and graduated from Montpellier University.

He also spent the year 1948–49 at the *École Normale Supérieure* in Paris. He received a doctoriate at the University of Nancy with his thesis *Produits tensoriels topologiques et espaces nucléaires* in 1954. He worked at the University of Sao Paulo and the University of Kansas before returning to France in 1956.

He was appointed to a chair at the *Institut des Hantes Études Scientifiques* (IHES) in 1959, where he stayed until 1970. IHES professors have no teaching responsibilities.

At IHES he provided unifying themes in geometry, number theory, topology and complex analysis. He received the Fields Medal in 1966 for building on the work of Weil and Zariski and effecting fundamental advances in algebraic geometry. He introduced the idea of K–theory (the Grothendieck groups and rings) and revolutionized homological algebra in his "Tohoku paper." Grothendieck influenced Hironaka at IHES in 1959.

Grothendieck was strongly against military built-up of the 1960s. He had several appointments at Orsay, *Collége de France*, Montpellier as well as *Centre National de la Recherché Scientifique* (CNRS) between 1970 and 1988.

In 1991, he left his home and disappeared. He is said to live in the South of France and decline to accept any visitors.

Figure 100.   Pierre R. Deligne

Pierre René Deligne

Pierre Deligne was born on 3 October 1944 in Etterbeek, Brussels, Belgium. He received his *Licence en mathematique* in 1966 at the *École Normale Supérieure* in Paris. Deligne was awarded a doctorate at the Free University of Brussels in November 1968 for his thesis *Théoreme de Lefschetz et criteres de dégénérescence de suites spectrales*.

He worked with Alexandre Grothendieck at IHES before he received his doctorate. After he received his doctorate he worked at IHES from 1970 to 1984. Deligne received the Fields Medal in 1978 for solving the three Weil conjectures concerning generalizations of the Riemann hypothesis to finite fields. His work did much to unify algebraic geometry and algebraic number theory. He also received the Wolf Prize in 2008.

In 1984 he was appointed a professor at the Institute for Advanced Study in Princeton. He was elected to the *Academie des Sciences* in 1978.

Henri Paul Cartan

Henri Paul Cartan was born in 8 July 1904 in Nancy, France, a son of Elie Cartan and Marie-Louise Bianconi. He studied at the *École Normale Supérieure* in Paris and received his *docteur és sciences mathematiques* in 1928 at the Sorbonne with his thesis *Sur les systems de fonctions holomorphes a varieties linéaires lacunaires et leurs applications.*

He taught at the University of Lille from 1929 to 1931 and at the University of Strasbourg from 1931 to 1940.

Cartan married Nicole Antoinette Weiss on 14 September 1935. They had two sons and three daughters.[3]

He was appointed professor at the University of Paris Sorbonne in 1940 and stayed there until 1969. He also taught at the Université de Paris-Sud at Orsay from 1970 until his retirement in 1975.

He was one of the founders of Bourbaki. Nicolas Bourbaki was a general under Napoleon. A group of young French mathematicians at the *École Normale Supérieure* in Paris formed the *Association des collaborateurs de Nicolas Bourbaki* and published papers under Bourbaki beginning 1935. The founding members were André Weil, Henri Cartan, Claude Chevalley, Jean Dieudonné, Szolem Mandelbrojt, Jean Coulomb, Jean Delsarté and René de Possel. Members were required to resign by age fifty.

Notations introduced by Bourbaki include the symbol ϕ for the empty set and the terms injective, surjective and bijective.

Cartan was active in human rights for dissident mathematicians through *Comité des Mathematicians* in the 1970s.

Figure 101. Henri P. Cartan

Figure 102. Jean-Pierre Serre

He was elected to the *Académie des Sciences* of Paris, the Royal Society of London (1971) and the National Academy of Sciences in Washington (1972). He received the Wolf Prize in 1980 and was made *Commandeur de la Légion d'Honneur* in 1989.

His important text *Homological Algebra* written with Samuel Eilenberg (1913–1998) was published in 1956.

Cartan died on 13 August 2008.

Jean-Pierre Serre

Jean-Pierre Serre was born on 15 September 1926 in Bages, France. He graduated from the *École Normale Supérieure* in Paris in 1948. In the same year he married Josiane Heulot

Serre received his doctorate in 1951 at the University of Paris Sorbonne for his thesis *Homologie singuliére des espaces fibrés* supervised by Henri Cartan.

After teaching for two years at the University of Nancy he was appointed in 1956 to the chair of Algebra and Geometry at the prestigious *Collège de France*. This position requires teaching only one course a year but it must be a different course every year.

He was awarded a Fields Medal in 1954 for achieving major results on the homotopy groups of spheres, especially in his use of the method of spectral sequences. He reformulated and extended some of the main results of complex variable theory in terms of sheaves.

He was elected a Fellow of the Royal Society of London in 1994. Serre received the Steele Prize in 1995 from the American Mathematical Society and the Wolf Prize in 2000. He was the first recipient of the Abel Prize by the Norwegian Academy of Science and Letters in 2003. He retired from the *Collège de France* in 1994.

René Thom

René Thom was born on 2 September 1923 in Montbéliard in France, 55 km west of Basel, Switzerland. He graduated from the *École Normale Supérieure* in Paris in 1946 and received his doctorate from the Sorbonne in 1951.

His dissertation was *Fibre spaces in spheres and Steenrod sequences* supervised by Henri Cartan.

Thom received his Fields Medal in 1958 for inventing and developing the theory of cobordism in algebraic topology. This classification of manifolds used homotopy theory in a fundamental way and became a prime example of a general cohomology theory.

Thom was very modest, claiming that John Milnor and Barry Mazur deserve the Fields Medal more than he did. John Milnor received it in 1962.

Thom taught at the University of Grenoble in 1953-54 and then at the University of Strasbourg from 1954 to 1963.

In 1964 he moved to the *Institut des Hautes Études Scientifique* (IHES).

He is known for his development of the catastrophe theory which deals with continuous action producing a discontinuous result.

Thom died on 25 October 2002 in Bures-sur-Yvette, France.

Jean Alexandre Eugéne Dieudonné

Jean Dieudonné was born on 1 July 1906 in Lille, France. He graduated from the École Normale Supérieure in Paris in 1927 and received his doctorate in 1931 with his thesis *Recherches sur quelques problems relatifs aux polynômese et aux fonctions bornées d'une variable complexe.*

He was *maître de conférences* (assistant professor) at the Faculty of Science at Nancy from 1937 to 1946. He was appointed professor of mathematics at Sao Paulo in Brazil (1946-1947) and Nancy (1948-1952). He also taught at Northwestern University in the United States from 1953 until 1959. He was a professor at the *Institut des Hautes Études Scientifiques* from 1959 until 1964. Then he was appointed to a chair at the University of Nice where he stayed until 1970.[3]

Dieudonné was a good piano player and had a fantastic memory but he did not enjoy teaching. He was a member of the Bourbaki which did broaden his mathematical outlook. In 1935 he married Odette Clavel and they had two children.

He wrote many well-known books on mathematics and the history of mathematics. He was elected to the *Académie des Sciences Paris* in 1968 and he received the Gaston Julia Prize in 1966.

Dieudonné died on 29 November 1992 in Paris, France.

Laurent Lafforgue

Laurent Lafforgue was born on 6 November 1966 in Antony, Paris metropolitan area, France. He graduated from the *École Normale Supérieure* in 1986. In 1994 he received his doctorate at the Université de Paris-Sud with his thesis *Caractéristiques d'Euler-Poincaré et sommes exponentielles* superzised by Gérard Laumon.

Lafforgue is a professor of mathematics at the *Institut des Hautes Études Scientifiques* (IHES) in Bures-sur-Yvette. He received a Fields Medal in 2002 at the International Congress of Mathematicians in Beijing for developing a new cohomology theory for algebraic varieties.

Figure 103. Laurent Lafforgue

Pierre-Louis Lions

Pierre-Louis Lions was born on 11 August 1956 in Grasse Alpes-Maritimes, France as the son of Jacques-Louis Lions and Andrée Olivier. Following in his father's footsteps, Pierre-Louis also studied at the *École Normale Supérieure*.

He received his *doctorate d'état es sciences* in 1979 at the University of Paris VI (Pierre and Marie Curie) with his thesis *Sur quelques classes d'équations aux derives partielles non lineaires et leur resolution numérique.*[10]

He was appointed a professor at the University of Paris-Dauphne in 1981 and Professor of Applied Mathematics at the *École Polytechnique* in 1992.

He married Lila Laurenti in December 1979 and they have one child. He was awarded the Fields Medal in 1994 for his work on the theory of nonlinear partial differential equations. He is a member of the *Académie des Sciences* and Chevalier of the Légion d'Honneur.

Figure 104. Pierre-Louis Lions

Gaston Maurice Julia

Gaston Julia was born on 3 February 1893 in Sidi Bel Abbés, Algeria. He was a second lieutenant in the French Army during World War I, and on 25 January 1915, he received a severe facial injury which resulted in the loss of his nose.

He had to wear a leather patch on his face for the rest of his life. He studied mathematics during his time in the hospital despite being in great pain. Given his injury, it is remarkable that in 1918 he published his 199 pages masterpiece *Mémoire sur l'iteration des fonction rationnelles* (Notes on the iteration of rational functions) in the *Journal de Mathematique Pures et Appliques*. He received the *Grand Prix de l'Académie des Sciences* for this work and made him famous.

In 1918 Julia married Marianne Chausson, one of the nurses who cared for him during his hospitalization and they had six children. He played both piano and violin in his leisure time.

H. Cramer wrote an essay on Julia's work which included the visualization of a "Julia set". He became a professor at the *École Polytechnique*.

His work was essentially forgotten until Benoit Mandelbrot brought it back to prominence in the 1970s through his fractal mathematics.

Julia died on 19 March 1978 in Paris, France.

Figure 105.   Gaston **Maurice** Julia   Figure 106.   Alain Connes

Alain Connes

Alain Connes was born on 1 April 1947 in Draguignan, 45 km west of Cannes, France. He graduated from the *École Normale Supérieure* in Paris in 1970. His doctoral thesis in 1973 was *A Classification of factors of type III* which was on von Neumann Operator algebras.

From 1976 to 1980 he was at the University of Paris VI. In 1979 he was appointed a professor at the *Institut des Hautes Études Scientifiques* and in 1984 professor of analysis and geometry at the prestigious *Collège de France*.

In 1982 he was awarded the Fields Medal for his contribution to the theory of operator algebras, particularly the general classification and structure theorem of factor of type III, classification of injective factors, and applications of the theory of C*-algebras to foliations and differential geometry in general.

He was elected to the *Académie des Sciences* in 1982 and received the Ampère Prize in 1980.

Jacques Salomon Hadamard

Jacques Hadamard was born on 8 December 1865 in Versailles, France. His ancestors on both sides were intellectuals of Jewish extraction. His father Amadée taught Latin at the *Lycée Charlemagne* and his mother was a noted music teacher. She taught him to play the violin at an early age.[4]

In 1884 he came first in the entrance examinations for the *École Normale Supérieure* and *École Polytechnique*. He chose to attend the former, as have many other mathematicians.

Hadamard received a doctorate in 1892 for his thesis *Essai sur l'etude des fonctions données par leur development de Taylor* (On the study of functions given by their Taylor expansion) while teaching at the secondary schools. In the same year he received the *Grand Prix des Sciences Mathématiques*.

In 1892 he married Louise-Anna Trenel, a childhood sweetheart with a similar background to his own.[4] They had five children. Their eldest sons Pierre and Etienne were killed in World War I and the youngest in World War II.

While he was working at the University of Bordeaux he proved the prime number theorem in 1896. Hadamard demonstrated that the number of primes less than a given number x was asymptotically equal to $x/\log x$, which was the most important result ever obtained in number theory.

It was independently proven by C.J. de la Vallée Poussin (1866-1962) using the methods of complex analysis.

Around 1898 Hadamard was appointed deputy professor at the *Collége de France* and in 1909 he became professor of analytic mechanics and celestrial mechanics. He remained there until 1937. Professors teach only one course per year but the lectures should contain new material.

In 1912 he succeeded Camille Jordan as professor of analysis at the *École Polytechnique*. In the same year he was elected to the *Académie des Sciences*. In 1920, he was appointed professor of mathematical analysis at the *École Centrale des Arts et Manufactures* where he retired in 1936. So he had three positions at the same time.

He was elected Fellow of the Royal Society in 1932. After the German invasion of France in 1940, he escaped to the United States in 1941, then to London, where he took part in operations research for the Royal Air Force. In 1945 he returned to France.

He was active in clearing the name of Alfred Dreyfus who was convicted of treason in 1894. In 1906, Dreyfus was fully rehabilitated, restored to the rank of major and decorated by the *Légion d'Honneur*. Hadamard became a communist sympathizer and an avowed pacifist.[4]

He died on 17 October 1963 at the age of ninety-seven in Paris, France.

Figure 107. Jacques S. Hadamard

Figure 108. Szolem Mandelbrojt

Szolem Mandelbrojt

Szolem Mandelbrojt was born on 10 January 1899 in Warsaw, Poland. He received a doctorate in 1923 at the University of Paris Sorbonne, supervised by Hadamard. He taught at the Rice Institute in Houston, Texas in 1926-27. After working as a *maître de conferences* at the University of Lille in 1928-29, he was appointed professor at the University of Clemont-Ferrand.

He was a member of the Bourbaki in its initial stage.

In 1938 he was appointed professor of mathematics and mechanics at the *Collége de France*. During World War II he went back to the Rice Institute, then returned in 1944 to his chair at the *Collége de France*.

Collaborating with Maurice Fréchet, Paul Levy and Laurent Schwartz, he published the works of Jacques Hadamard in four volumes in 1968.[3] He retired in 1972. He was the uncle of Benoit Mandlbrot.

Szolem Mandelbrojt died in 1983 in Paris, France.

Paul Pierre Lévy

Paul Lévy was born on 15 September 1886 in Paris, France. His father and grandfather were also mathematicians. He was placed first for the entrance examination to the *École Normale Supérieur* and second to the *École Polytechnique*.

He chose to attend the *École Polytechnique* where his father was an examiner. After graduating in first place, he also graduated the *École des Mines* in 1910. While he was studying at the *École des Mines* he attended lectures at the Sorbonne by Darboux and Picard as well as lectures by Georges Humbert and Hadamard at the *Collége de France*.

Levy received *Docteur és Sciences* at the University of Paris in 1912 supervizsed by Hadamard. Since *Grandes Écoles* do not award degrees, students at *Grandes Écoles* receive their degrees at a University.

Levy was a professor at *École des Mines* in Paris from 1913 to 1920. In 1920 he became professor of analysis at the *École Polytechnique* where he remained until his retirement in 1959.[3]

His main research fields were probability, functional analysis and partial differential equations. He is known for Lévy C curve, Lévy process, Lévy measure, Lévy constant, and Lévy distribution.

He wrote several books including *Lecons d'analyse fonctionnelle* (1922), Calcul des probabilitiés (1925) and *Processus stochastiques et mouvement brownien* (1948). He was elected to the *Académie des Sciences* in 1964.

Lévy died on 15 December 1971 in Paris, France.

Benoit Mandelbrot

Benoit Mandelbrot was born on 20 November 1924 in Warsaw, Poland. His family emigrated to France in 1936 and his uncle Szolem Mandelbrojt, professor of mathematics and mechanics at the *College de France* (1938-1972), took responsibility for Benoit's education.

He graduated from the *École Polytechnique* in 1947 where he was influenced by Paul Lévy. Mandelbrot went on to the California Institute of technology, receiving his MS in 1948 and returned to Paris.

He was awarded *Docteur es Sciences* at the University of Paris in 1952 for a thesis *Contribution a la théorie mathématique des communications*. He claims that he did not have any supervisor for his dissertation.[94] Between 1949 and 1957 he was a staff member at the *Centre National de la Recherche Scientifiques* (CNRS), Paris.

During that time he spent a year at the Institute for Advanced Study at Princeton in 1953 until 1957 and two years as assistant professor at the University of Geneva from 1955 until 1957. In 1958 he became a research staff member at the IBM Thomas J.

Watson Research Center and an IBM Fellow in 1974. In 1987 he became the Abraham Robinson Professor of Mathematical Sciences at Yale University.[5]

At IBM Watson Research Center he was able to show how Gaston Julia's work is a source of the most beautiful fractals known today.

Fractal geometry has been the unifying theme in all of Mandelbrot's work. Clouds, coatlines, and fluid turbulence are examples of natural phenomena which are best described in the language of fractal geometry. He published many books including *Fractals: Form, Chance, and Dimension* (1977) and *The Fractal Geometry of Nature* (1982).

Mandelbrot received the *Légion d'Honneur* in 1989, the Wolf Prize for physics in 1993 and the Japan Prize for Science and Technology in 2003.

Figure 109.   Maurice René Fréchet

Figure 110.   Benoit Mandelbrot

Maurice René Fréchet

Maurice Fréchet was born on 2 September 1878 in Maligny, Bourgogone, France. At the Lycée Buffon in Paris, he was taught mathematics by Hadamard. He entered the *École Normale Supérieure* in 1900 and he was awarded his *Agrégation des Sciences Mathematiques* in 1903.

He submitted his doctoral dissertation *Sur quelques points du calcul functional* in 1906. It concerns 'functional operations' and 'functional calculus' and was developed from the ideas of Hadamard and Volterra.[3] He married Suzanne Carrive in 1908, and they had four children.

During World War I, he served for about two and a half years as an interpreter the British Army. He was a professor of mechanics at the Faculty of Science in Poitiers from 1910 until 1919 and at Strasbourg from 1919 until 1927.

He moved to Paris and became a professor at the Sorbonne and the *École Normale Supérieure*. He was elected to the *Académie des Sciences* in 1956. Fréchet made major contributions to the topology of point sets, statistics, probability and calculus, and defined and founded the theory of abstract spaces.

He died on 4 June 1973 in Paris, France.

Figure 111. Sergei N. Bernstein

Figure 112. André Weil

Sergei Natanoich Bernstein

Sergei Natanoich Bernstein was born on 5 March 1880 in Odessa, Ukraine. After graduating from high school in 1898 he went to France. His doctoral dissertation which contained a general result about a class of elliptical partial differential equations was submitted to the Sorbonne in 1904.

He defended his thesis before Poincaré, Hadamard and Picard. When he returned to Russia in 1905 he had to receive a Candidate degree in Russia in order to teach at Russian universities. He received a Candidate degree in St. Petersburg in 1906 and another one in Kharkov University in 1908. In 1913 he received a Doctor of Science degree solving Hilbert's twentieth problem which demonstrates the existence of solutions of partial differential equations with given boundary values and generalization of regular variational problems.

He taught at Kharkov University for 25 years beginning in 1907. Beginning in 1933 he lectured at the University of Leningrad and also at the Leningrad Polytechnic Institute.

In 1943 he moved to the University of Moscow. He greatly extended Chebyshev's 1854 works. In 1911, he introduced the Bernstein polynomials to give a constructive proof of the Weierstrass theorem that a continuous function on a finite subinterval of the real line can be uniformly approximated as closely as we wish by a polynomial.[3]

He generalized Lyapunov's conditions for the central limit theorem and worked on Markov's processes and stochastic processes.

He is not to be confused with Felix Bernstein (1878–1956) who was an extraordinary professor at Göttingen from 1901 to 1933.

Sergei Bernstein died on 26 October 1968 in Moscow, USSR.

André Weil

André Weil was born to Jewish parents on 6 May 1906 in Paris, France. He graduated from the *École Normale Supérieure* in 1925 and received his doctorate in 1928 with his thesis on Diophantine Equations at the Sorbonne.[10]

He was a professor at the Aligarh Muslim University from 1930 until 1932, lecturer at Marseilles from 1932 until 1933, and at Strasbourg from 1933 until 1939.

In 1935, Weil and his colleagues introduced the fictitious Nicholas Bourbaki as a member of the Royal Academy of Poldevia. Bourbaki was even said to have a daughter named Betti. His colleagues were Henri Cartan, Jean Delsarte, Claude Chevally and Jean Dieudonne.

In November 1939 Weil was arrested by the Finnish Police in Helsinki and they tried to have him executed as a spy.

Finland was already in war with Soviet Union at that time.

Finnish mathematician Rolf Nevanlinna heard of the situation and arranged it so that Weil was deported to Sweden. He was sent to Scotland, then London, then Southampton and finally to Le Havre jail in France.

He was officially a French soldier and was tried in May 1940 for failing to report for duty. He received five year suspended sentence and was discharged from the Army.

He arrived in New York on 3 March 1941 and received a Rockefeller Foundation grant to work at the Haverford College. From 1945 to 1947 he was at the University of Sao Paulo in Brazil.

In 1947 Weil was invited to join the University of Chicago by Marshall Stone, and he stayed there until 1958. His colleagues were Marshall Stone, S.S. Chern, Antoni

Zygmund and Saunders MacLane. The University of Chicago became one of the great mathematics research centers in the world.

Weil moved to the Institute for Advanced Study at Princeton in 1958, and he retired from there in 1976.

In 1949 he proposed the Weil Conjectures about the congruence Zeta function of algebraic varieties over finite fields. In 1978 Deligne was awarded a Field Medal for solving the Weil Conjectures.

Weil was elected a Fellow of the Royal Society of London in 1966. He was also elected to the *Académie des Sciences* in Paris and to the National Academy of Sciences in the United States.

In 1979 he was awarded the Wolf Prize, the Steele Prize in 1980, and the Kyoto Prize in 1994.

Weil died on 6 August 1998 in Princeton, New Jersey.

Figure 113. Jean-Christophe Yoccoz

Figure 114. Wendelin Werner

Jean-Christophe Yoccoz

Jean-Christophe Yoccoz was born on 29 May 1957 in France. He placed first in the entrance examinations at the *École Normale Supérieure* and the *École Polytechniquein* 1975 as Hadamard, Borel and Darboux had many year before. He studied at the *École Normale* and received his *Agrégation de Mathématiques* in 1977. He received a doctorate in 1985 at the University of Paris-Sud Orsay where he was then appointed a professor.

In 1994, he was awarded the Fields Medal for his work on many important results in dynamical systems.

Yoccoz has developed a method of combinational study of Julia sets and Mandelbrot sets-called Yoccoz puzzles.

He was appointed to the chair of differential equations and dynamical systems at the *Collége de France* in 1995.[19]

Wendelin Werner

Wendelin Werner was born on 23 September 1968 in Germany. He became a French national in 1977 and graduated from the *École Normale Supérieure* in 1991. His 1993 doctorate was awarded at the University of Paris VI (Pierre-et-Marie Curie), supervised by Jean-Francois Le Gall.

Werner was awarded the Fields Medal in 2006 for introducing new ideas and concepts combining probability theory and complex analysis for understanding critical phenomena in certain phase transitions such as the transition from liquid to gas.

He also received the European Mathematical Society Prize in 2000, the Fermat Prize in 2003 and the Polya Prize in 2006.

He has also been an actor, appearing in the 1982 French film *La Passante du Sans-Souci.*

Jean Leray

Jean Leray was born on 7 November 1906 in Chantenay, near Nantes France. He graduated from *the École Normale Supérieure* in 1929 and he received his docotorate in 1933 at Sorbonne supervised by Henri Villat. He was a professor at the University of Nancy in 1938 and 1939.

During World War II, he served as an officer in the French Army. He was captured in 1940 and was placed in a prisoner of war camp in Austria until the war ended in 1945.

While at the camp Leray and some of his fellow prisoners organized a *"Université en captivité"* and Leray became its director. Although he was an expert in hydrodynamics he claimed to be a topologist so that the Germans would not force him to do other work for them.

After his release in 1945 Leray published a three-part book Algebraic topology he developed in captivity. He was a professor at Sorbonne from 1945 until 1947, when he was appointed to the chair of the differential and functional equations at the *Collége de France.* He retired in 1978.

Leray combined energy estimaties for partial differential equations with ideas from algebraic topology (such a fixed point theorems) which cracked open the toughest problems.[3]

He married Marquerite Trunier on 20 October 1932, and they had three children.

Leray was elected to the *Académie des Sciences* in 1953 and to the National Academy of Sciences in the United States in 1965. He was also a Fellow of the Royal Society of London and the USSR Academy of Sciences.

He received the Wolf Prize in 1979 and the Lomonosov Gold Medal in 1988. He was also made *Commandeur de la Légion d'Honneur.*[3]

He is best known for the Leray spectral sequences, the Leray Cover, the Leray's theorem and the Leray-Hirsch theorem. He died on 10 November 1998 in La Baule, Loire-Atlantique, France.

Figure 115.   Jean Leray

Figure 116.   Armand Borel

Armand Borel

Armand Borel was born on 21 May 1923 in La Chaux-de-Fonds, Switzerland. He graduated from the *Eidgenossische Technische Hochschule Zürich* in 1947 after completing his military service. Borel worked for two years at ETH as a teaching assistant. He spent a year in France an exchange grant from the French *Centre National de la Recherche Scientifique* (CNRS). While in Paris, where he met Henri Cartan, Jean Dieudonné, Laurent Schwartz and Jean Leray.

He returned to Geneva and wrote his doctoral dissertation on the cohomology with integer coefficients of Lie groups which he defended at the Sorbonne in 1952. Jean Leray was his thesis supervisor.

In 1952 Borel married Gabrille (Gaby) Aline Pittet and they had two daughters. After working at the Institute for Advanced Study at Princeton and at the University of Chicago, he was appointed professor of mathematics at ETH in 1955. Then in 1957 he accepted a permanent professorship at the Institute for Advanced Study at Princeton where he spent most of the remainder of his career.

He worked on Lie groups and their actions, as well as on algebraic and arithmetic groups and opened core questions regarding many different areas; algebraic topology, differential geometry, algebraic geometry and number theory.[3]

Borel was elected to the *Académie des Sciences* and the National Academy of Sciences in the United States in 1987. He received the Steele Prize in 1991.

He died on 11 August 2003 in Princeton, New Jersey.

Jacques Charles François Sturm

Jacques Sturm was born on 29 September 1803 in Geneva, Switzerland. When he was 16 years old, his father died and his family became very poor. Sturm became a tutor to the youngest son of Mme de Staël. The Staël family spent six months in Paris in 1823 and Sturm naturally accompanied them.[3] Sturm met many distinguished mathematicians there.

Sturm lived at François J. D. Arago's (1786–1853) house as a tutor to his son. Arago later helped Sturm to be appointed as professor of mathematics at the Collège Rollin. Sturm became a French citizen in 1833 and was elected to the *Académie des Sciences* in 1836.[3]

Sturm and Liouville published papers in 1837–37 known today as the Sturm–Liourille problem, an eigenvalue and eigenfunction expansions in second-order linear differential equations.

Sturm became a professor of analysis and mechanics at the *École Polytechnique* as well as professor of mechanics in the *Faculté des Sciences*, Paris succeeding Poisson in 1840.[3]

Sturm died on 18 December 1855 in Paris after a long illness.

Russian School

Martin Bartels   ⟶   Nikolai I. Lobachevsky (1792–1856)

(1769–1833)   1811 Kazan Univ

               1822 Prof Kazan

               1827 Rector Kazan

Nikolai D. Brashman ⟶ Pafnuty L. Chebyshev ⟶ Dmitri A. Grave ⟶ Boris N. Delone

(1796–1866)   (1821–1894)   (1863–1939)   (1890–1980)

             1849 D.Sc. St. Petersburg   1896 St. Petersburg   1913 Kiev

             5 students

                                 ⟶ Aleksandr Lyapunov

                                   (1857–1918)

                                   D.Sc. 1892 Moscow

                                 ⟶ Andrei A. Markov

                                   (1856–1922)

                                   1884 D.Sc. St. Petersburg

                                   4 students

D.A. Grave   ⟶   Nikolai G. Chebotaryov   ⟶   Mark G. Krein

               (1894–1947)   (1907–1989)

               1928 Prof Kazan   1929 Ph.D. Odessa State U.

                                     1938 D.Sc. Moscow State U.

                                     1982 Wolf Prize

                                     50 students

B.N. Delone ⟶ Igor R. Shafarevich ⟶ Yuri Manin ⟶ Vladimir G. Drinfeld

               (1923–   )   (1937–   )   (1954–   )

               1946 Steklov Inst.   1961 Ph.D. Moscow   1978 Ph.D. Moscow

               NAS member   45 students   1988 D.Sc. Steklov

                                           1990 Fields Medal

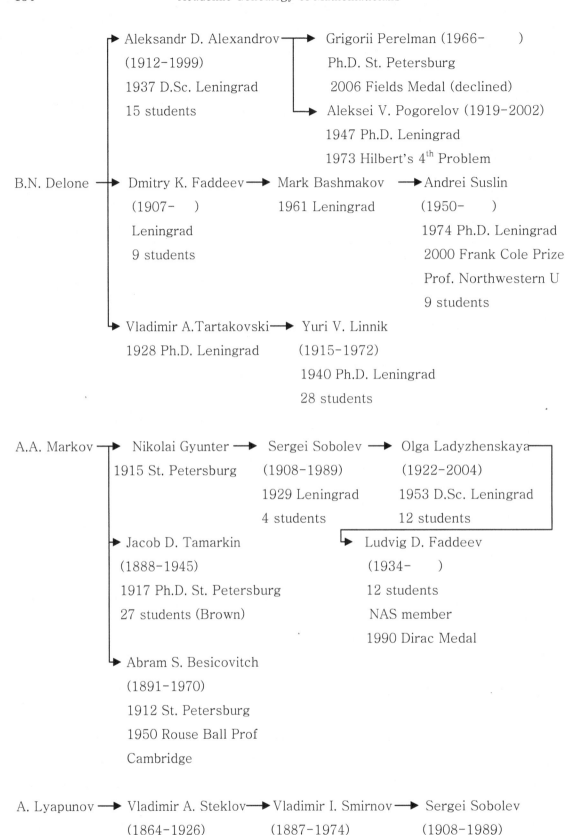

Aleksandr D. Alexandrov → Grigorii Perelman (1966-   )
(1912-1999)   Ph.D. St. Petersburg
1937 D.Sc. Leningrad   2006 Fields Medal (declined)
15 students   Aleksei V. Pogorelov (1919-2002)
                          1947 Ph.D. Leningrad
                          1973 Hilbert's 4th Problem

B.N. Delone → Dmitry K. Faddeev → Mark Bashmakov → Andrei Suslin
              (1907-   )   1961 Leningrad   (1950-   )
              Leningrad   1974 Ph.D. Leningrad
              9 students   2000 Frank Cole Prize
                                                     Prof. Northwestern U
                                                     9 students

Vladimir A.Tartakovski → Yuri V. Linnik
1928 Ph.D. Leningrad   (1915-1972)
                           1940 Ph.D. Leningrad
                           28 students

A.A. Markov → Nikolai Gyunter → Sergei Sobolev → Olga Ladyzhenskaya
              1915 St. Petersburg (1908-1989)   (1922-2004)
                                   1929 Leningrad   1953 D.Sc. Leningrad
                                   4 students   12 students

Jacob D. Tamarkin   Ludvig D. Faddeev
(1888-1945)   (1934-   )
1917 Ph.D. St. Petersburg   12 students
27 students (Brown)   NAS member
                                   1990 Dirac Medal

Abram S. Besicovitch
(1891-1970)
1912 St. Petersburg
1950 Rouse Ball Prof
Cambridge

A. Lyapunov → Vladimir A. Steklov → Vladimir I. Smirnov → Sergei Sobolev
              (1864-1926)   (1887-1974)   (1908-1989)
              1901 D.Sc. Kharkov   1936 D.Sc. Leningrad

Sergei Yu Maslov ——————→ Yuri V. Matiyasevich
(1939–1982)   (1947–   )
Car Accident   1970 Ph.D. Steklov
                 Hilbert's 10th Problem

Eduard Kummer ——→ Nicolai Bugaev ——→ Dmitri F. Egorov ——————→ Nicolai N. Luzin
              (1837–1903)   (1869–1931)   (1883–1950)
              1863 Ph.D. Moscow 1901 D.Sc. Moscow 1915 D.Sc. Moscow
              1866 D.Sc. Moscow 12 students   14 students

                                           → Ivan G. Petrovsky
                                           (1901–1973)
                                           1951–73 Rector MSU

                                           → Pavel S. Aleksandrov
                                         (1896–1982)
                                         1921 Ph.D. Moscow
                                         1927 D.Sc.
                                         14 students

N.N. Luzin ——————→ Andrei N. Kolmogorov (1903–1987)
               1927 Ph.D. Moscow
               1980 Wolf Prize
               79 students

             → Aleksandr Khinchin ——→ Aleksandr O. Gelfond ——→ I.I. Piatetski–Shapiro
               (1894–1959)   (1906–1968)   (1929–   )
               1927 Prof. Moscow   1935 D.Sc. Moscow   1954 Ph.D.
                                                 1990 Wolf Prize

                                  → Boris V. Gnedenko
                                    (1912–1995)
                                    1937 Ph.D. Moscow
                                    1942 D.Sc. Moscow
                                    7 students

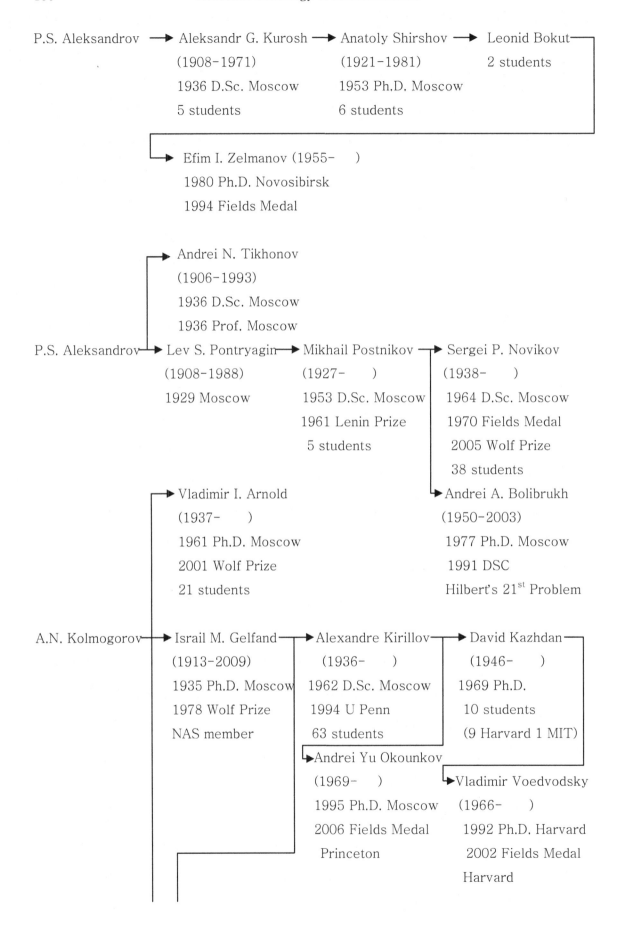

P.S. Aleksandrov ⟶ Aleksandr G. Kurosh ⟶ Anatoly Shirshov ⟶ Leonid Bokut

(1908-1971) (1921-1981) 2 students

1936 D.Sc. Moscow 1953 Ph.D. Moscow

5 students 6 students

Efim I. Zelmanov (1955-   )

1980 Ph.D. Novosibirsk

1994 Fields Medal

Andrei N. Tikhonov

(1906-1993)

1936 D.Sc. Moscow

1936 Prof. Moscow

P.S. Aleksandrov ⟶ Lev S. Pontryagin ⟶ Mikhail Postnikov ⟶ Sergei P. Novikov

(1908-1988) (1927-   ) (1938-   )

1929 Moscow 1953 D.Sc. Moscow 1964 D.Sc. Moscow

1961 Lenin Prize 1970 Fields Medal

5 students 2005 Wolf Prize

38 students

Vladimir I. Arnold Andrei A. Bolibrukh

(1937-   ) (1950-2003)

1961 Ph.D. Moscow 1977 Ph.D. Moscow

2001 Wolf Prize 1991 DSC

21 students Hilbert's 21st Problem

A.N. Kolmogorov ⟶ Israil M. Gelfand ⟶ Alexandre Kirillov ⟶ David Kazhdan

(1913-2009) (1936-   ) (1946-   )

1935 Ph.D. Moscow 1962 D.Sc. Moscow 1969 Ph.D.

1978 Wolf Prize 1994 U Penn 10 students

NAS member 63 students (9 Harvard 1 MIT)

Andrei Yu Okounkov

(1969-   )

1995 Ph.D. Moscow Vladimir Voedvodsky

2006 Fields Medal (1966-   )

Princeton 1992 Ph.D. Harvard

2002 Fields Medal

Harvard

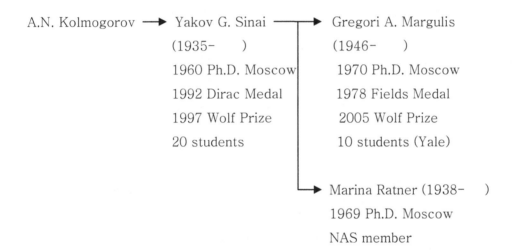

G.E. Shilov → A.K. Kostinchenko → V.A. Sadovnichy
(1939-   )
1974 D.Sc. Moscow
Rector MSU

Eugene B. Dynkin (1924-   )
1948 Ph.D. Moscow
NAS member

Vladimir A. Rokhlin → Mikhail Gromov
(1919-1984)   (1943-   )
1951 Ph.D   1968 Ph.D. Leningrad
12 students   1993 Wolf Prize
                     2009 Abel Prize
                     5 students

A.N. Kolmogorov → Yakov G. Sinai → Gregori A. Margulis
                  (1935-   )   (1946-   )
                  1960 Ph.D. Moscow 1970 Ph.D. Moscow
                  1992 Dirac Medal   1978 Fields Medal
                  1997 Wolf Prize   2005 Wolf Prize
                  20 students   10 students (Yale)

                                     Marina Ratner (1938-   )
                                     1969 Ph.D. Moscow
                                     NAS member

Russian School

The Russian academic degree system differs from the systems in Western Europe and the United States. The first degree granted by Russian universities, the Diplom, comes after five years of study. The Diplom is comparable to a Master's degree. Graduate school (Aspirantura) award the first advanced degree, the Candidate of Science (Kandidat Nauk) after three years of study and research with mandatory publications in peer reviewed journals and public defense of the thesis with open scientific discussions. Russians believe the Candidate degree is equivalent to the Ph.D.

The final and highest academic degree is the Doctor of Science (Doktor Nauk). This degree is granted to those who made a substantial contribution to research in their area of study, after public defense of a thesis and a number of publications including monographs.

Only holders of Doctor of Science qualify for the academic title "Professor" awarded by the Supreme Certifying Committee of the Government. The degree is similar to German habilitation.

Russian Universities have a Faculty of Mathematics and Mechanics because, traditionally, the study of mathematics includes mechanics.

Nikolai Ivanovich Lobachevsky

Nikolai Lobachevsky was born on 1 December 1792 in Nizhny Novgorod (was Gorky from 1932-1990), Russia. In 1807 Lobachevsky graduated from Kazan Gymnasium and entered Kazan University which have been founded only three years before. Lobachevsky met and worked with J.C. Martin Bartels (1769-1833), Professor of Mathematics from Germany and friend of Gauss.[3]

Lobachevsky received a Diplom in physics and mathematics in 1811. In 1814, Kazan University appointed him to a lectureship. Two year later he became an extraordinary professor and in 1822 he became a full professor.

In 1827, Lobachevsky became Rector of Kazan University holding the position until 1846. The University of Kazan flourished under his rectorship. Under Lobachevsky's leadership the university embarked on a vigorous program of construction-a library, an

---

Russian middle name represents the patronymic, father's first name. In case of male it ends with "vich" and in case of female, it ends with "vna".

astronomical observatory, medical facilities and physics, chemistry and anatomy laboratories.

Despite the heavy administrative load, he managed to teach many courses such as mechanics, hydrodynamics, integration, differential equations, the calculus of variation and mathematical physics.[3]

In 1832, he married Varvara Alexivna Moisieva and they had seven children. At the time of his marriage he was forty years old while the bride was a young girl.

His idea on non-euclidean geometry was first reported on 23 February 1826 at the University of Kazan and he published the work on non-euclidean geometry in 1829. The non-Euclidean geometry that Lobachevsky developed was called hyperbolic geometry. Euclid's parallel postulate states that, in two-dimensional space, given a point and a separate line, only one line can be drawn through the point that does not intersect the separate line. In hyperbolic geometry (Wolfram Math World),

For any infinite straight line L and any point P not on it, there are many other infinitely extending straight lines that pass through P which do not intersect L.

A famous consequence is that the sum of angles in a triangle must be less than 180 degrees.

Janos Bolyai independently discovered hyperbolic geometry and published in 1832. Even before that, Gauss described it in a 1824 letter to Taurinus but did not publish his work.

Figure 117. Nikolai I. Lobachevsky

Lobachevsky also developed a method for the approximation of the roots of algebraic equations.

Soon after he was dismissed by the University of Kazan, for reasons that remain obscure, his favorite eldest son died. Lobachevsky was hit hard by this tragedy. His health deteriorated and he became blind. His great mathematical achievements were not recognized in his lifetime and he died without having any notion of the fame that his work would achieve.[3]

Lobachevsky died on 24 February 1856 in Kazan, Russia.

Nikolai Dmetrievich Brashman

Nikolai Brashman was born on 14 June 1796 in Rassnova near Brno in what is now the Czech Republic. He studied at the University of Vienna and Vienna Polytechnic Institute. After which he worked at the Kazan University. In 1834, the Moscow State University selected him as a professor of applied mathematics.

Brashman wrote research articles on the Principle of Least Action. He wrote one of the best analytic geometry texts of his time, for which he was awarded the Demidov Prize by the Russian Academy of Sciences in 1836.[3]

While at Moscow State University, Brashman founded the Moscow Mathematical Society and its journal.

Brashman died on 13 May 1866 in Moscow, Russia.

Pafnuty Lvovich Chebyshev

Pafnuty Chebyshev (Tchebycheff) was born on 16 May 1821 in Okatovo, west of Moscow, Russia. Chebyshev entered Moscow State University in 1837 which was founded in 1755. He graduated in 1841 and continued to study for his Candidate degree under Brashmann's supervision.

Chebyshev defended his Doctor of Science thesis on 27 May 1849. This work received a prize from the Russian Academy of Sciences. He was promoted to extraordinary professor at the University of St. Petersburg in 1850 and full professor in 1860.

His famous Chebyshev polynomials appeared in his 1854 paper *Théorie des mécanismes connus sous le nom de paralélogrammes*. Chebyshev learned French as a child from his cousin. The Chebyshev polynomials are widely used in electrical filter design.

In 1852 he visited France, London and Berlin meeting Liouville, Hermite, Serret, Lebesgue, Poncelet, Cayley, Sylvester and Dirichlet.

He never married but he had a daughter whom he refused to officially acknowledge.[3] He retired from the University of St. Petersburg in 1882.

Chebyshev was elected to the Berlin Academy of Sciences in 1871, the Royal Society of London in 1877. He was awarded the French *Légion d'Honneur*.

He died after a heart attack on 8 December 1894 in St. Petersburg, Russia.

Figure 118.   Pafnuty L. Chebyshev

Dmitry Aleksandrovich Grave

Dmitry Grave was born on 6 September 1863 in Kirillov, Russia. He received the Candidate degree from the University of St. Petersburg in 1889 and the Doctor of Science degree in 1896. His doctoral thesis was on map projections which were built on ideas of Euler, Lagrange and Chebyshev.

Grave became professor at Kharkov in 1897 and, from 1902, he was appointed professor at the University of Kiev, where he founded the Kiev school of algebra.[3]

At Kiev, he studied algebra and number theory. After the Russian Revolution in 1917, the new government required mathematical research to have more pragmatic application. As a result, Grave abandoned algebra and instead elected to teach applied mathematics.

Grave served as the first director of the Institute of Mathematics of the Academy of Sciences in Kiev in 1934 simultaneously holding the professorship at the Kiev University. He remained in both positions until his death in 1939.

He was elected to the Academy of Sciences of the USSR in 1929. He died on 19 December 1939 in Kiev, Ukraine.

Boris Nikolaevich Delone

Boris Delone (sometimes spelled Delaunay) was born on 15 March 1890 in St. Petersburg, Russia. Delone got his surname from an Irish ancestor named Deloney, who was among the mercenaries left in Russia after Napoleon's invasion of 1812. By 1913 Delone had become one of the top three mountain climbers in Russia. The Soviet government recognized his expertise by naming him a Master of Mountain Climbing of the USSR.

He graduated from Kiev University in 1913 probably receiving his Candidate degree. After teaching at Kiev he moved to Petrograd (former St. Petersburg later changed to Leningrad in 1924) in 1920.

He worked as a professor at the Leningrad University from 1922 to 1935. When he accepted an appointment at the Moscow State University where he stayed until 1958.

Delone also led the Algebra Department of the Steklov Mathematical Institute which moved from Leningrad to Moscow in 1935. During that period, he invented Delone triangulation. He died on 17 July 1980 in Moscow, Russia.

Figure 119.   Boris N. Delone

Figure 120.   Aleksandr M. Lyapunov

Aleksandr Mikhailovich Lyapunov

Aleksandr Lyapunov was born on 6 June 1857 in Jaroslavl, Russia. He graduate the Gymnasium in Gorky (Nizhny Novgorod) in 1876 where he befriended Andrei Markov. They both entered the St. Petersburg University where Lyapunov graduated in 1880.

He received his Candidate of Science degree at St. Petersburg in 1885 with his thesis *On the stability of ellipsoidal forms of equilibrium of a rotating liquid*

He became a lecturer (dozent) at Kharkov University where he continued research for his doctoral thesis. Lyapunov was awarded the Doctor of Science degree at Moscow State University in 1892 with his thesis *The general problem of the stability of motion.*

In 1893 he was appointed professor at Kharkov University where he remained until 1902. He married his cousin Natalia Rafailovna Sechenov in 1886.

From 1902 to 1917 Lyapunov worked at the Russian Academy of Sciences in St. Petersburg devoting himself completely to scientific research.

He was elected to the French Academy of Science in 1916. The second method of Lyapunov is still widely used in the study of nonlinear system stability. If the system has an asymptotically stable equilibrium state, then the stored energy of the system displaced within the domain of attraction decays with increasing time until it finally assumes its minimum value at the equilibrium state. For purely mathematical systems, it is very hard to define an "energy function."

Lyapunov also introduced the Lyapunov function, a fictitious energy function. Any scalar function satisfying the hypotheses of Lyapunov stability theorems can serve as Lyapunov functions.[21]

In 1917 he moved to the University in Odessa, where his wife Natalia's health deteriorated rapidly. She died of tuberculosis on 31 October 1918. On the same day he shot himself and died on 3 November 1918 in a hospital in Odessa.

Andrei Andreyevich Markov

Andrei Markov was born on 14 June 1856 in Ryazan, 330 km southwest of Gorky. He entered St. Petersburg University in 1874 two years before Lyapunov and graduated in 1878 receiving the gold medal for the best essay. He was awarded the Candidate degree in 1880 for his thesis on the binary quadratic forms with positive determinant.

While teaching as a privatdozent he received his Doctor of Science degree in 1884 for his dissertation on certain applications of continued fractions.[3]

Figure 121. Andrei Andreyevich Markov

In 1883 Markov married Maria Ivanova Valvatyeva. He became an extraordinary professor in 1886 and ordinary professor in 1893 at St. Petersburg University.

Russian Academy of Sciences elected him as a corresponding member in 1890 and a full member 1896. His son of the same name was born on 9 September 1903 and became a renowned mathematician in his own right.

Markov was against the Romanov dynasty which had ruled since 1613. He volunteered to teach mathematics at a secondary school in Zarajsk, 130 km southeast of Moscow. He returned to St. Petersburg and continued to lecture on probability at the university in 1921. He died on 20 July 1922 in St. Petersburg.

The Markov matrix A is an $n \times n$ nonnegative matrix such that the sum of it's entries in each column equal to 1. It is interesting that the columns of A^r approaches the eigenvectors of a dominant eigenvalue of A for large r.

For example

$$A= \begin{bmatrix} 0.8 & 0.3 & 0.2 \\ 0.1 & 0.5 & 0.4 \\ 0.1 & 0.2 & 0.4 \end{bmatrix}$$

The eigenvalues of A are 1, 0.5303 and 0.1697. The eigenvector for the dominant eigenvalue 1 is $(0.8744, 0.3975, 0.2782)^T$. And A^{50} is given by

$$A^{50} = \begin{bmatrix} 0.5641 & 0.5641 & 0.5641 \\ 0.2564 & 0.2564 & 0.2564 \\ 0.1795 & 0.1795 & 0.1795 \end{bmatrix}$$

Each column of A^{50} and all higher powers of A is actually a multiple of (0.8744 0.3975 0.2782)^T.

Markov is best remembered for his study of Markov Chains, sequences of random variables in which the future variable is determined by the present variable but is independent of the way in which the present state arose from its predecessors.[3] This work launched the theory of stochastic process which is very important in modern communication theory.

Figure 122. Mark G. Krein

Figure 123. Igor R. Shafarevich

Mark Grigorievich Krein

Mark Krein was born into a Jewish family on 3 April 1907 in Kiev, Ukraine. He studied at Odessa University under Nikolai Chebotaryov and received his Candidate degree in 1929. He built up one of the most important centres for functional analysis research in the world. He was awarded Doctor of Science degree in 1938 at Moscow State University without defense of his thesis. In those days, doctorates were awarded this way occasionally.

In 1941 he left Odessa when the university was evacuated due to the German invasion. When he returned to Odessa in 1944 he was dismissed and the functional analysis group was closed down presumably for having too many Jewish students.

From 1944 he held the chair of theoretical mechanics at Odessa Marine Engineering Institute. From 1944 to 1952 he also held a part-time head of the functional analysis and algebra department at the Mathematical Institute of the Ukrainian Academy of Sciences in Kiev.[3]

From 1954 until his retirement in 1974 Krein was professor of theoretical mechanics at the Odessa Civil Engineering Institute.

Despite the persecution he suffered, Krein published more than 270 papers and monographs. The National Academy of Sciences in the United States elected him a Foreign Member in 1979 and he received the Wolf Prize in 1982. But the Soviet government prevented him from attending the award ceremony.

He died on 17 October 1989 in Odessa, USSR.

Igor Rostislavovich Shafarevich

Igor Shafarevich was born on 3 June 1923 in Zhytomyr, Russia. He was awarded Candidate degree in 1942 and Doctor of Science degree in 1946 at Moscow State University. He became an associate at the Steklov Mathematical Institute of the USSR Academy of Sciences and also taught at Moscow State University (1946-1974).

He publicly supported Andrei Sakharov's Human Rights Committee from 1970 and became an active participant in supporting Solzhenitsyn. As a consequence the Moscow State University dismissed him in 1975.

His most noteworthy work includes the theory of the Tate-Shafarevich group in Galois cohomology and the Golod-Shafarevich theorem on class field theory.

He was awarded the Lenin Prize in 1959 and elected to the USSR Academy of Sciences in 1991. Shafarevich was president of the Mathematics Society of Moscow from 1970 to 1974.

The US National Academy of Science recognized his work by electing him as a foreign member in 1974.

Yuri Ivanovich Manin

Yuri Manin was born on 16 February 1937 in Simferopol in the Crimean peninsula near the Black Sea, USSR. He graduated from Moscow State University in 1958 and received his Candidate degree in 1960 at which time he was appointed as Principal Researcher at the Steklov Mathematical Institute in Moscow.

Figure 124. Yuri I. Manin

Figure 125. Vladimir G. Drinfeld

In 1963 he received the Doctor of Science degree at the Steklov Mathematical Institute and two years later. Moscow State University appointed him as professor of algebra.

He spent two years at Columbia University and Massachusetts Institute of Technology in 1991−93. He went to the Max Planck Institute, Bonn in 1993 and retired in 2005.

He wrote many books on mathematics and physics.

Manin received many honors including the Lenin Prize in 1967 and the Georg Cantor medal of the German Mathematical Society in 2002.

Vladimir Gershonovich Drinfeld

Vladimir Drinfeld was born on 4 February 1954 in Kharkov, Ukraine. He won a gold medal in the International Mathematics Olympiad in 1969 and graduated from Moscow State University in 1974.

Drinfeld received his Candidate degree in 1978 at Moscow State University and he was awarded Doctor of Science degree in 1988 at Steklov Mathematics Institute in Moscow.

In 1985 Drinfeld introduced the idea of quantum groups independently with Japanese mathematician Jimbo. He also proved the Langlands conjecture for GL(2) over a functional field.

In 1990 he was awarded a Fields Medal at the International Congress of Mathematicians in Kyoto, Japan for his work on quantum groups and for his work in number theory.

He was elected a member of the Academy of Sciences of the Ukraine in 1992. He currently serves as the Harry Pratt Judson Distinguished Service Professor at the University of Chicago.

Figure 126. Aleksandr D. Aleksandrov

Figure 127. Grigori Y. Perelman

Aleksandr Danilovich Aleksandrov (Alexandrov)

Aleksandr Aleksandrov was born on 4 August 1912 in Volyn, Ryazan (Rázan), Russia, but he lived in St. Petersburg (Petrograd then Leningrad) from a very young age. He studied physics under V.A. Fok and mathematics under B.N. Delone at the Leningrad State University where he graduated with Diplom in theoretical physics in 1933.

He received his Candidate degree in 1935 with his thesis on the topic of mixed volumes of convex bodies. Aleksandrov was awarded Doctor of Science degree in 1937 for a thesis on additive set functions and the geometrical theory of weak convergence.[3]

In 1952 he became Rector of the Leningrad State University. In 1964 he was appointed to chair of geometry at the University of Novosibirsk and at the Mathematical Institute of the Siberian Branch of the USSR Academy of Sciences.

His love of mountaineering originated from by his teacher Delone. On his fiftieth birthday he climbed to 6222m in the Shakhdarinsk ridge, Pamir.

He received the Lobachevsky Prize in 1951. Aleksandrov died on 27 July 1999 in St. Petersburg.

Grigori Yakovlevich Perelman

Grigori Perelman was born on 13 June 1966 in Leningrad (now St. Petersburg), Russia. He is known as Grisha Perelman. He won a gold medal in the International Mathematical Olympiad in 1982.

He received his Candidate degree at the Leningrad State University with his thesis Saddle Surfaces in Euclidean Spaces.

In the late 1980's and early 90's, he held posts at several universities in the United States returning to the St. Petersburg Department of the Steklov Institute of Mathematics in 1996.

Richard Hamilton of Princeton University had been investigating an approach to solve the Poincaré Conjecture using the Ricci Flow which describes the behavior of a tensorial quantity, the Ricci curvature tensor.

Hamilton's hope was that under the Ricci flow, concentration of large curvature will spread out until a uniform curvature is achieved over the entire three-manifold.[22]

After many years of studying Hamilton's work and investigating the concept of entropy, Perelman wrote an article that combined with Hamilton's work would provide a proof of William Thurston's Geometrisation Conjecture and thus the Poincare Conjecture.[22]

Perelman was publicly offered the Fields Medal on 22 August 2006 for his contributions to geometry and his revolutionary insights into the analytical and geometric structure of the Ricci flow, but he declined to accept the medal.

According to some sources, in the spring of 2003, the faculty of the Steklov Institute allegedly declined to re-elect Perelman as a member which was a great disappointment for him. Some faculty members apparently had continuing doubt over his claims regarding the geometrization conjecture.

He also resented when the Fields Medalist Shing-Tung Yau's downplayed Perelman's role in the work of Huai-Dong Cao and Xi-Ping Zhu regarding the complete proof of the Poincaré conjecture.[22]

As of the spring of 2003 Perelman no longer works at the Steklov Institute. He is currently jobless, living with his mother in St. Petersburg.

Perelman is a talented violinist and play table tennis.

Yuri Vladimirovich Linnik

Yuri Linnik was born on 21 January 1915 in Belaya Tserkov, Ukraine. He graduated from Leningrad State University in 1938 and began his postgraduate studies at the same university. In the winter of 1939–1940 he was called up for military service and served as a platoon leader in the Soviet Army. In the spring of 1940 he defended his Doctor of Science degree, surpassing Candidate degree.

In June 1941 Linnik joined the People's Volunteer Corps to take part in fighting the Nazis German troops. In the autumn of 1941, being inside the besieged Leningrad, he suffered from dystrophia. The siege lasted until January 1944.

From 1944 until his death in 1972 he worked at the Leningrad Department of the Steklov Mathematical Institute of the USSR Academy of Sciences while also working as professor of mathematics at Leningrad State University. His main research interests were number theory, probability theory and mathematical statistics.

In 1970 he received the Lenin Prize and was elected to the USSR Academy of Sciences as a full member in 1964. He served as president of the Leningrad Mathematical Society from 1959 to 1965.

He spoke seven languages fluently and wrote poetry in Russian, English, German and French.[23] Linnik died on 30 June 1972 in Leningrad, Russia.

Figure 128.   Yuri V. Linnik

Sergei Lvovich Sobolev

Sergei Sobolev was born on 6 October 1908 in St. Petersburg. He graduated from the Leningrad State University in 1929. He taught at the Leningrad Electrotechnical Institute (now St. Petersburg Electrotechnical University) in 1930–31. Working with Vladimir Smirnov studied functionally invariant solutions of the wave equation. The methods led them to a complete theory of Rayleigh surface waves and Sobolev also solved a problem on diffraction.

Because of this work USSR Academy of Sciences elected him as corresponding member in 1933 at the age of 25. Six years later he became a full member of the Academy.

When the Steklov Mathematical Institute moved to Moscow in 1934 he also moved and was appointed Head of the Department of the Theory of Differential Equations. During World War II the Institute moved to Kazan where Sobolev was appointed the Director of the Institute in 1943. He also served as a professor at Moscow State University from 1935 to 1957, during which time he was appointed the chair of computational mathematics in 1952.

From 1960 to 1978 Sobolev was a professor at the Novosibirsk University and Director of the Institute of Mathematics. The *Académie des Sciences* in Paris elected him as a member and he received the 1988 M V Lomonosov Gold Medal. Sobolev introduced the Generalized functions later known as distributions in 1935 which were further developed by Laurent Schwartz in 1948.

Sobolev is remembered for Sobolev space, Sobolev inequality, Sobolev conjugate and Sobolev generalised derivative. He died on 3 January 1989 in Leningrad, Russia.

Figure 129. Olga A. Ladyzhenskaya

Olga Alexandrovna Ladyzhenskaya

Olga Ladyzhenskaya was born on 7 March 1922 in Kologriv near picturesque river Unzha Russia. Stalinist authorities executed her father, a mathamtics teacher, in 1937, because his familial links to Russian nobility. In 1943 she became a student at Moscow

State University studied under Petrovsky, Stepanov, Gelfand, Delone and Tikhonov before she graduated in 1947.

She moved to Leningrad and received her Candidate degree in 1949 at Leningrad State University for a thesis on the development of finite differences method for linear and quasilinear hyperbolic systems of partial differential equations. Officially, Sobolev was her supervisor, but in practice, Smirnov was her adviser.

In 1951 she completed her thesis for Doctor of Science degree but did not receive it until the death of Stalin in 1953. In 1954 she taught at Leningrad State University and became a researcher at the Steklov Mathematical Institute.

From the mid-1950's Ladyzhenskaya and her students made important advances in the study of boundary-value problems for quasilinear elliptic and parabolic equations. One result gave the solution to Hilbert's 19th problem for a second-order equation.[3]

She was elected a full member of the USSR Academy of Sciences in 1990 and awarded the Sofia Kovalevskaya Prize in 1992 as well as the Lomonosov Gold Medal. She served as President of the St. Petersburg Mathematical Society from 1990 to 1998.

She died unexpectedly in her sleep on 12 January 2004 in St. Petersburg.

Abram Samoilovitch Besicovitch

Abram Besicovitch was born on 24 January 1891 in Berdyansk, Russia. He graduated from the University of St. Petersburg in 1912 having been influenced by A.A. Markov. In 1916 Besicovitch married Valentina Vietalievna. When he escaped from Russia in 1924, his wife remained in Russia and the marriage was dissolved in 1928. In 1930 he married Valentina Alexandrovna, the sixteen-year-old daughter of a family friend.

From 1917 to 1920 Besicovitch served as a professor at the University of Perm where Alexander Friedmann was a colleague. In 1921 he moved to Leningrad University (at that time it was called Petrograd). He left Petrograd for Copenhagen in 1924 where he worked with Harald Bohr supported by the Rockefeller Fellowship.

The next year, G.H. Hardy saw the mathematical genius in Besicovitch (known Bessy) and found a post for him in Liverpool. In 1927 Besicovitch moved to University of Cambridge where he stayed until his death in 1970. While at Cambridge he succeeded Littlewood to the Rouse Ball Chair of Mathematics (1950-1958) after which he stayed at Trinity College as a Fellow.

Around 1930 he extended Hausdorff's density property of sets to those of finite Hausdorff measure.

Hausdorff and Besicovitch were the two mathematical pioneers on whose work Mandelbrot's development of fractals is based.[3]

The Royal Society elected Besicovitch a Fellow in 1934 and he received the Sylvester Medal from the Society in recognition of his outstanding work on almost-periodic functions, the theory of measure and many other topics of the theory of functions. In 1950 The London Mathematical Society awarded him the De Morgan Medal.

He spoke tolerably good English when he arrived in Britain in the 1920's, but he always ignored definite articles when speaking.

At a dinner at Cambridge University, Besicovitch asked the name of the tasty food that they were eating. When he heard the reply, he exclaimed, "In Russia we are not allowed to eat the peasants."[35]

He used to say "Mathematician reputation rests on number of his bad proofs."

Besicovitch died on 2 November 1970 in Cambridge, England.

Figure 130. Abram S. Besicovitch

Figure 131. Jacob D. Tamarkin

Jacob David Tamarkin

Jacob Tamarkin (originally Yakov Davydovich Tamarkin) was born into a wealthy family on 11 July 1888 in Chernigov, Ukraine. He attended the second Gymnasium in St. Petersburg with Alexander Friedmann where the two became very close friends. Smirnov attended the same school but he was one year older than them.

In 1905 Tamarkin and Friedmann wrote a paper on Bernoulli numbers which they submitted to Hilbert in Göttingen for publication. The paper was accepted and published in *Mathematische Annalen* in 1906.[3]

Tamarkin graduated from the Second Gymnasium in 1906 with a gold medal and entered the University of St. Petersburg. Tamarkin and Friedmann published another paper in Crelle's Journal in 1909 and both graduated in 1910.

Tamarkin received his Candidate degree in 1917 and his thesis was published in English in *Mathematische Zeitschrift* in 1928 as *Some general problems of the theory of ordinary linear differential equations and expansion of an arbitrary function in series of fundamental functions.*

In 1919 he married Helen Weichart and they had one son. The composer Shostakovitch and the son of Rimsky-Korsakov and Tamarkins used to play string quartets at the house of Tamarkins. From 1919 to 1922 he was professor at the University of Perm as a colleague of Friedmann and Besicovitch. In 1925 he emigrated to the United States where he taught at Dartmouth College. Two years later he was appointed to Brown University where he remained for the rest of his career. He served as editor of the Transactions of the American Mathematical Society (1927-1936) and the Annals of Mathematics (1928-1939). He later served as Vice President of the American Mathematical Society (1942-1943).

After a heart attack he died on 18 November 1945 in Washington, D.C.

Vladimir Andreevich Steklov

Vladimir Steklov was born on 9 January 1864 in Nizhny Novgorod (was Gorky from 1932-1990), Russia. Inheriting his parents talents Steklov had considerable musical and literary ability.

He studied at Moscow State University and Kharkov University where he graduated in 1887. In 1893, he received Candidate degree with his thesis on the equations of a solid body moving in an ideal non-viscous fluid. In 1902, he was awarded Doctor of Science degree for a dissertation on boundary-value problems of Dirichlet type using rigorous mathematical analysis.

In the same year Karkhov University appointed him to the Chair of Applied Mathematics succeeding Lyapunov who had moved to St. Petersburg.

In 1906, Steklov also moved to St. Petersburg where he taught Friedmann, Smirnov and Tamarkin. Four years later the Russian Academy of Sciences elected him as a member. In 1921, Steklov founded the Institute of Physics and Mathematics and served as its director until his death in 1926. In 1934, the Institute split into an Institute of Physics and a separate Institute of Mathematics. The latter was named the V.A. Steklov

Mathematical Institute and moved to Moscow, but a branch remained in Leningrad (St. Petersburg).

Steklov kept a diary for about 20 years, documenting in detail the events of his life. He gave details of letters he had received and letters he had written. He also made entries about those he visited and those who visited him.[3]

Steklov died on 30 May 1926 in Gaspra, Crimea, USSR.

Figure 132.   Vladimir A. Steklov

Figure 133.   Vladimir I. Smirnov

Vladimir Ivanovich Smirnov

Vladimir Smirnov was born on 10 June 1887 in St. Petersburg, Russia. He graduated the Second Gymnasium in St. Petersburg winning the gold medal. He was friendly with Tamarkin and Friedmann at the Gymnasium although he was one year older than them. They also studied together at St. Petersburg University. Smirnov published a joint paper with Friedmann in 1913 in the Journal of the Russian Physics-Chemical Society.

Smirnov wrote a five volume book *A Course in Higher Mathematics*, the first volume jointly with Tamarkin. From 1912 he taught at the St. Petersburg Institute of Railway Engineering and from 1919 to 1922 at Simferopol University in southern Ukraine.

Smirnov received his Doctor of Science degree in 1936 at Leningrad and was appointed to the Chair of Mathematics and Mechanics. He was also elected to the USSR Academy of Sciences.[3]

In 1959 he restarted the Leningrad Mathematical Society which was disbanded in the late 1920's. Sergei Sobolev was one of his students. Smirnov died on 11 February 1974 in Leningrad, USSR.

Figure 134. Nicolai V. Bugaev

Figure 135. Dmitri F. Egorov

Nicolai Vasilievich Bugaev

Nicolai Bugaev was born on 14 September 1837 in Dusheti near Tiflis now Tbilisi, Gruzia (Georgia). He graduated from Moscow University in 1859 and received his Candidate degree in 1863 at the same university. He also studied under Kummer and Weierstrass in Berlin and Liouville in Paris for two and half years. In 1866 Bugaev was awarded a Doctor of Science degree with his dissertation on numerical identities associated with the number e.

In 1867, Moscow State University appointed Bugaev as a professor. His work there eventually led to the creation of the Moscow School of the theory of functions of a real variable by Dmitri Egorov, one of his students. In 1911,[3] Bugaev was a member of the Moscow Mathematical Society and served as its president from 1891.

He delivered a paper on the philosophy of mathematics at the International Congress of Mathematics in Zurich in 1897.

He died on 11 June 1903 in Moscow, Russia.

Dmitri Fedorovich Egorov

Dmitri Egorov was born on 22 December 1869 in Moscow, Russia. He entered Moscow University in 1887 where he was most influenced by Bugaev. Egorov taught at Moscow

University from 1894 and received his Doctor of Science degree there in 1901. Moscow State University appointed him as a professor in 1903.

Egorov influenced Gaston Darboux's work on differential geometry and mathematical analysis. A theorem in real analysis, Egorov's theorem, is named in his honor.

Among his students N.N. Luzin, I.G. Petrovsky and P.S. Aleksandrov are well known. In 1922, he became president of the Moscow Mathematical Society. One year later he became director of the Institute for Mechanics and Mathematics at Moscow State University. Egorov was a deeply religious man and tried to prevent the imposition of Marxist methodology on scientists. In 1929, he was dismissed as director of the Institute for Mechanics and Mathematics because of his political views.

In 1930, he was arrested and imprisoned for being a religious sectarian. Not long thereafter the Moscow Mathematical Society expelled him denouncing him as a reactionary and a churchman.[3]

Egorov went on hunger strike in prison and was taken to the prison hospital in Kazan. Coincidentally, Mathematician Chebotaryov's wife was a doctor at the prison hospital. Chebotaryov and his wife took pity on Egorov, and with the prison's permission, allowed Egorov to stay at their home in Kazan where Egorov died on 10 September 1931.

Nikolai Nikolaevich Luzin

Nikolai Luzin was born on 9 December 1883 in Irkutsk, Russia. His family moved to Moscow in 1901 and he entered Moscow University. But his father lost all their savings in stock dealings which forced Luzin to move into a room owned by the widow of a doctor. In 1908 Luzin married the widow's daughter.

He graduated Moscow State University in 1905 but he was not sure whether to devote himself to mathematics, medicine or theology all of which he had studied.

In 1910 he started teaching mathematics at Moscow State University just before the Revolution. He was awarded Doctor of Science degree in 1915 for his thesis *The integral and trigonometric series* at Moscow originally intended for a candidate degree. Over the next ten years Luzin and Egorov built up an impressive research group at Moscow State University. He proved an important result in 1919 on the invariance of sets of boundary points under conformal mappings.[3] Ten years later The USSR Academy of Sciences elected Luzin as a full member.

From 1935 he also headed the Department of the Theory of Functions of Real Variables at the Steklov Mathematics Institute. After Egorov died in 1931 in despair and misery due to political persecution, Luzin became the victim of a violent political campaign through the newspaper Pravda. He was accused of publishing all of his important results abroad and only minor papers in Soviet Journals.

He fortunately escaped from a tragic fate as the Soviet authorities may have feared the international criticism for attacking a world renowned scientist.

From this moment on, Soviet mathematicians began to publish exclusively in Soviet journals in Russian.

Luzin died on 25 February 1950 in Moscow.

Figure 136. Nikolai N. Luzin   **Figure 137. Pavel S. Aleksandrov**

Pavel Sergeevich Aleksandrov (Alexandroff)

Pavel Aleksandrov was born on 7 May 1896 in Bogorodsk (Noginsk), Russia where his father worked as a doctor in the local hospital. Aleksandrov learned French and German from his mother. His home was always filled with music.

Aleksandrov entered Moscow State University in 1913 and became a student of Luzin, where he proved his first important result in 1915 that every non-denumerable Borel set contains a perfect subset.

In 1921 he received his Candidate degree and was appointed a lecturer at Moscow State University. In the summers of 1923 and 1924 Aleksandrov and Pavel S. Urysohn (1898–1924) visited Western Europe to meet Emmy Noether, Richard Courant and David Hilbert in Göttingen. In Bonn they met Hausdorff and Brouwer in Holland. The mathematicians in Göttingen were impressed with Aleksandrov and Urysohn's findings that a topological space is metrisable.

In Bonn the two of them swam across the Rhine every day which provoked Hausdorff's worry. On 17 August 1924 Urysohn drowned while swimming in the Atlantic near Bourg de Batz in Brittany.

From 1926 Aleksandrov and Heinz Hopf became close friends working together. They spent the academic year 1927–28 at Princeton in the United States collaborating with Lefschetz, Veblen and Alexander.

In 1927 Aleksandrov was awarded the Doctor of Science degree at Moscow and appointed Professor of Mathematics two years later. Aleksandrov and Kolmogorov, another Luzin student, became good friends and bought a house together in 1935 in Komarovska, a small village outside Moscow.[3]

He was elected a corresponding member in 1929 and a full member in 1953 of the USSR Academy of Sciences. He served as president of the Moscow Mathematical Society from 1932 to 1964.

Aleksandrov was an active participant in the political persecution of Luzin, his former teacher.[24] Luzin blocked Aleksandrov's election to a full member until Luzin's death. However, Aleksandrov always included Luzin and Egorov in Moscow, Emmy Noether and Hilbert in Göttingen and Brouwer in Amsterdam among his teachers.

Aleksandrov was the first to use the term 'Kernel of homomorphism.' In 1935 he published his book *Topology* which he originally planned to write with Heinz Hopf.

Aleksandrov died on 16 November 1982 in Moscow. In the last three years of his life he had been totally blind.

Ivan Georgievich Petrovsky

Ivan Petrovsky was born on 18 January 1901 in Sevsk, Russia. He studied at Moscow State University and became a professor there in 1943. His knowledge was encyclopaedic, including a thorough understanding of modern science and all its interconnections.

His main mathematical work was on the theory of partial differential equations, and the topology of algebraic curves and surfaces. He made remarkable progress in the solutions of two Hillbert problems: the 19th (analyticity of solutions of variational equations), and the first half of the 16th problem (topology of real algebraic curves and surfaces).[12]

He received the Doctor of Science degree in 1935 at Moscow without defense. He was elected to the USSR Academy of Science in 1946 and was the Rector (President) of Moscow State University (1951–1973).

He led the International Congress of Mathematicians in Moscow 1966 which opened Soviet mathematics to the international community after thirty years behind the Iron Curtain.

He selected the university faculty not by ideological, but by overall scientific and human criteria during his Presidency of Moscow State University. He helped many people, who otherwise were destined to be broken by the communist totalitarian system to realize their talents. Olga Ladyzheuskaya was one of them.

Petrovsky died on 15 January 1973 in Moscow. He is buried in the Novodevich Cemetery in Moscow.

Figure 138. Ivan G. Petrovsky

Figure 139. Andrey N. Kolmogorov

Andrey Nikolaevich Kolmogorov

Andrey Kolmogorov was born on 25 April 1903 in Tambov, Russia. His parents, who were not married, both died when Kolmogorov was young. His father, Nikolai Kataev, was the son of a priest and an agronomist by trade who was killed during the Civil War in 1919. His mother died while giving birth to Kolmogorov. His maternal aunt, Vera Yakovlena, adopted him and he always had the deepest affection for her.

Kolmogorov's name came from his maternal grandfather not from his father. His father's given name Nikolai appears as the patronymic. In 1920 Kolomogorov entered Moscow State University where lectures of V.V. Stepanov (1889–1950) on trigonometric series impressed him the most. He first achieved international renown in 1922 when he had constructed a Fourier series that diverges almost everywhere.

Kolmogorov graduated from Moscow State University in 1925 when he published eight papers.[3] In 1935 he received a Doctor of Science degree. In the summer of 1931 Kolmogorov and Pavel Aleksandrov, seven years senior to Kolmogorov, visited Berlin, Göttingen, München and Paris. In the same year he was appointed a professor at Moscow State University. Kolmogorov and Aleksandrov bought a house together in Komarovka, a small village outside Moscow in 1935 where many famous mathematicians visited.

On Sunday Kolmogorov invited his students on long walks to discuss the current problems of mathematics and the progress of culture in areas such as painting, architecture and literature.

In 1938 he was appointed as Head of the Department of Probability and Statistics at the Steklov Mathematical Institute of the USSR Academy of Sciences while retaining his position at the university. During his career, Kolmogorov achieved numerous significant accomplishments.

Kolmogorov made a major contribution to solving Hilbert's sixth problem and completely solved Hilbert's 13th problem in 1957.[12] His name is honored in Kolmogorov axioms, Kolmogorov continuity theorem, Kolmogorov extension theorem, Landau-Kolmogorov inequality, Kolmogorov space, Kolmogorov-Arnold-Moser (KAM) theorem, Borel-Kolmogorov paradox, Hahn-Kolmogorov theorem, etc. He also received many accolades for his work.

In 1939 he was elected to the USSR Academy of Sciences, He was also elected to the Royal Society of London in 1964, the National Academy of Science in 1968. He received the Lenin Prize in 1965 and the Lobachevsley Prize in 1987. Kolmogorov had 79 doctoral students[10], four more than Hilbert.

In 1942 he married Anna Dmietrievna Egorova. He enjoyed excellent health until about 1980 when he was found to be suffering from Parkinson's disease and he lost the power of speech. He became almost completely blind due to glaucoma.[4]

He died on 20 October 1987 in Moscow and was buried in the Moscow Novodevich Cemetery.

Aleksandr Yakovlevich Khinchin

Aleksandr Khinchin was born on 19 July 1894 in Kondrovo, Russia. He graduated from Moscow State University in 1916. Around 1922 he began to study the theory of numbers and probability theory.

In 1927 Khinchin was appointed a professor at Moscow State University and published *Basic Law of Probability Theory*. He also published *Mathematical Principles of Statistical Mechanics* in 1943 and *Mathematical Foundations of Quantum Statistics* which many considers equal to John von Neumann's *Mathematical Foundations of Quantum Mechanics*.[3]

His book *Mathematical Foundations of Information Theory* was translated into English from the original Russian in 1957. He was elected to the USSR Academy of Sciences in 1939 and received a State Prize in 1940.

Most electrical engineers remember the Wiener-Khinchin theorem which states that the power spectral density of a wide-sense-stationary random process is the Fourier transform of the corresponding autocorrelation function.

He died on 18 November 1959 in Moscow.

Figure 140. Aleksandr Y. Khinchin

Figure 141. Aleksandr O. Gelfond

Aleksandr Osipovich Gelfond

Aleksandr Gelfond was born on 24 October 1906 in St. Petersburg. He entered Moscow State University in 1924 and received his Candidate degree in 1930 under supervision of Aleksandr Khinchin and Vyacheslaw Stepanov.

Gelfond developed basic techniques in the study of transcendental numbers that are not the solution of an algebraic equation with rational coefficients. From 1931 he taught mathematics at Moscow State University.

In 1934 Gelfond solved Hilbert's seventh problem in full generality. Building on the method of his 1929 result on the function $2^{\sqrt{-2}z}$ and using an important lemma from

Siegel's 1929 paper, he added several new ideas and created a new method that was used to solve many other problems.[12] Theodor Schneider (1911–1988) independently solved the seventh problem in 1935. Gelfond received his Doctor of Science degree in 1935 without defense.

In 1933 he was appointed to the faculty of the Steklov Mathematics Institute in addition to his university position. He addressed the Second All-Union Mathematics Congress in Leningrad in 1934 on transcendental numbers. In 1929 Gelfond proposed an extension of the theorem known as the Gelfond's Conjecture that was proved by Alan Baker in 1966.

$2^{\sqrt{2}} = 2.6651441\cdots\cdots$ is called the Gelfond-Schneider constant and e^{π} is known as Gelfond's constant. He was an expert at chess, literature, mineralogy and the history of science.[3]

Ivan Petrovsky said that Gelfond is the only Soviet mathematician who solved all the problems in Pólya and Szegös book *Aufgaben und Lehrsätze aus der Analysis* (Poblems and Theorems in Analysis) of 1925.[12]

Gelfond died on 7 November 1968 in Moscow.

Boris Vladimirovich Gnedenko

Boris Gnedenko was born in Simbirsk on 1 January 1912. University of Saratov admitted Gnedenko at the age of fifteen after he petitioned the Soviet Ministr of Education. The government allowed him to graduate in 1930 after just three years of study rather than the normal five years because he was an excellent student.

In 1934, he began his postgraduate study at Moscow State University under Khinchin and Kolmogorov. In 1937 he was awarded Candidate of Science degree for his thesis on the theory of infinitely divisible distributions. Then he was appointed as an assistant researcher in the Mathematics Institute at Moscow State University.

When Khinchin left in 1935 to spend two years at Saratov University, Kolmogorov took over as Gnedenko's supervisor. During the summer of 1937 Gnedenko went on a hiking expedition to the Caucusus Mountains with Kolmogorov and some fellow researchers. The hikers discussed a range of topics, including politics and mathematics, and during at least one conversation, Gnedenko stated his distate for Soviet policies.

That winter, on 1 December 1937, he was conscripted into the Soviet Army and arrested on 5 December. He had been denounced by one of the hikers.

After six months in prison he was released and reinstated to his post of Assistant Researcher with strong support from Khinchin and Kolmogorov. He was not allowed to join the Soviet Army in 1941 when German forces invaded.

Gnedenko married Natalia Konstantinova in 1939 and they had two sons. He was awarded Doctor of Science degree in 1942.

On the recommendation of Kolmogorov the Ukrainian Academy of Sciences elected him a member in 1945. In 1949, he became Director of the Kiev Institute of Mathematics. After returning to Moscow in 1960 he was appointed to the chair of probability theory in 1966. He held this post for 29 years until his death.

One of Gnedenko's most famous books is *Course in the Theory of Probability* which was published in 1950 and has gone through six editions. It has also been translated into English, German, Polish and Arabic. He also published *Outline of the History of Mathematics* in Russia in 1946.

Gnedenko enjoyed classical music and had a large collection of records. One of his student was Domokos Szasz, the Hungarian mathematician.

Gnedenko died on 27 December 1995 in Moscow.

Figure 142. Boris V. Gnedenko   Figure 143. Efim I. Zelmanov

Efim Issakovich Zelmanov

Efim Zelmanov was born on 7 September 1955 in Khabarovsk, USSR. He graduated Novosibirsk State University in 1977 and received his Candidate degree in 1980 at Novosibirsk with his thesis on nonassociative algebra. He was awarded Doctor of Science degree in 1985 at Leningrad State University.

Zelmanov described his work on Jordan algebras in his invited lecture to the International Congress of Mathematicians at Warsaw in 1983.

In 1987 Zelmanov solved one of the big open questions in the theory of Lie algebras. He proved that the Engel identity implies that the algebra is necessarily nilpotent in the case of infinite dimension.

In 1994 he was awarded a Fields Medal at the International Congress of Mathematicians in Zürich for his solution of the restricted Burnside problem.

In 1990 he moved to the United States becoming a professor at the University of Wisconsin-Madison, the University of Chicago, Yale University and in 2002 the University of California, San Diego.

He also received the *Collége de France* Medal in 1992.

Lev Semenovich Pontryagin

Lev Pontryagin was born on 3 September 1908 in Moscow. The family was very poor and his mother, Tatyana Andreevna Pontryagina, used her sewing skills to help out the family finances. At the age of 14 Pontryagin lost his eyesight in a portable stove explosion. Nonetheless, Pontryagin managed to excel at mathematics thanks to the assistance of his mother, who acted as his secretary, reading scientific works aloud to him, writing in the formulas in his manuscripts, correcting his work and so on. In order to do this she learnt to read foreign languages.[3] Incredibly, Tatyana Andreevna had no mathematical training and could only describe mathematical symbols by their appearance.

In 1925 he entered Moscow State University and was strongly influenced by Pavel Aleksandrov's courses. He graduated in 1929.

In 1934 he proved the duality between the homology groups of founded closed sets in Euclidean space and the homology groups Hilbert's fifth problem for abelian groups using the theory of characters on locally compact abelian groups which he had introduced. In 1934 he joined the Steklov Mathematics Institute and one year later he became head of the Department of Topology and Functional Analysis there.

In 1952 he began to study differential equations and control theory. In 1961 he published *The Mathematical Theory of Optimal Processes* with three former students. In 1962 he received the Lenin Prize for his book. His maximum principle is fundamental to the modern control theory of optimization. He also introduced the idea of a bang-bang control principle to describe situations where either the maximum steer should be applied to a system, or none at all. Small heating systems are usually run this way, with the gas flame either fully on or fully off.

The USSR Academy of Sciences elected Pontryagin a corresponding member in 1939 and a full member in 1959. Politically, Pontryagin sided with Soviet regime and participated in some of their campaign against "enemies of the State" including Nicolai Luzin.

He is remembered by Pontryagin duality, Pontryagin classes, Pontryagin Spaces other than the maximum principle.

He died on 3 May 1988 in Moscow.

Figure 144.   Parents of Pontryagin

(Courtesy of Aleksandr Logginov)

Figure 145.   Lev S. Pontryagin   **Figure 146.   Andrei N. Tikhonov**

Andrei Nikolasvich Tikhonov (Tychonoff)

Andrei Tikhonov was born on 30 October 1906 in Gzhatska, Smolensk, Russia. He entered Moscow State University in 1922. By 1926 he had discovered the topological construction which is today named after him the Tikhonov topology defined on the product of topological spaces. He proved that the product of any set of compact topological spaces is compact. He graduated Moscow State University in 1927 and became a research student.

In 1936 he was awarded Doctor of Science degree with his dissertation *Functional Equations of Volterra Type and Their Applications to Mathematical Physics.* In the same year Moscow State University appointed him as a professor. Three years later he was elected as a Corresponding Member of the USSR Academy of Sciences.

Tikhonov's work led from topology to functional analysis with his famous fixed point theorem for continuous maps from convex compact subsets of locally convex topological space in 1935.

One of the most outstanding achievements in computational mathematics is the theory of homogeneous difference schemes which he developed in collaboration with Samarskii.[3] He was awarded the Lenin Prize in 1966. In the same year the USSR Academy of Sciences elected him as a Full Member.

Tikhonov founded the Faculty of Computing and Cybernetics at Moscow State University and served as Dean from 1970 to 1990. He died on 7 October 1993.

Figure 147. Israil M. Gelfand

Figure 148. Vladimir I. Arnold

Israil Moiseevic Gelfand

Israil Gelfand was born on 2 September 1913 in Krasnye Okny, Odessa in Ukraine. He went to Moscow in 1930 before completing his secondary education. He began teaching mathematics in evening classes and attended lectures at Moscow State University.

In 1932 Gelfand was admitted as a research student under Kolmogorov's supervision. In 1935 he was awarded Candidate degree with his thesis *Abstract Functions and Linear Operators*.

In 1938 he was awarded Doctor of Science degree with his thesis on the theory of commutative normed rings. In 1941 Moscow State University appointed him as a professor.

He also worked on the theory of representations of non-compact groups and the inverse Sturm-Liouville problem. Between 1968 and 1972 he produced a series of important papers on the Cohomology of infinite dimensional Lie algebras.

He was president of the Moscow Mathematical Society in 1968-70. In 1990 he emigrated to the United State where he became Distinguished Visiting Professor at Rutgers University. He published over 500 papers in mathematics and biology. He established a correspondence school in mathematics in Russia and the United States.

He died on 5 October 2009 in New Brunswick, New Jersey.

Vladimir Igorevich Arnold

Vladimir Arnold was born on 12 June 1937 Odessa, USSR (now Ukraine). He graduated from Moscow State University in 1959 receiving his Diplom degree with a thesis *On Mappings Of A Circle To Itself*. He was awarded his Candidate degree in 1961 with his thesis *On the representation of continuous functions of 3 variables by the superpositions of continuous functions of 2 variables*. This thesis contained a solution to Hilbert's 13th problem.

In the late 1950's and early 1960's the Faculty of Mathematics and Mechanics at Moscow State University had the constellation of great mathematicians: Kolmogorov, I.M. Gelfand, A.O. Gelfond, Petrovsky (Rector), Pontryagin, P.S. Novikov (father of Sergei Novikov), Markov, Khinchin and Pavel Aleksandrov. Students include Yuri Manin, Yakov Sinai, Sergei Novikov, Aleksandr Kirillov and Vladimir Arnold. Previously, the Russian center of mathematics had been in Leningrad, but the concentration of those great minds shifted the balance to Moscow.

In 1963 Arnold was awarded Doctor of Science degree for the thesis *Small Denominators And Celestial Mechanics* at the Institute of Applied Mathematics in Moscow. In 1965 he was appointed professor at Moscow State University, a position he held until 1986.

He was a professor at the University of Paris-Dauphine from 1993 to 2005. Arnold married Voronina Elionora Aleksandrova in 1976 and they had one son.

His work on Hamiltonian dynamics which includes co-creation of KAM (Kolmogorov-Arnild-Moser) theory and the discovery of "Arnold diffusion" made him world famous at an early age.

He received many honors: the Lenin Prize with Kolmogorov in 1965, election to the National Academy of Science in the United States (1983), the Academy of Sciences of Paris (1984), the Royal Society of London (1988), the USSR Academy of Sciences (1990), the Lobachevsky Prize (1992) and the Wolf Prize in Mathematics (2001).

Arnold is an excellent sportsman enjoying marathon, biking, swimming and skiing. When he has difficulties in proving some mathematical theorems, he used to ski 40 to 60 km.

Figure 149.   Sergei P. Novikov

Sergei Petrovich Novikov

Sergei Novikov was born on 20 March 1938 in Gorky (now Nizhny Novogorod), Russia. His father, Peter Sergeevich Novikov (1901-1975) was a professor of mathematics at Moscow State University and his mother, Ludmila V. Keldush was also a professor of mathematics. His uncle Mstislav Keldish (1911–1978) was also a mathematician and served as president of the USSR Academy of Sciences (1961–1975).

Novikov graduated from Moscow State University in 1960, but his first important publication *Cohomology of the steenrod algebra* appeared in 1959. He received Candidate degree in 1964 with his thesis *Differentiable sphere bundles* and a Doctor of Science degree in 1965 at the Steklov Institute of Mathematics under Postnikov's supervision. The USSR Academy of Sciences elected him a Corresponding Member in 1966 and a Full Member in 1981.

Novikov served as president of the Moscow Mathematical Society from 1985 to 1996. Since September 1996 he has been professor of mathematics at University of Maryland. In 1965 he proved his famous theorem on the invariance of Pontryagin Classes and stated the conjecture, now known as the Novikov conjecture, concerning the homotopy invariance of certain polynomials in the Pontryagin classes of a manifold, arising from the fundamental group. It is one of the most fundamental problems in topology.[3]

Novikov discussed his conjecture in a lecture given at the 1970 International Congress of Mathematicians in Nice where he received his Fields Medal.

After 1971 he became interested in mathematical physics and dynamical models, the theory of solitons, the spectral theory of linear operators, quantum field theory and string theory.[3]

In 1981 he received the Lobachevsky Prize of the USSR Academy of Sciences. Then in 2005 he was awarded the Wolf Prize for his contributions to algebraic topology, differential topology and mathematical physics. He became one of twelve mathematicians who received both the Fields Medal and the Wolf Prize.

He has been married since 1962 and has one son and two daughters.

Eugene (Evgenii) Borisovich Dynkin

Evgenii Dynkin was born into a family of Jewish origin on 11 May 1924. His father disappeared in the Gulag in 1937 and perished under Stalin. Despite his father being "people's enemy" he graduated Moscow State University in 1945. His poor eyesight saved from military service during World War II.

He discovered the "Dynkin diagram" approach to the classification of the semisimple Lie algebras. In 1948 he received his Candidate of Science degree and in 1951 Doctor of Science degree at Moscow State University.

After Stalin's death in 1953 he was appointed to a chair at Moscow State University. He published two books on the theory of Markov processes. In 1968 his post at Moscow

State University was compulsorily interrupted due to a political reason and he became a senior scientific worker at the Central Economics and Mathematics Institute at the USSR Academy of Sciences.

In 1976 Dynkin emigrated to the United States where he began to work at Cornell University in Ithaca, New York. Dynkin has obtained exciting results in the theory of "superprocesses", a class of measure-valued Markov processes which can be used to give probabilistic solutions to certain nonlinear partial differential equations.[3]

Yakov Grigorevich Sinai

Yakov Sinai was born on 21 September 1935 in Moscow. His maternal grandfather was Veniamin Kagan (1869-1953), a famous Russian geometer. Sinai received his Candidate degree in 1960 at Moscow State University under Kolmogorov's supervision.

He has made major contributions to the modern Metric Theory of Dynamical Systems. His work has also linked deterministic dynamical systems with probabilistic systems.

In 1971 he became a professor at Moscow State University and a senior research at the Landau Institute of Theoretical physics. Since 1993 he was been a Professor of Mathematics at Princeton University. He is member of the United States National Academy of Sciences, and the Russian Academy of Sciences. He received the Dirac Medal in 1992 and the Wolf Prize in Mathematics in 1997.

He is known by Kolmogorov-Sinai entropy, Sinai's billiards, Sinai's random walk, Sinai-Ruelle-Bowen measures, Pirogov-Sinai theory etc.[26]

Gregori Aleksandrovich Margulis

Gregori Margulis was born on 24 February 1946 in Moscow. After graduating Moscow State University in 1967 he received Candidate of Science degree in 1970 with his thesis *On Some Problems in the Theory of U-Systems*.

He received a Fields Medal in 1978 at the International Congress of Mathematicians in Helsinki for proving innovative analysis of the structure of Lie groups. But the Soviet government did not allow him to attend the award ceremony. His work belongs to combinatorics, differential geometry, ergodic theory, dynamical systems, and Lie groups.

Margulis proved the full conjecture of Oppenheim in 1986, which was given by Sir Alexander Oppenheim (1903-1997) in 1929.

In 1990 Margulis immigrated to the United States. From 1991 he has been a professor at Yale University.

Figure 150.   Yakov G. Sinai   **Figure 151.   Gregori A. Margulis**

Marina Ratner

Marina Ratner was born in 1938 into a Jewish family in USSR. She received her Diplom degree at Moscow State University in 1961 and taught at a high school for gifted students in Moscow.

She was awarded the Candidate degree in 1969 at Moscow State University for a thesis *Geodesic Flows on Unit Tangent Bundles of Compact Surfaces of Negative Curvature* under Sinai's supervision. After living in Israel in 1971-75, she began to work at the University of California, Berkeley where she became a full professor in 1982. Her main area is ergodic theory.

She was elected to the National Academy of Science in 1993 and in the same year she received the 1993 Ostrowski Prize.[27]

She is known by Ratner's theorems on Unipotent Flows.

Vladimir Abramovich Rokhlin

Vladimir Rokhlin was born on 23 August 1919 into a Jewish family in Baku. He entered Moscow State University in 1935. When the Germans invaded the Soviet Union in volunteered for the Soviet Army but was captured and confined as a prisoner of war in a German concentration camp. Victory by the Allies did not bring immediate freedom as Rokhlin spent two years in a Soviet filtration camp.

During 1947–49 he made fundamental contributions to measure theory and ergodic theory. He obtained his famous results in topology (homotopy groups of spheres, cobordisms of low dimensional manifolds, signature of 4-manifolds, characteristic classes).

From 1959 to 1984 he was a professor at Leningrad State University. He died on 3 December 1984 in Leningrad. His son, Vladimir Rokhlin is a professor of computer science, mathematics and physics at Yale University, a winner of the Steele Prize.

Mikhail Leonidovich Gromov

Mikhail (Misha) Gromov was born on 23 December 1943. He received a Candidate degree in 1968 and Doctor of Science degree in 1973 at Leningrad State University under Rokhlin's supervision.

His impact has been felt heavily in geometric group theory, where he characterized groups of polynomial growth and created the notion of hyperbolic group; symplectic topology, where he introduced pseudoholomorphic curves, and in Riemann geometry. His h-principle on differential relations is the basis for a geometric theory of partial differential equations.[28]

Gromov received many prizes including Oswald Veblen Prize in Geometry (1981), Wolf Prize in Mathematics (1993), Lebachevsky medal (1997) and Bolyai Prize (2005).

He is a permanent member of the *Institut Hautes et Etudes Scientilique* (IHES) and Jay Gould Professor of Mathematics at New York University.

Alexander Kirillov

Alexander Kirillov was born in 1936. He received Doctor of Science degree in 1962 at Moscow State University for a thesis *Unitary Representation of Nilpotent Lie Groups* which was originally intended for a Candidate degree. This is quite unusual. Nikolai Luzin is another mathematician who received a Doctor of Science degree in 1915 for a thesis for a Candidate degree. At that time Kirillov was the youngest Doctor of Science in the Soviet Union. His thesis adviser was Israil Gelfand.

He was a professor at Moscow State University until 1994 when the University of Pennsylvania appointed him as Francis J. Carey Professor of Mathematics.

His son, Alexander Kirillov, Jr is a mathematician at State University of New York at Stony Brook.

Andrei Yu Okounkov

Andrei Okounkov was born in 1969 in Moscow. He received a Candidate of Science degree in 1995 at Moscow State University under Kirillov's supervision. His thesis was *Admissible Representation of Gelfand Pairs Associated with the Infinite Symmetric Group.*

He has worked on the representation theory of infinite symmetric groups, the statistics of plane partitions, and the quantum cohomology of the Hilbert scheme of points in the complex plane.

At the 24th International Congress of Mathematicians in Madrid in 2006, he received the Fields Medal for his contributions to bridging probability, representation theory and algebraic geometry.[29]

He was a faculty member at the University of Chicago, the University of California, Berkeley and since 2002 he has been a professor at Princeton University.

Figure 152.   Andrei Y. Okounkov

Mikhail Mikhailovich Postnikov

Mikhail Postnikov was born on 27 September 1927 in Shatura near Moscow. His father, an electrical power engineer, was arrested and executed in 1937, but he was later rehabilitated. Postnikov received Candidate degree in 1949 at Moscow State University for a thesis on classifying the maps from a 3-dimensional polyhedron into a simply-connected space. He also received Doctor of Science degree in 1953. From 1954 to 1960 he lectured on algebraic topology at Moscow State University.

In 1961 he received the Lenin Prize for his work in homotopy theory.

Andrei Andreevich Bolibrukh

Andrei Bolibrukh was born on 30 January 1950 in Moscow. His father served as a lieutenant general in the Soviet Army. Bolibrukh graduated from Moscow State University in 1972 and received his Candidate of Science degree in 1977 with his thesis *On the Fundamental Matrix of A Pfaffian System of Fuchs Type*, which stated the analytic theory of differential equations on multidimensional complex domains.

Bolibrukh solved Hilbert's 21st problem at the end of 1980's.

The problem has to do with constructing in a complex domain a system of linear ordinary differential equations with, roughly speaking, a prescribed character of branching of the solution at singular points and with a relatively moderate character of the singularities at those points.[30]

He was awarded the Doctor of Science degree in 1991 at the Steklov Institute of Mathematics. From 1992 to 1997 he taught at the Moscow Institute of Physics and Technology (Fiztekh). In 1996 he became a professor at Moscow State University.

He was elected a full member of the Russian Academy of Sciences in 1997. He was awarded the Lyapunov Prize in 1995 and in 2001 he received the State Prize of the Russian Federation. He had temporary positions in Nice, Strasbourg and Bonn in the late 1990's.

Bolibrukh died on 11 November 2003. At the time of his death he was Deputy Director of the Steklov Institute of Mathematics, Chairman of the mathematics section of the Department of Mathematical Sciences of the Russian Academy of Sciences.

Aleksei Vasilevich Pogorelov

Aleksei Pogorelov was born on 3 March 1919 in Korocha (now Belgorod Oblast), Russia. He graduated Kharkov University in 1941 and N.E. Zhukowsky Air Force Academy in 1945. He attended external post-graduate courses defending his Candidate thesis in 1947 and Doctor of Science thesis in 1948 at Moscow State University.

He was elected a Corresponding Member of the USSR Academy of Sciences in 1960 and a Full Member in 1976. He was a professor at Kharkov State University (1950-1959) and at the geometry department at B. Verkin Institute for Low Temperature Physics and Engineering, Academy of Sciences of Ukraine (1960-2002).

In 1973 he published a paper titled *A Complete Solution of Hilbert's Fourth Problem* in Soviet Math Doklady. In this paper he considered only the two-dimensional case. In 1975 he published a monograph entitled *The Fourth Hilbert Problem* in which the three-dimensional case was considered.[12] Pogorelov received the Lobachevsky Prize in 1959 and Lenin Prize in 1982.

He wrote more than 200 publications including about 40 monographs and textbooks. He died in 2002.

Yuri Vladimirovich Matiyasevich

Yuri Matiyasevich was born on 2 March 1947 in Leningrad. He graduated from Leningrad State University in 1969. He won a gold medal at the Sixth International Mathematical Olympiad in Moscow in 1964. In 1970, he received Candidate of Science degree and two years later a Doctor of Science degree at the Leningrad Department of Stektov Institute of Mathematics (LOMI).

His doctoral thesis was devoted to the negative solution of Hilbert's tenth problem. His book about the tenth problem was published under the title *Hilbert's Tenth Problem* in 1993 by the MIT Press.[31]

He was elected Corresponding Member of the Russian Academy of Sciences in 1997. Since 1998 Matiyasevich has been a Vice-President of St. Petersburg Mathematical Society.

British School

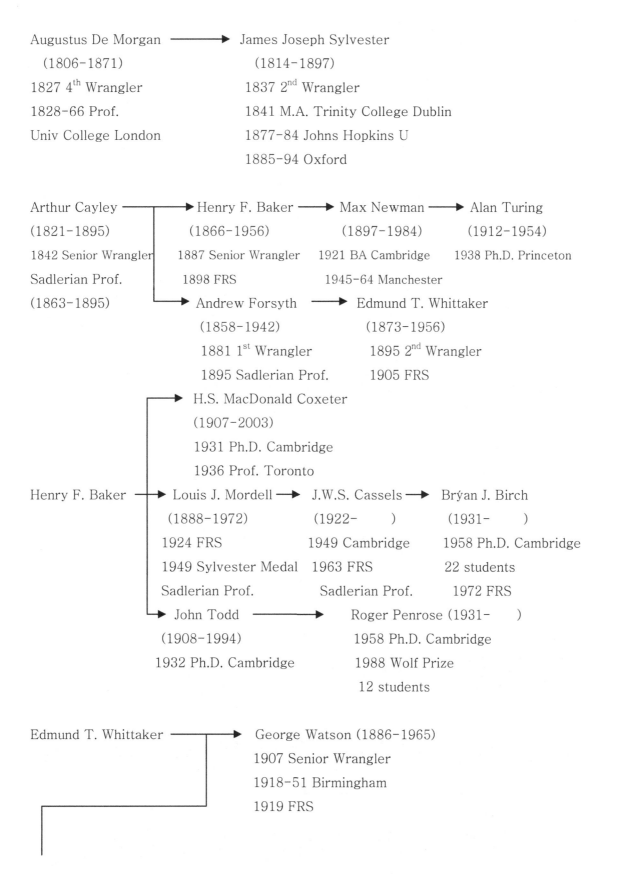

Augustus De Morgan ⟶ James Joseph Sylvester
(1806-1871)   (1814-1897)
1827 4th Wrangler   1837 2nd Wrangler
1828-66 Prof.   1841 M.A. Trinity College Dublin
Univ College London   1877-84 Johns Hopkins U
                                          1885-94 Oxford

Arthur Cayley ⟶ Henry F. Baker ⟶ Max Newman ⟶ Alan Turing
(1821-1895)   (1866-1956)   (1897-1984)   (1912-1954)
1842 Senior Wrangler 1887 Senior Wrangler 1921 BA Cambridge 1938 Ph.D. Princeton
Sadlerian Prof.   1898 FRS   1945-64 Manchester
(1863-1895)   Andrew Forsyth ⟶ Edmund T. Whittaker
                           (1858-1942)   (1873-1956)
                           1881 1st Wrangler   1895 2nd Wrangler
                           1895 Sadlerian Prof.   1905 FRS

                      H.S. MacDonald Coxeter
                      (1907-2003)
                      1931 Ph.D. Cambridge
                      1936 Prof. Toronto

Henry F. Baker ⟶ Louis J. Mordell ⟶ J.W.S. Cassels ⟶ Brýan J. Birch
                     (1888-1972)   (1922-   )   (1931-   )
                     1924 FRS   1949 Cambridge   1958 Ph.D. Cambridge
                     1949 Sylvester Medal 1963 FRS   22 students
                     Sadlerian Prof.   Sadlerian Prof.   1972 FRS
                  John Todd ⟶ Roger Penrose (1931-   )
                     (1908-1994)   1958 Ph.D. Cambridge
                     1932 Ph.D. Cambridge 1988 Wolf Prize
                                                  12 students

Edmund T. Whittaker ⟶ George Watson (1886-1965)
                                    1907 Senior Wrangler
                                    1918-51 Birmingham
                                    1919 FRS

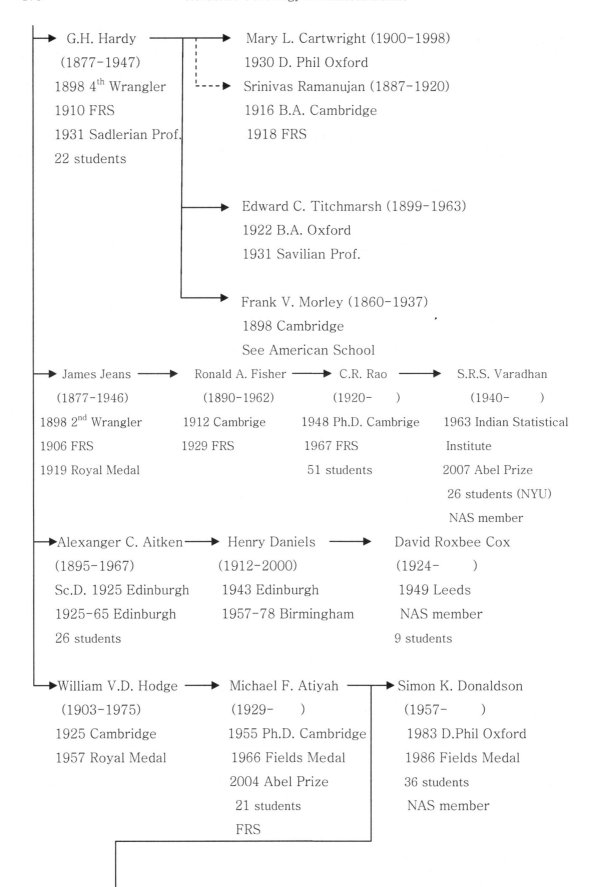

G.H. Hardy
(1877–1947)
1898 4th Wrangler
1910 FRS
1931 Sadlerian Prof.
22 students

Mary L. Cartwright (1900–1998)
1930 D. Phil Oxford

Srinivas Ramanujan (1887–1920)
1916 B.A. Cambridge
1918 FRS

Edward C. Titchmarsh (1899–1963)
1922 B.A. Oxford
1931 Savilian Prof.

Frank V. Morley (1860–1937)
1898 Cambridge
See American School

James Jeans
(1877–1946)
1898 2nd Wrangler
1906 FRS
1919 Royal Medal

Ronald A. Fisher
(1890–1962)
1912 Cambrige
1929 FRS

C.R. Rao
(1920–)
1948 Ph.D. Cambrige
1967 FRS
51 students

S.R.S. Varadhan
(1940–)
1963 Indian Statistical
Institute
2007 Abel Prize
26 students (NYU)
NAS member

Alexanger C. Aitken
(1895–1967)
Sc.D. 1925 Edinburgh
1925–65 Edinburgh
26 students

Henry Daniels
(1912–2000)
1943 Edinburgh
1957–78 Birmingham

David Roxbee Cox
(1924–)
1949 Leeds
NAS member
9 students

William V.D. Hodge
(1903–1975)
1925 Cambridge
1957 Royal Medal

Michael F. Atiyah
(1929–)
1955 Ph.D. Cambridge
1966 Fields Medal
2004 Abel Prize
21 students
FRS

Simon K. Donaldson
(1957–)
1983 D.Phil Oxford
1986 Fields Medal
36 students
NAS member

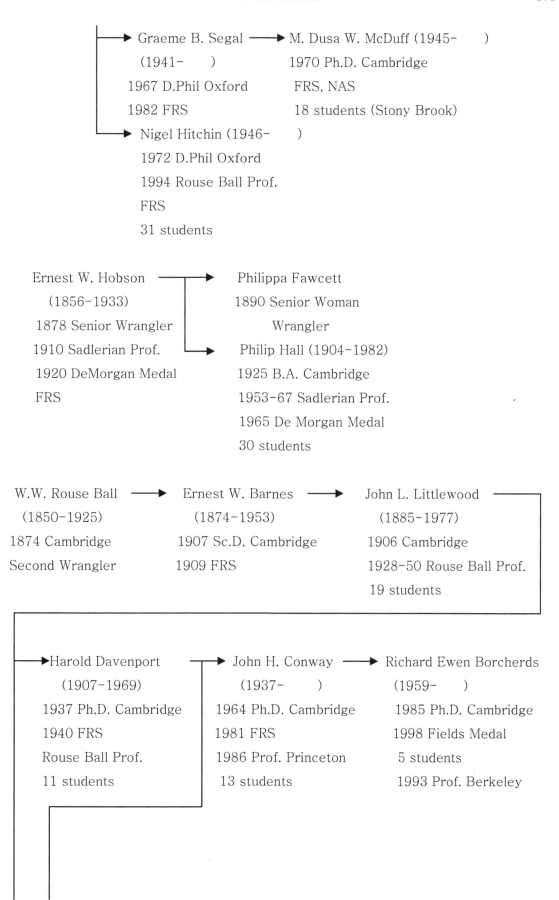

Graeme B. Segal ⟶ M. Dusa W. McDuff (1945-   )
(1941-   )   1970 Ph.D. Cambridge
1967 D.Phil Oxford   FRS, NAS
1982 FRS   18 students (Stony Brook)
Nigel Hitchin (1946-   )
1972 D.Phil Oxford
1994 Rouse Ball Prof.
FRS
31 students

Ernest W. Hobson ⟶   Philippa Fawcett
(1856-1933)   1890 Senior Woman
1878 Senior Wrangler   Wrangler
1910 Sadlerian Prof.   Philip Hall (1904-1982)
1920 DeMorgan Medal   1925 B.A. Cambridge
FRS   1953-67 Sadlerian Prof.
                       1965 De Morgan Medal
                       30 students

W.W. Rouse Ball ⟶   Ernest W. Barnes ⟶   John L. Littlewood
(1850-1925)   (1874-1953)   (1885-1977)
1874 Cambridge   1907 Sc.D. Cambridge   1906 Cambridge
Second Wrangler   1909 FRS   1928-50 Rouse Ball Prof.
                                             19 students

Harold Davenport ⟶   John H. Conway ⟶   Richard Ewen Borcherds
(1907-1969)   (1937-   )   (1959-   )
1937 Ph.D. Cambridge   1964 Ph.D. Cambridge   1985 Ph.D. Cambridge
1940 FRS   1981 FRS   1998 Fields Medal
Rouse Ball Prof.   1986 Prof. Princeton   5 students
11 students   13 students   1993 Prof. Berkeley

Alan Baker ⟶ John H. Coates ⟶ Andrew John Wiles

(1939-   )   (1945-   )   (1953-   )

1964 Ph.D. Cambridge   1969 Ph.D. Cambridge   1980 Ph.D. Cambridge

1970 Fields Medal   1985 FRS   1995 Wolf Prize

1973 FRS   1986 Sadlerian Prof.   1997 Frank Cole Prize

6 students   19 students   15 students (14 Princeton)

⟶ Donald C. Spencer

(1912-2001)

1939 Ph.D. Cambridge

H.P.F. Swinnerton-Dyer   1948 Bocher Prize

(1927-   )   27 students

1967 FRS   See American School

2006 Sylvester Medal

Theodor Eastermann ⟶ Klaus F. Roth (1925-   )

1925 Ph.D. Hamburg   1950 Ph.D. London

3 students   1958 Fields Medal

Ralph Fowler ⟶ Paul A.M. Dirac ⟶ Harish-Chandra

(1859-1944)   (1902-1984)   (1923-1983)

1915 Cambridge   1926 Ph.D. Cambridge   1947 Ph.D. Cambridge

                   1933 Nobel Prize   1954 Frank Cole Prize

Francis Galton ⟶ Karl Pearson

(1822-1911)   (1857-1936)

1856 FRS   1879 Cambridge

                   Univ College London

John Burkill ⟶ Frederick W. Gehring

(1900-1993)   1952 Ph.D. Cambridge

1921 B.A. Cambridge   29 Students

1953 FRS   NAS member

Shaun Wylie ⟶ J. Frank Adams ⟶ Béla Bollóbas ⟶ William T. Gowers

(1913-   )   (1930-1989)   (1943-   )   (1963-   )

1937 Ph.D.   1955 Ph.D. Cambridge   1972 Ph.D. Cambridge   1990 Ph.D. Cambridge

Princeton   Sylvester Medal   1984 Sc.D. Cambridge   1998 Fields Medal

5 students   27 Students   1998 Rouse Ball Prof.

                                 (24 Cambridge)   1999 FRS

⟶ William Tutte

(1917-2002)

1948 Ph.D. Cambridge

1948-62 Univ. of Toronto

1962-85 Univ. of Waterloo

⟶ Erik Christopher Zeeman

(1925-   )

1953 Ph.D. Cambridge

1964 Prof. Warwick

30 students

British School

Number of B.A. graduates from the University of Cambridge in 1901–40 are shown below.[33]

	BA Math
1901–05	538
1906–10	482
1911–15	394
1916–20	174
1921–25	395
1926–30	467
1931–35	462
1936–40	441
total	3353

Average 84 per year

The tripos was distinctive written examinations of undergraduate student at the University of Cambridge. From about 1780 to 1909, the "Old Tripos" system published an order of merit of successful candidates. In descending order of merit, a Class I candidate was called Wrangler, Class II called Senior Optime, and Class III Junior Optime. The student with the highest grade was called the Senior Wrangler, which was a great honor.

All degree Candidates were expected to show at least competence in mathematics. The level of technique required of candidate was high, and the time pressure in the examinations acute.

A long process of development of coaching went hand-in-hand with gradual increase in the difficulty of the most testing questions asked.[34] Tuition for coaches should be paid by the candidate. Among many coaches Edward Routh (1831–1907) was the most outstanding. He was the Senior Wrangler in 1854 while J.C. Maxwell was just below him.

Emphasizing the tripos system actually was detrimental in mathematical research at Cambridge compared to Germany and France. Mathematical research in Britain was behind the continent in the nineteenth century partly due to the tripos system. The objective of tripos training was simply to achieve the best possible marks by learning all the tricks of the trade. Coaches were not interested in the subject of mathematics.

Consequently, G.H. Hardy and E.W. Hobson helped to abolish the old tripos system in 1909. There are no longer senior wrangler, second wrangler, junior optime, etc.

Students are given Class I, Class II Division i, Class II Division ii, and Class III, ratings. Students in Class I receive B.A. degree (Honor).

In Oxford and Cambridge, college teachers are called Fellows who teach students individually or in a very small group. At Cambridge Fellows were not allowed to get married and required them to live at a College until 1860. College fellows should submit a thesis to be appointed in the past. Senior Wranglers were almost automatically elected a fellow. Election to fellowship was roughly equivalent to gaining doctorates in Germany.

The M.A. (Master of Arts) degree at Oxford and Cambridge is not an earned degree. Three years after receiving B.A. degree, M.A. degree is awarded upon request. The first research degree is called M.Phil (Master of Philosophy) in the United Kingdom.

In 1917 British Universities resolved to create a Ph.D. degree. The first Oxford D.Phil in mathematics was awarded in 1921. The first Cambridge Ph.D. in mathematics was awarded in 1924 to an Australian Thomas Cherry.

Before the doctor of philosophy (Ph.D. or D.Phil) degree was introduced in the British universities, these were doctor of divinity (D.D.), doctor of medicine (M.D.) and doctor of law (LL.D.), awarded by the Faculty of Theology, Faculty of Medicine and Faculty of Law respectively.

Everything other than these three areas belonged to the Faculty of Philosophy which include history, philosophy, social science, mathematics and natural philosophy.

The natural philosophy was renamed natural science and the Faculty of Natural Science was created later. Engineering belonged to natural science and eventually separated as the Faculty of Engineering. That is why the Ph.D. degrees are conferred to candidates studying in humanities, social science, natural science, mathematics and engineering.

In 1961-2, 72% of university teachers in science and mathematics had a doctorate of some kind, including a higher doctorate.

In the United Kingdom the meaning of word can be quite different from that in the United States. The "Public School" in the U.K. would be considered a private school in America and a "State School" in the U.K. would be considered a public school in America. Even more confusing is that the secondary schools are called "College" in some regions.

For example, Chesterton Community College in Cambridge is a secondary school. Most students finish their education after 10th grade. Students preparing to enter a university study two more years at a Sixth-form College which is 11th and 12th grade in the United States.

Augustus De Morgan

Augustus De Morgan was born on 27 June 1806 in Madura (now Madurai, Tamil Nadu), India, the son of a Lieutenant-Colonel in the Indian Army. He graduated from Trinity College, Cambridge with a B.A. in 1826. He did not take M.A. degree which required a theological test at that time. Without his M.A. he was not eligible for a Fellowship at Cambridge.

De Morgan entered Lincoln's Inn to study for the Bar in 1826. In 1827, he applied for the chair of mathematics in the newly founded University College London. Despite having no mathematical publications, he received an appointment in 1828.[3] He remained at the University College until 1866 except five years from 1831 to 1836. In 1866, he was a co-founder of the London Mathematical Society and became its first president.[3]

De Morgan never became a Fellow of the Royal Society as he refused to let his name be put forward. After De Morgan's death on 18 March 1871 in London, Lord Overstone bought his library of more than 3000 books and presented them to the University of London. Together with George Boole, De Morgan can be credited with stimulating the upsurge of interest in logic that took place in the mid-19th century.[5]

Figure 153. Augustus De Morgan

Figure 154. James J. Sylvester

James Joseph Sylvester

James Joseph Sylvester was born into a Jewish family on 3 September 1814 in London England. His father was Abraham Joseph but his son added the surname Sylvester before he began his university studies.

Sylvester studied only five months in 1828 at University College London but had to withdraw because a fellow student accused Sylvester of threatening him with a knife. He matriculated at St. John's College, Cambridge in 1831 and took the mathematics tripos examination in 1837 achieving the Second Wrangler.[102]

At that time students were expected to accept the Thirty-Nine Articles of the Church of England before graduating but Sylvester refused to take the necessary oath. Because of this he was not eligible for a Smith's prize or for a Fellowship.

From 1838 to 1841 he worked as professor of natural philosophy at the University College, London. During that time the Royal Society of London elected him as a Fellow in 1839. He was awarded a B.A. and an M.A. by Trinity College, Dublin in 1841.

He went to the United States in 1841 to become a professor of mathematics at the University of Virginia in Charlottesville. There, a student insulted him so Sylvester struck him with a sword stick. The student collapsed in shock and Sylvester thought that he had killed the student, but he was not seriously injured. Sylvester fled to New York where one of his elder brothers lived. He stayed in New York until November 1843 trying unsuccessfully for academic positions.

On his return to England in 1846 he entered the Inner Temple, and was called to the Bar in 1850. In the course of his legal studies he became friends with the mathematician Arthur Cayley who was also studying law in the nearby Lincoln's Inn.[105]

Sylvester worked as an actuary and lawyer at the Equity Law and Life Assurance Company until 1855. Sylvester and Cayley discussed mathematics as they walked around the courts.

Sylvester became professor of mathematics at the Royal Military Academy at Woolwich in 1855 where he resigned in 1870.

He was elected to the Paris Academy of Sciences in 1863 and became the second president of the London Mathematical Society in 1866 succeeding De Morgan. In 1876 Sylvester accepted a Chair at Johns Hopkins University in Baltimore, Maryland for a salary $5000 to be paid in gold[105] and one year later he founded the American Journal of Mathematics, the first mathematical journal in the United States. He supervised the doctorates of nine students at Johns Hopkins. Sylvester received an honorary doctorate of civil law from Oxford in 1880.[105]

Sylvester succeeded Henry Smith (1826-1883) as Savilian Professor of Geometry at Oxford in 1883. College tutors at Oxford delivered lectures open to members of other colleges. While highly popular to the students, this tended to make professors' lectures seem less relevant. He liked to lecture on his own research but students did not appreciate it because they only wanted information relevant to their examinations. In

1892, he returned to London. By this time he was partially blind and suffering from loss of memory.

He is one of the founders of modern invariant algebra. He wrote two long memoirs on the nature of roots in quintic equations and worked on the theory of numbers, especially in partitions and Diophantine analysis. He coined the term "matrix" in 1850.

Sylvester received the Royal Medal in 1861, the Copley Medal in 1880 and the De Morgan Medal in 1887.

He loved poetry and was fluent in French, German, Italian and Greek. Sylvester died on 15 March 1897 in London, England.

Arthur Cayley

Arthur Cayley was born on 16 August 1821 in Richmond, Surrey, England. His father Henry Cayley was a merchant working in St. Petersburg, Russia. Arthur was born when his parents returned to England for a visit. Arthur spent the first eight years of his childhood in St. Petersburg.

Cayley graduated as Senior Wrangler and won the Smith's prize in 1842 at Trinity College, Cambridge. He became the youngest Fellow at Trinity College which imposed a light teaching load allowing time for research. During four years as a fellow, he published 28 papers in the Cambridge Mathematical Journal.

When his fellowship came to an end he decided to prepare for the bar. In April 1846 Cayley was admitted to Lincoln's Inn, where he became a pupil of a celebrated conveyancing counsel named J.H. Christie.[4] Among the friends Cayley made at Lincoln's Inn was James Joseph Sylvester (1814-1897) who studied at the nearby Inner Temple.

Cayley was called to the Bar in 1849, thereby becoming a fully fledged barrister. He continued to mathematical pursue his mathematical work when not engaged in conveyancing. He wrote 300 mathematical papers from 1849 to 1863 at the bar. He was tired of routine legal works, and he chose a professorship with less pay. Cayley was appointed to a new chair, Sadlerian Professor of Mathematics in 1863 which he held thirty two years. In September 1863 Cayley married Susan Moline and they had one son and one daughter.

He delivered one course per year and later two courses. Since each course only had a few students, he dispensed with the use of chalk and blackboard. He brought the lecture written out on paper and went through it, making comments.[4] He lectured mostly on his current research which was unusual for the time, but is common in today's graduate school lectures. Unfortunately, students found him to be a poor lecturer. Nonethless, Cayley's work was highly respected in Germany and France. Cayley developed the theory of algebraic invariance and he united projective geometry and metrical geometry which is dependent on size of angles and lengths of lines. In matrix theory, he is remembered in Cayley-Hamilton theorem.

He wrote over 900 papers and notes on nearly every aspect of modern mathematics. Cayley was awarded the Royal Medal (1859) and the Copley Medal (1881) of the Royal Society. He died on 26 January 1895 in Cambridge, England.

Figure 155.   Arthur Cayley

Figure 156.   Henry F. Baker

Henry Frederick Baker

Henry Baker was born on 3 July 1866 in Cambridge in 1884. In 1887 he was Senior Wrangler in the Mathematical Tripos a position he shared with two other students. He was elected a Fellow of St. John's College in 1889 and in the same year he won the Smith's Prize.[3]

He was a College Lecturer from 1890 to 1895, then as a University Lecturer until 1914. At that time he became Lowndean Professor of Astronomy and Geometry which was created in 1749. He retired from that post in 1936.

Baker's most important contribution was a six volume masterpiece *Principles of Geometry* published between 1922 and 1925. He also edited Sylvester's collected papers in four volumes between 1904 and 1912.

Baker was elected a Fellow of the Royal Society in 1898 and he won the Sylvester Medal in 1910 as well as the De Morgan Medal in 1905. He died on 17 March 1956 in Cambridge, England.

Harold Scott MacDonald Coxeter

His original name was Harold MacDonald Scott Coxeter (HMS Coxeter) which sound like a ship. So it was changed to Harold Scott MacDonald Coxeter, known as Donald Coxeter.[3]

He was born on 9 February 1907 in London, England. Coxeter received his B.A. in 1929 and a Ph.D. in 1931 at the University of Cambridge under the supervision of Henry F. Baker. His thesis was *Some contributions to the theory of regular polytopes.*[10]

He was appointed to a faculty position at the University of Toronto in 1936 where he remained until his death.[3] He had supervised 17 Ph.D. students. Coxeter was a Fellow of the Royal Society of London and a Fellow of the Royal Society of Canada. Coxeter wrote 12 books including The *Real Projective Plane* (1955), *Introduction to Geometry* (1961), *Regular Polytopes* (1963) and *Non-euclidean Geometry* (1965). He was made a Companion of the Order of Canada in 1997.

He died on 31 March 2003 in Toronto, Canada.

Maxwell Herman Alexander Newman

Maxwell Newman was born on 7 February 1897 in Chelsea, London. His father, Herman Alexander Neumann, was German. Max changed his last name to Newman in 1916. He entered St. John's College, Cambridge in 1915, but undertook war related work during World War I. He graduated St. John's College in 1921 and became a Fellow of St. John's College in 1923. From 1927 he was a lecturer at Cambridge at the same time holding his Fellowship.[3]

He married Lyn Irvine in 1934 and they had two sons. Alan Turing took his course on the foundations of mathematics introducing to the concept of decidability. In 1939 Newman was elected a Fellow of the Royal Society of London and received the Sylvester Medal in 1958. During World War II he worked at the Government Code and Cipher School at Bletchley Park.

In 1945 he succeeded Mordell as Fielden Professor of Mathematics at Manchester and in 1948 he appointed Turing as a Reader in his department.[3] Newman retired in 1964.

His field was combinatorial topology and theoretical computer science. He wrote one book *Elements of the Topology of Plane Sets of Points* in 1939.[3]

Newman served as President of the London Mathematical Society in 1949-1951 and received its DeMorgan Medal in 1962. Newman was a gifted pianist and a good chess player. His wife Lynn died in 1973 and he remarried Margaret Penrose, the window of Lionel Sharples Penrose.

Newman died on 22 February 1984 in Comberton, near Cambridge, England.

Alan Mathison Turing

Alan Turing was born on 23 June 1912 at Paddington, London, England. While studying at a public school (private school) he read Einstein's papers on relativity and Eddington's *The Nature of the Physical World*. When his friend Christopher Morcom died in 1930, Turing had a shattering experience. Turing graduated King's College, Cambridge in 1934 where he was interested in mathematical logic, mathematical philosophy and quantum mechanics.

He was elected a Fellow of King's College, Cambridge, in 1935 for a dissertation *On the Gaussian error function*. In 1936 he won a Smith Prize. In the same year he introduced the Turing machine in a paper *On Computable Numbers, with an Application to the Entscheidungs Problem*.

He went to the United States in 1936 and received a Ph.D. in 1938 at Princeton University for a dissertation *Systems of Logic Based on Ordinals* under Church's supervision.

When World War II started in 1939 Turing moved to work full time at the Government Code and Cypher School at Bletchley Park. By the middle of 1941 Turing's statistical approach combined with captured information had led to the German Enigma signals being decoded at Bletchley. In 1945 Turing was awarded the O.B.E. (Officer of the Order of British Empire) for his vital contribution to the war effort.[3]

In 1948 he was appointed as a Reader at the University of Manchester. He was elected a Fellow of the Royal Society of London in 1951 mainly for his work on Turing machines.

He was a very good athlete and he once was placed fifth in a full marathon in 1947. Turing was arrested for violation of British homosexuality statues in 1952 and found guilty on 31 March 1952. After his conviction, his security clearance was withdrawn.

He died of potassium cyanide poisoning on 7 June 1954 in Wilmslow, Cheshire, England. The Cyanide was found on a half eaten apple beside him.[3]

Figure 157. Andrew Russell Forsyth   **Figure 158. Alan Mathison Turing**

Andrew Russell Forsyth

Andrew Forsyth was born on 18 June 1858 in Glasgow, Scotland. He entered Trinity College, Cambridge in 1877 where he studied under Cayley. In 1881 he was First Wrangler in the mathematical Tripos. In the same year he was appointed a Fellow at Trinity College with his thesis on double theta functions.

He was appointed to the Chair of Mathematics at the University of Liverpool in 1882 at the age of 24. Two years later he came back to Cambridge accepting a lectureship. In 1886 the Royal Society of London elected him as a Fellow. In 1897 he received the Royal Medal from that Society.

Forsyth was appointed Sadlerian Professor of Pure Mathematics in 1895 succeeding Cayley. In 1910 he had to resign his chair at Cambridge due to his love affair with Marion Amelia Boys, wife of Sir Charles Vernon Boys (1855–1944), physicist and inventor of a fused quartz fibre suspension and other instruments. Forsyth married Marion Boys and moved to Calcutta.[5] He was appointed to the chair in Imperial College London in 1913 where he retired in 1923.

Forsyth had unusually good skills as a linguist. He is known by the Smith–Forsyth theorem which generalized a large number of identities between double theta functions.[5] His wrote many books that critics on the continent panned as lacking in creativity. Forsyth died on 2 June 1942 in London, England.

Figure 159.   Edmund T. Whittaker

Figure 160. Louis J. Mordell

Edmund Taylor Whittaker

Edmund Whittaker was born on 24 October 1873 in Southport, Lancashire, England. He graduated from Trinity College, Cambridge in 1895 as Second Wrangler in the Mathematical Tripos. In 1896 he was elected as a Fellow of Trinity College and he became a Smith's prizeman in 1897. In 1901 he married Mary Ferguson McNagten Boyd and they had three sons and two daughters. One of his sons became a mathematician.

Among those who attended his lectures at Cambridge were H.M. Bateman, A.S. Eddington, G.H. Hardy, J.H. Jeans, J.E. Littlewood and G.N. Watson. He was appointed Royal Astronomer of Ireland in 1906.

He succeeded George Chrystal to the Chair of Mathematics in Edinburgh in 1911 and retired in 1946. He is best known for his book *A Course of Modern Analysis* written in 1915 with George Watson.

He was elected a Fellow of the Royal Society in 1905, he won the Sylvester Medal (1931), De Morgan (1935) and the Copley Medal (1954). He was knighted in 1945. Whittaker died on 24 March 1956 in Edinburgh, Scotland.

Louis Joel Mordell

Louis Mordell was born on 28 January 1888 in Philadelphia, Pennsylvania, USA. His parents were Jewish and emigrated from Lithuania to the United States in 1881. When Mordell was about 13 years old, he began buying second hand mathematics books to teach himself the subject. These books presented many examples taken from the Cambridge Tripos examinations. Through his studies, he developed an ambition to study mathematics at Cambridge.[3]

After graduating from the Central High School in Philadelphia he traveled to England. He placed first in the Cambridge Scholarship Examination and entered St. John's College. Mordell graduated as Third Wrangler in the Mathematical Tripos but failed to be appointed a Fellow at St. John's College.

He was appointed as a lecturer at Birkbeck College, London in 1913. Three years later, he married Mabel Elizabeth Cambridge. During World War I he worked on statistics at the Ministry of Munitions.

In 1917 he proved Ramanujan's conjecture on the tau-function. This work later extended by Erich Hecke is now fundamental in the theory of the Hecke operator.[3] He moved to the Manchester College of Technology in 1920 and stayed there two years.

During this time he discovered the finite basis theorem which proved a *conjecture of Poincaré*. Mordell theorized that there are only finitely many rational points on any curve of genus greater than one. This became known as the Mordell Conjecture. In 1983 Gerd Faltings proved the Mordell Conjecture to be true.

From 1922 to 1945 he taught at the University of Manchester. He then succeeded Hardy as the Sadlerian Professor at the University of Cambridge and was elected a Fellow at his alma mater, St. John's College. In 1953 he retired from the Sadlerian Chair, but he continued working. In fact, almost half of his 270 publications appeared after his retirement.

He was an enthusiastic mountaineer and swimmer. Mordell was a generous man to his younger colleagues.

Figure 161.   John W.S. Cassels

Figure 162.   John A. Todd

He was elected a Fellow of the Royal Society of London in 1924 and became a British subject in 1929. He received the Sylvester Medal in 1949, and the De Morgan Medal in 1941. Mordell died on 12 March 1972 in Cambridge, England.

John William Scott Cassels

John Cassels was born on 11 July 1922 in Durham, England. He graduated from Edinburgh University with an M.A. in 1943 and received a Ph.D. in 1949 at the University of Cambridge. In the same year he was elected a fellow of Trinity College, Cambridge.

From 1950 to 1963 Cassels was a lecturer and in 1963 he was appointed Reader in Arithmetic at Cambridge. In 1963, the Royal Society of London elected him as a Fellow. During that time, he served as president of the London Mathematical Society (1976-78). In 1967, Cambridge appointed him as the Sadlerian Professor of Pure Mathematics. He retired from Cambridge in 1984.

He wrote *Introduction to Diophantine Approximation* in 1957 and *An Introduction to the Geometry of Numbers* in 1959.[3]

John Arthur Todd

John Todd was born on 23 August 1908 in Liverpool, England. He graduated from Trinity College, Cambridge in 1928 and received a Ph.D. in 1932 under H.F. Baker's supervision. Although he was awarded the highly prestigious Smith's prize, he failed to win a Research Fellowship at Trinity College. He spent the 1933-34 academic year at Princeton University on a Rockefeller Scholarship. During that year, he expanded his study of geometry beyond the classical approach he had learnt under H.F. Baker.[3] In 1937, Todd was appointed a lecturer at Cambridge.

In 1948, Todd was elected a Fellow of the Royal Society of London, but because some members of Trinity College perceived him to be socially gauche, he never became a Fellow there. In 1958, he left Trinity and he was appointed a Fellow at Downing College, Cambridge where he spent a happier period in his life.[3]

Two years later, he became a Reader. He retired in 1973.

The Todd Class in the higher dimensional Riemann-Roch theorem is an example of a characteristic Class (a reciprocal of one) that was discovered by Todd in his work published in 1937. The Todd-Coxeter process for coset enumeration is a major method of computational algebra. In 1953 Todd and H.S.M. Coxeter discovered the Coxeter-Todd lattice. He died on 22 December 1994 in Croydon, England.

Figure 163.   Roger Penrose

Roger Penrose

Poger Penrose was born on 8 August 1931 in Colchester, Essex, England. His mother was a doctor while his father was a medical geneticist who became Professor of Human Genetics at University College London in 1945.

Penrose graduated from University College London with a B.Sc. First Class Honours. Then he went to Cambridge for his Ph.D. attending lectures by Hermann Bondi, Paul Dirac and Dennis Sciama.

After receiving a Ph.D., he became a Research Fellow at St. John's College, Cambridge and he married Joan Isabel Wedge in 1959. He spent the years 1959-61 in the United States on NATO Research Fellowship and the years 1961-63 as a Research Associate at King's College, London.

Then the University of Texas at Austin appointed him a Visiting Associate Professor (1963-1964). From 1964 to 1973 he worked at Birkbeck College, London. In 1973, he was appointed Rouse Ball Professor of Mathematics at the University of Oxford where he retired in 1998.

In 1965 Penrose proved an important theorem which, under conditions of a trapped surface, a singularity must occur in a gravitational collapse using topological methods. Under these conditions space-time cannot be continued and the classical general relativity breaks down.[3]

One of his major breakthroughs was his introduction of twistor theory in an attempt to unify relativity and quantum theory. Together with physicist Wolfgang Rindler, Penrose published two volumes of *Spinors and space-time* in 1984 and 1986. He wrote many other books on cosmology, consciousness, and computers. The author likes his book "The Road to Reality" most.

Penrose was elected a Fellow of the Royal Society of London in 1972, the US National Academy of Sciences in 1998. He received the Royal Medal of the Royal Society in 1985 and the Wolf Prize for Physics jointly with Stephen Hawking in 1988. Penrose also received the Dirac Medal and Prize of the British Institute of Physics in 1989. In 1994 he was knighted for services to science.

George Neville Watson

George Watson was born on 31 January 1886 in Westward Ho!, Devon, England. His father, George Wentworth Watson, was a famous genealogist.

When Watson entered Trinity College, Cambridge in 1904, three young fellows of Trinity influenced Watson's mathematics: Whittaker, Barnes, and Hardy. Watson graduated as Senior Wrangler in 1907 in the Mathematical Tripos. He won a Smith's Prize in 1909 and became a Fellow of Trinity College in 1910.[3]

Trinity College is the most prestigious and affluent College in Cambridge where 32 Nobel laureates had been associated with. But becoming a Fellow and a professor at the university are different thing. Watson stayed four years as Trinity Fellow but could not be appointed a professor at the University of Cambridge.

Consequently, Watson left Cambridge to become Mason Professor of Pure Mathematics at the University of Birmingham from 1918 to 1951. He is best known as a joint author with Whittaker of *A Course of Modern Analysis* published in 1915. He also published *The Theory Bessel Functions* in 1922.

In 1902, Oliver Heaviside had predicted that there was an conducting layer in the atmosphere which allowed low frequency radio waves to follow the Earth's curvature. In 1923 Heaviside conjecture was experimentally proved correct. Watson obtained mathematical solutions to the problem in 1918 making a contribution to wave propagation theory in earth-ionosphere waveguide.

Watson carefully examined Ramanujan's (1887-1920) notebook and wrote twenty-five papers relating to Ramanujan's results. He was elected a Fellow of the Royal Society of

London in 1919 and received the Sylvester Medal in 1946 and the De Morgan Medal in 1947.

Watson died on 2 February 1965 in Leamington Spa, Warwickshire, England.

Figure 164. G. Harold Hardy

Godfrey Harold Hardy

G. H. Hardy was born on 7 February 1877 in Cranleigh, Surrey, England. His father was bursar and an art master at a school and his mother was a teacher. Hardy graduated from Winchester College (Secondary School) in 1896 and entered Trinity College, Cambridge winning an open scholarship.

Professor August Love (1863–1940) advised Hardy to read Camille Jordan's *Cours d'Analyse* from where Hardy learned for the first time what mathematics really meant.[51] Hardy was placed as Fourth Wrangler in Part I of the Mathematical Tripos of 1898, a result annoyed him despite feeling that the system was very silly. Hardy graduated in 1899 and the following year took Part II of the Tripos which is more challenging. He scored first in Part II.[36] He was elected a Fellow of Trinity College in 1900 and awarded a Smith's Prize jointly with J.H. Jeans.

Hardy began his collaboration with J.E. Littlewood in 1911. The two men worked together for 25 years. Littlewood was eight years younger than Hardy and the former became a Fellow of Trinity College.

In early 1913 Hardy received a letter and manuscripts from Ramanujan whose genius was quickly spotted by Hardy. He brought Ramanujan to Cambridge where they wrote five remarkable papers together.

From 1919 to 1931 Hardy was Savilian Professor of Geometry at Oxford after which he succeed E.W. Hobson as Sadlerian Professor at Cambridge. Hardy retired in 1942.

His only passion other than mathematics was cricket. He had remained youthful in both body and mind until he had a heart attack in 1939 at the age of 62.

Although he was a handsome man, he could not endure having his photograph taken and he hated mirrors. His first action on entering any hotel room was to cover any mirror with a towel.[3] He hated telephone, watches and fountain pens. He loved cats, hated dogs. He loved The London Times crossword puzzles.[36]

He was elected a Fellow of the Royal Society of London in 1910 and received the Royal Medal in 1920, De Morgan Medal in 1929 and the Sylvester Medal in 1940. He was twice president of the London Mathematical Society. He wrote *A Mathematicians Apology* in 1940. It offers a brilliant and engaging account of mathematics as very much more than a science.

It is said that Hardy had four ambitions in life:

To prove the Riemann hypothesis

To score the winning play in an important game of Cricket

To murder Mussolini

To prove the non-existence of God.

Hardy did not achieve any of these goals.[35]

By 1946 he could only get around by taking a taxi, a few steps would make him short of breath. He died on 1 December 1947 in Cambridge, England. In 1966, 36 years after Hardy had asked for a Mathematics Institute at Cambridge, a new purpose-designed building was completed.[105]

Figure 165.   Edward C. Titchmarsh

Edward Charles Titchmarsh

Edward Titchmarsh was born on 1 June 1899 in Newbury, Berkshire, England. He entered Balliol College, Oxford in October 1917. He joined the Royal Engineers as a dispatch rider, first on horseback and then on a motorcycle during World War I.

He graduated Balliol College with a First Class degree in 1922. In 1923 he was appointed as a Senior Lecturer at the University College London. He became a Fellow at Magdalen College in 1923, so he had duties at both London and Oxford.[3] He married Kathleen Blomfield in 1925 and they had three daughters.

Titchmarsh held the chair of pure mathematics at Liverpool from 1929 to 1931, then he succeeded to Hardy's Savilian Chair at Oxford in 1932. Titchmarsh held the Savilian Chair of Geometry for 31 years without obligation to lecture on geometry.[102]

He was elected to the Royal Society of London in 1931 and received the Sylvester Medal in 1955. Titchmarsh served as President of the London Mathematical Society in 1945-47 and received its De Morgan Medal in 1953.

He enjoyed listening to music and watching cricket. Titchmarsh wrote many five books including *The Theory of Functions* (1932), *The Theory of the Riemann Zeta-Function* (1951) and *Eigenfunction Expansions Associated with Second-Order Differential Equations* (1946, 1958).

He died on 18 January 1963 in Oxford, England.

Augustus Edward Hough Love

Augustus Love was born on 17 April 1863 in Weston-Super-Mare near Bristol, England. In 1882 he entered St. John's College, Cambridge winning a scholarship. He attained the rank of Second Wrangler in the Mathematical Tripos after which he became a fellow at St. John's College in 1886 and won the Smith's Prize.

The Royal Society elected Love a Fellow in 1894. Oxford appointed him to the Sedleian Chair of Natural Philosophy in 1899.[3]

He wrote the two volume work *A Treatise on the Mathematical Theory of Elasticity* in 1892-93 and his later book, *Some Problems in Geodynamics,* won him the Adams Prize at Cambridge in 1911.

He received the Royal Medal in 1909, the Sylvester Medal in 1937 and the De Morgan Medal in 1926. He was President of the London Mathematical Society in 1912-13. He never married so one of his sisters kept house for him. He died on 5 June 1940 in Oxford, England.

Figure 166.   Augustus E. H. Love

Figure 167.   Srinivasa A. Ramanujan

Srinivasa Aiyangar Ramanujan

Srinivasa Ramanujan was born on 22 December 1887 in Erode, Tamil Nadu State, India. His father, K. Srinivasa Iyengar worked as a clerk in a cloth merchant's shop. Ramanujan did not have a family name. Many Indians do not have a family name. He was born at the place of residence of his maternal grandparents.[36]

When he was sixteen, Ramanujan came across George S. Carr's book, *A synopsis of elementary results in pure and applied mathematics*, which features a collection of 5000 theorems, and introduced Ramanujan to the world of mathematics. In 1906, he entered Pachaiyappa's College in Madras. He passed mathematics but failed all other subjects and therefore failed his examination.

He married to S. Janaki Ammal, a ten year old girl on 14 July 1909. This was not untypical in that region at that time. He did not live with his wife until she was twelve years old.[3] After publication of his paper on Bernoulli numbers in 1911 in the *Journal of Indian Mathematical Society* he gained recognition for his work in India.

He wrote to E.W. Hobson and H.F. Baker at Cambridge trying to interest them in his results but he never got any replies. In January 1913 he wrote to G.H. Hardy and on 8 February 1913, Hardy replied to Ramanujan.[36]

Ramanujan arrived in London on 14 April 1914 after almost one month of traveling. After two weeks, he moved to a room in Trinity College. Since he was a strict vegetarian, he had problems maintaining adequate diet in the meat-rich Trinity dining hall, which created health problems.

On 16 March 1916 he graduated from Cambridge with a Bachelor of Arts by Research. The degree was called a Ph.D. from 1920. His dissertation was on highly composite numbers which consisted of seven papers he published in England.

Ramanujan compiled nearly 3900 results mostly identities and equations. Most of his claims have now been proved to be correct. He was elected a Fellow of the Royal Society of London in 1918 and a Fellow of Trinity College in the same year.[36]

Ramanujan worked out the Riemann series, the elliptic integrals, hypergeometric series and functional equations of the zeta function. But he had only a vague idea of mathematical proof.[3]

He sailed back to India on 27 February 1919 in very poor health. Ramanujan died on 26 April 1920 in Kumbackonam, Tamil Nadu State, India.

Mary Lucy Cartwright

Mary Cartwright was born on 17 December 1900 in Aynho, Northamptonshire, England. Her father was the vicar at Aynho at the time of Mary's birth.

In 1919 she entered St. Hugh's College in Oxford to study mathematics. She was awarded a first class degree in Final Honours and graduated from Oxford in 1923. After teaching four years in secondary schools she came back to Oxford for her doctoral studies. Cartwright received a D.Phil with her thesis on zeros of integral functions in 1930 under supervision of Hardy and Titchmarsh. She was examined by Littlewood as an external examiner.[105]

In 1930 she went to Girton College, Cambridge, a women's college, on a Yarrow Research Fellowship. She was appointed an assistant lecturer at the University of Cambridge in 1934 and two years later she became Director of Studies in Mathematics a position that required her to supervise other fellows teaching students.

In 1947, the Royal Society of London made Cartwright the first woman mathematician in the Society. In 1948, she was appointed Mistress (Principal) of Girton College. She acted as a Reader in the theory of functions at Cambridge from 1959 to 1968, but she was never appointed full professor at Cambridge.

Cartwright received the Sylvester Medal in 1964 and the De Morgan Medal in 1968. In 1969 she became Dame Mary Cartwright, Commander of the British Empire. She was the first woman president of the London Mathematical Society in 1961-62.[41] She died on 3 April 1998 at the age of 97 in Cambridge, England.

Figure 168.   Mary L. Cartwright

Walter William Rouse Ball

Walter Rouse Ball was born on 14 August 1850 in Hampstead, London, England. After graduating from University College, London in 1871, he entered Trinity College, Cambridge. In the 19th Century and early 20th Century, many college graduates matriculated at Cambridge as a freshman.

In 1874 he graduated Trinity College as Second Wrangler in the Mathematical Tripos. In 1875 he was elected a fellow of Trinity College, Cambridge. In 1876 he was called to the Bar as a barrister by the Inner Temple in London. He practiced a short time as an equity draftsman and conveyance.[3]

In 1878, Trinity College appointed Rouse Ball as a lecturer and two years later as assistant tutor. In 1891, he was appointed Director of Mathematical Studies at Trinity, but he did not hold a professorship at Cambridge.

He is best known for his book *A Short Account of the History of Mathematics* published in 1888 and *The History of Mathematical Studies at Cambridge* in 1889. Rouse Ball died on 4 April 1925 in Elmside, Cambridge. The Rouse Ball professorship of mathematics were founded in 1927 by a bequest from him. Rouse Ball left 25,000 pounds to Oxford and a rather larger amount to Cambridge. It is one of the senior Chair in the Mathematics Departments at the University of Cambridge and the University of Oxford. The Rouse Ball Professor could lecture on topics of his own choice rather than courses in the regular curriculum.

The holders of Rouse Ball Professorship are as follows:

Rouse Ball Professors: Cambridge
 1. John Edensor Littlewood (1928–1950)
 2. Abram Samoilovitch Besicovitch (1950–1958)
 3. Harold Davenport (1958–1969)
 4. John G. Thompson (1971–1993)
 5. Nigel Hutchin (1994–1997)
 6. William Timothy Gowers (1998–present)

Rouse Ball Professors: Oxford
 1. E.A. Milne (1929–1950)
 2. Charles Coulson (1952–1972)
 3. Roger Penrose (1973–present)
 4. Philip Candelas (1999–present)

Ernest William Barnes

Ernest Barnes was born on 1 April 1874 in Birmingham, England. He was a Second Wrangler in 1896 and placed in the first division of the first class in Part II of the Mathematical Tripos in 1897 at the University of Cambridge.

In 1898 he received a Smith Prize and was elected a Fellow at Trinity College, Cambridge. In 1902 he was appointed a lecturer in mathematics and a tutor in 1908. The Royal Society elected him a Fellow in 1909.

He was ordained in 1903. By 1915, when he was made Master of the Temple, he had established a reputation for outspoken and provocative preaching, which he maintained as canon of Westminster (1918–24). Appointed bishop of Birmingham in 1924, he attacked ritualistic practices and in 1929 evoked much protest by his refusal to install an Anglo-Catholic priest. An uncompromising pacifist, he refused to take part in national days of prayer during World War II.[37]

He is remembered by Barnes G-function and Mellin–Barnes integrals. Barnes died on 29 November 1953 in Sussex, England.

Figure 169.   John E. Littlewood

John Edensor Littlewood

John Littlewood was born on 9 June 1885 in Rochester, Kent, England. His father Edward Thornton Littlewood was Ninth Wrangler in the Mathematical Tripos at Cambridge in 1882. He was appointed Headmaster at a new school in South Africa in 1892 and the family moved to South Africa. In 1900 Littlewood returned to England to attend St. Paul's School in London. He entered Trinity College in October 1903 and graduated as Senior Wrangler (equal first) in 1906. His tutor E.W. Barnes gave him the Riemann hypothesis as his research problem.[3]

He was elected a fellow of Trinity College in 1908. He lectured as Richardson Lecturer at the University of Manchester from 1907 to 1910. Around 1911 he began a 35 year long collaboration with G.H. Hardy.

Littlewood was appointed the first Rouse Ball Professor of Mathematics at Cambridge in 1918, a post he held until 1950. He could lecture on topics of his own choice and he no longer had to take part in routine teaching. Beginning 1938 Littlewood, working jointly with Mary Cartwright, spent 20 years working on nonlinear differential equations such as Van der Pol's equation which are important for radio engineers.

Littlewood was elected a Fellow of the Royal Society of London in 1915 and received the Royal Medal in 1929, the Sylvester Medal in 1943, the De Morgan Medal in 1938 and the Copley Medal in 1958.

He had an entertaining sense of humor and was full of fascinating stories.[40] He had been one of the best gymnasts and a hard hitting batsman at school. He was good in rock climbing and skiing. On most days he walked many miles in the country.

Littlewood never married and lived at Trinity College. But he fathered a daughter with a married woman. He did not publicly acknowledge his child until near his death.[36,40] Littlewood suffered from depression for most of his life beginning while he was at school in South Africa. But there is no evidence that it impaired his ability to do mathematics.

Although he was President of the London Mathematical Society in 1941-43 he never chaired a meeting, the Vice-President took the chair.[3] His last paper was published in 1972, when he was 87. He died on 6 September 1977 in Cambridge, England.

Harold Davenport

Harold Davenport was born on 30 October 1907 in Huncoat, Lancashire, England. He graduated from Manchester University with first class honors in mathematics in 1927. He entered Trinity College, Cambridge in the same year to take another bachelor degree which was a common thing at that time.

He was awarded a Rayleigh Prize in 1930 but failed to get a Smith Prize. He was elected a Fellow of Trinity College in 1932. Davenport Collaborated with Helmut Hasse in Germany during which time Davenport taught Hasse to speak English. In turn, Davenport became fluent in German during his years at Göttingen. He received a Ph.D. at Cambridge in 1937. From 1937 to 1941 he was an assistant lecturer at the University of Manchester at the invitation of Mordell. From 1941 to 1945 he held the Chair of Mathematics at the University College of North Wales at Bangor. And from 1945 to 1958 he was Astor Professor of Mathematics at the University College, London.[3]

In 1958 Davenport succeeded Besicovitch as Rouse Ball Professor of Mathematics at Cambridge.

In 1944, he married Anne Lofthouse and they had two sons. He spent the academic year 1947-48 at Stanford University where he started a life-long friendship with George Pólya and Gabor Szego.

Davenport was elected a Fellow of the Royal Society in 1940 while he was an assistant lecturer at Manchester. He received the Sylvester Medal in 1967 and was President of the London Mathematical Society during 1957-59.

Davenport's conservative outlook on life led him to say, "All change is for the worse". Always a heavy smoker and Davenport died of lung cancer at the age of 61 on 9 June 1969 in Cambridge.

Figure 170. Harold Davenport

John Horton Conway

John Conway was born on 26 December 1937 in Liverpool, England. He graduated from Gonville and Caius College, Cambridge in 1959 and he received a Ph.D. at Cambridge in 1964. In the same year he became a Lecturer in Pure Mathematics and a Fellow at Sidney Sussex College.

In 1970, he was elected a Fellow at Gonville and Caius College and three years later, he was appointed Reader in Pure Mathematics and Mathematical Statistics.

The Royal Society of London elected Conway as a Fellow in 1981 and two years later Cambridge appointed him professor of mathematics.[94]

In 1986 he left Cambridge for Princeton University to become John von Neumann Chair of Mathematics.

Figure 171. Alan Baker

Figure 172. John H. Coates

He wrote many books. Conway is active in the theory of finite groups, knot theory, number theory, combinatorial game theory and coding theory. He was inventor of the Game of Life (the cellular automaton, not the board game). He was the first recipient of the Pólya Prize of the London Mathematical Society in 1987 and Leroy P. Steele Prize by the American Mathematical Society in 2000.

Alan Baker

Alan Baker was born on 19 August 1939 in London, England. He graduated from the University College, London in 1961 receiving his B.Sc. He then did his graduate work at Cambridge and was awarded a Ph.D. in 1964 under Davenport's supervision.

He was elected a Fellow of Trinity College in the same year and remained there for the next ten years, as a research fellow (1964-1968) and as director of studies in mathematics (1968-1974). He was elected a Fellow of the Royal Society of London in 1973 and appointed professor of pure mathematics at the University of Cambridge in 1974.[5]

In 1970 at the International Congress of Mathematicians at Nice he was awarded a Fields Medal for generalizing the Gelfond-Schneider theorem (the solution to Hilbert's seventh problem). From this work he generated transcendental numbers not previously identified. Two years later, he received the University of Cambridge's Adams Prize, which commemorates John Couch Adam's discovery of Neptune in 1846.

Hilbert's 7th problem asked whether or not a^q was transcendental when a and q are algebraic. Baker's most important publication is *Transcendental Number Theory* published in 1975. He is one of the most important number theorists.

John Henry Coates

John Coates was born on 26 January 1945 in New South Wales, Australia. He graduated from the Australian National University receiving a B.Sc. degree and also studied at the *École Normale Superieure in Paris*. He was awarded a Ph.D. in 1969 at the University of Cambridge under Alan Baker's supervision.

He remained at Cambridge where he was appointed a lecturer and a Fellow of Emmanuel College. He supervised Andrew Wile's Ph.D. thesis, and together they proved a partial case of the Birch and Swinnerton-Dyer Conjecture for elliptic curves with complex multiplication.[38]

From 1978 to 1985 Coates was a professor at the University of Paris XI at Orsay. In 1985 he became a professor and director of mathematics at the *École Normale Superieure in Paris*.

In 1986 he was appointed as Sadlerian Professor of Pure Mathematics at Cambridge succeeding John Cassels. Coates was elected a Fellow of the Royal Society of London in 1985 and was President of the London Mathematical Society (1988-1990). Later, he became Vice President of the International Mathematical Union (1991-1995).

He married Julie Turner in 1966 and they have three sons. His hobby is collecting oriental porcelains.

Andrew John Wiles

Andrew Wiles was born on 11 April 1953 in Cambridge, England. He graduated from Merton College, Oxford in 1974 and received a Ph.D. at Cambridge in 1980 under Coates's supervision. His thesis was *Reciprocity laws and the conjecture of Birch and Swinnerton-Dyer*. He was a Junior Research Fellow at Clare College, Cambridge from 1977 to 1980.

After about one year at the Institute for Advanced Study in Princeton, he was appointed professor of mathematics at Princeton University in 1982. During 1985-1986 he visited the *Institut des Hautes Études Scientifique* in Paris on a Guggenheim Fellowship. From 1988 to 1990 Wiles was a Royal Society Research Professor at Oxford. He returned to Princeton in 1990 and he was appointed Eugene Higgins Professor in 1994.

Beginning around 1987 Wiles abandoned all his other research and concentrated solely on attempting to prove the Shimura-Taniyama-Weil conjecture because in 1986 Ken Ribet proved that this conjecture will lead to Fermat's Last Theorem.

With some help from Richard Taylor, a former student, Wiles finally proved Fermat's Last Theorem in September 1994. His paper *Modular Elliptic Curves and Fermat's Last Theorem* appeared in the *Annals of Mathematics* in 1995.

He was elected a Fellow of the Royal Society of London in 1989 and received the Royal Medal in 1996. He received the Wolf Prize in 1995 and AMS Cole Prize in 1997. Wiles was knighted in 2000.

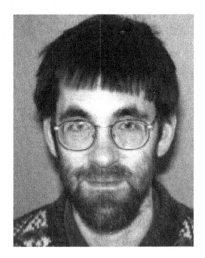

Figure 173. Richard E. Borcherds   **Figure 174. Andrew J. Wiles**

Richard Ewen Borcherds

Richard Borcherds was born on 29 November 1959. He graduated from Trinity College, Cambridge and received his Ph.D. in 1985 for a thesis *The leech lattice and other lattices* under John Conway's supervision.

He was a research fellow at Trinity (1983–1987), the Royal Society University Research Fellow (1988–1992), and a lecturer (1992–1993). Then he was appointed professor of mathematics at the University of California in Berkeley (1993–1996). From 1996 to 1999 he was the Royal Society Professor of Mathematics at Cambridge. Since that time he has been back at UC Berkeley.

Borcherds was elected a Fellow of the Royal Society of London in 1994 and awarded a Fields Medal in 1998 for his proof of the so-called Moonshine conjecture. This conjecture was formulated at the end of the 1970s by John Conway and Simon Norton and presents two mathematical structures in such an unexpected relationship that the experts gave it the name "Moonshine". The Moonshine conjecture provides an interrelationship between the so-called "monster group" and elliptic functions.

Henry Peter Francis Swinnerton-Dyer

Peter Swinnerton-Dyer was born on 2 August 1927. He graduated from Trinity College, Cambridge. He was a Research Fellow (1950–1954), Fellow of Trinity College (1955–1973), Master of St. Catharine's College (1973–1983), university lecturer (1960–1971), Professor of mathematics (1971–1988) and Vice-Chancellor (1979–1981) at the University of Cambridge.

He was elected a Fellow of the Royal Society in 1967 and knighted in 1987. In the early 1960s, B. Birch and Swinnerton-Dyer conjectured that if a given elliptic curve has an infinite number of solutions, then the associated L-series has value zero at a certain fixed point.

In 1976, Coates and Wiles showed that elliptic curves with complex multiplication having an infinite number of solutions have L-series that are zero at the relevant fixed point (Coates-Wiles theorem).[39] Swinnerton-Dyer received the Sylvester Medal in 2006.

Figure 175.   H. Peter F. Swinnerton-Dyer   Figure 176.   James H. Jeans

James Hopwood Jeans

James Jeans was born on 11 September 1877 in Ormskirk, Lancashire, England. When he entered Trinity College, Cambridge in 1896 he was a fellow student with G.H. Hardy. Jeans was Second Wrangler in the Mathematical Tripos examination of 1898 and he graduated in 1900. In 1901 he was elected a Fellow at Trinity College. He was appointed a lecturer in mathematics at Cambridge in 1904. During 1905-1909 he was professor of Applied Mathematics at Princeton University.[3]

In 1907 he married an American, Charlotte Tiffany Mitchell and in the same year he was elected a Fellow of the Royal Society of London. In 1909 Jeans became Stokes Lecturer in Applied Mathematics at Cambridge, but he gave up this post three years later to devote himself for research and writing books. In 1917 he won the Adams Prize from University of Cambridge, the Royal Medal in 1919, and was knighted in 1928.

Initially he was against Max Planck's theory on black-body radiation and he disagreed Eddington's ideas on the energy creation in stars. Jeans wrote many books on dynamic theory of gas, mathematical theory of electricity and magnetism, Cosmogony and Stellar dynamics as well as many popular books.

Jean's wife died in 1934 and he remarried one year later to Suzanne Hock, an accomplished musician.[3] Jeans himself was a good organ player. He also had an outstanding memory. He died of a heart attack on 16 September 1946 in Dorking, Surrey, England.

Ronald Aylmer Fisher

Ronald Fisher was born on 17 February 1890 in London, England. He graduated from Gonville and Caius College, Cambridge in 1912. He continued his studies on the theory of errors reading George B. Airy's manual the Theory of Errors which eventually led him to investigate statistical problems.

He taught mathematics and physics at several schools between 1915 and 1919. He married Ruth Eileen Guinnes, 17 year old girl on 26 April 1917.[3] In 1919 he accepted the post of statistician at the Rothamsted Agricultural Experiment Station where he made contributions to concept of likelihood and published *Statistical Methods for Research Workers* in 1925. Five years later he published *The Genetical Theory of Natural Selection*.

He formulated a methodology in which the analysis of results obtained using small samples produced interpretations that were objective and valid overall.[5] In 1933 Fisher was appointed as Galton Professor of Eugenics at University College London succeeding Karl Pearson. Ten years later he was appointed as Arthur Balfour professor of genetics at the University of Cambridge where he retired in 1959. He then moved to the University of Adelaide in Australia.

Fisher was elected a Fellow of the Royal Society in 1929, received the Royal Medal in 1938 and was knighted in 1952. In 1955, he was awarded the Copley Medal for his contributions to developing the theory and application of statistics in the field of biology.

He died on 29 July 1962 in Adelaide, Australia.

Figure 177. Alexander C. Aitken

Alexander Craig Aitken

Alec Aitken was born on 1 April 1895 in Dunedin, New Zealand. He fought in the New Zealand Expeditionary Force during World War I. He was wounded at the battle of the Somme, France. His war experiences troubled him for the rest of his life, affecting his health. Aitken graduated from the University of Otago, Dunedin, New Zealand in 1920. In the same year he married Mary Winifred Betts and they had two children.

In 1923, he went to the University of Edinburgh to seek a Ph.D. under Whittaker. His Ph.D. thesis *Smoothing of data* was considered so outstanding that he was awarded Doctor of Science degree in 1926. The Doctor of Science degree is normally awarded to a person with a Ph.D. and substantial research output requiring no additional dissertation. It is called the Higher Doctorate in Britain. He was appointed to Edinburgh University as a lecturer in 1925, a Reader in 1936 and was appointed to Whittaker's Chair in 1946.[3]

Aitken had an incredible memory and played the violin and composed music to a very high standard. He was elected a Fellow of the Royal Society in 1936. Aitken wrote a famous book *The Theory of Canonical Matrices* in 1932 with Herbert Turnbull (1885–1961).

Aitken's mathematical field was in statistics, algebra and numerical analysis. He died on 3 November 1967 in Edinburgh, Scotland.

Figure 178. William V. D. Hodge

Figure 179. Michael F. Atiyah

William Vallance Douglas Hodge

William Hodge was born on 17 June 1903 in Edinburgh, Scotland. After graduating with First Class Honors in mathematics from Edinburgh in 1923, he entered St. John's College, Cambridge. It was customary to get another bachelor degree at Cambridge even though the student already have a bachelor degree elsewhere.

Hodge was appointed to an assistant lectureship at the University of Bristol in 1926 and spent five years there.[3] He married Kathleen Anne Cameron in Bristol and they had one son and one daughter.

In 1930 he was elected a Fellow at St. John's College, Cambridge. In 1933 he was appointed as a university lecturer and two years later elected a Fellow of Pembroke College, Cambridge.

He won the Adams Prize in 1937 for his work on the theory of harmonic integrals. In 1936 Hodge was appointed as Lowndean Professor of Astronomy and Geometry, succeeding Henry Baker.

Hodge was elected a Fellow of the Royal Society of London in 1938 and received the Royal Medal in 1957. He also received the De Morgan Medal in 1959 and the Copley Medal in 1974. He was knighted in 1959. In 1958 he was appointed as Master of Pembroke College and he retired from university life in 1970.

Hodge was modest and unassuming and he got on well with his colleagues and students. Hodge died on 7 July 1975 in Cambridge, England.

Michael Francis Atiyah

Michael Atiyah was born on 22 April 1929 in London, England. His father Edward Atiyah was Lebanese and his mother Jean Levens was Scottish.

He graduated from Trinity College, Cambridge and received his doctorate in 1955 for a thesis *Some applications of topological methods in algebraic geometry* under Hodge's supervision. He became a Fellow of Trinity College in 1954 and a college lecturer in 1957.

In 1961 he became a Reader at the University of Oxford and two years later he was appointed Savilian Professor of Geometry at Oxford, holding this chair until 1969. For the next three years Atiyah worked at the Institute for Advanced Study in Princeton. From 1973 to 1990 he was a Royal Society Research Professor at Oxford. In 1990 he became Master of Trinity College and Director of the newly opened Isaac Newton Institute for Mathematical Sciences in Cambridge.

In 1966 he was awarded a Fields Medal at the International Congress of Mathematicians for his joint work with Hirzebruch in K-theory; proving jointly with I.M. Singer the index theorem of elliptic operators on complex manifolds; working in collaboration with Raoul Bott to prove a fixed point theorem related to the "Lefschetz formula".

Atiyah was elected a Fellow of the Royal Society of London in 1962, received the Royal Medal in 1968 and the Copley Medal in 1988. He was President of the Royal Society from 1990 to 1995. He was President of the London Mathematical Society in 1974–1976 and received the De Morgan Medal in 1980. He was knighted in 1983. Atiyah received 25 honorary degrees. He also received the Abel Prize in 2004.

He married to Lily Brown in 1955 and they have one son. His younger brother Patrick Atiyah is a professor of English law at Oxford University.[42]

Figure 180. Ernest W. Hobson

Figure 181. Simon K. Donaldson

Ernest William Hobson

Ernest Hobson was born on 27 October 1856 in Derby, England. He entered Christ's College, Cambridge in 1874 and he graduated as Senior Wrangler in the Mathematical Tripos of 1878. He was appointed a Fellow of Christ College in 1879.

He married Selina Rosa Knüsli in 1882. From 1879 to 1904 he taught only at Christ College in private coaching and lecturing. In 1904 he was appointed to the Stoke's lectureship so that he could teach at the University.

He wrote *Theory of Functions of a Real Variable* in 1907 which was the first English book on the measure and integration developed by Borel and Lebesgue. Hobson was elected Sadlerian Professor in 1910 after Forsyth's forced resignation.

Hobson was elected a Fellow of the Royal Society of London in 1893 and received the Royal Medal in 1907. He was President of the London Mathematical Society (1900–1902) and received the De Morgan Medal in 1920. Hobson and Hardy were instrumental in reforming the Mathematical Tripos system in 1909. Since 1909 students have not been ranked individually by test score. Instead, they are grouped with other students in a tiered ranking system by Class I, Class II Division i, Class II Division ii and Class III. And there are no more Senior Wranglers, Senior Optime and Junior Optime. Hobson died on 19 April 1933 in Evelyn nursing home in Cambridge, England.

Simon Kirwan Donaldson

Simon Donaldson was born on 20 August 1957 in Cambridge, England. He graduated from Pembroke College, Cambridge in 1979. One of his tutors at Pembroke felt he was not the top student in his year probably because he would always come to his tutorials carrying a violin case.[3] Tutorials are given at Colleges and lectures are given at the University in Cambridge.

Donaldson was awarded D.Phil degree in 1983 at Oxford with his thesis *The Yang-Mills equations on Kähler manifolds* under Atiyah and Hitchin's supervision. Donaldson's doctoral thesis appeared in the Bulletin of American Mathematical Society in 1983 as self-dual connections and the topology of smooth 4-manifolds. Atiyah described that Donaldson proved a result that stunned the mathematical world.[3]

Donaldson spent the academic year 1983-1984 at the Institute for Advanced Study in Princeton. After returning to Oxford he was appointed Wallis Professor of Mathematics in 1985. In 1999, he moved to Imperial College London. In 1986 he was elected a Fellow of the Royal Society of London and in the same year he received a Fields Medal at the International Congress of Mathematicians at Berkeley. His Fields Medal was awarded for his work on topology of four-manifolds, especially for showing that there is a differential structure on euclidean four-space that is different from the usual structure. In February 2006 he was awarded the King Faisal International Prize for Science for his work in linking pure mathematics to physics, which has helped the understanding of matter at a sub nuclear level.[43]

He is married to Ana Nora Hurtado since 1986 and have two sons. His hobby is sailing.[42]

Margaret Dusa Waddington McDuff

Margaret Dusa McDuff (née Waddington) was born on 18 October 1945 in London, England. She graduated from the University of Edinburgh in 1967 and she married David W. Mc Duff in 1968. They have one daughter but the marriage was dissolved in 1978.

She received a Ph.D. at Girton College, Cambridge in 1971 supervised by George A. Reid. She solved a difficult problem on von Neumann algebras, constructing infinitely many different factors of type II_1, and published the work in the *Annals of Mathematics*.[3]

In early 70s, McDuff stayed six month in Moscow where Israil Gelfand gave her a deeper appreciation of mathematics. She was appointed a lecturer at the University of York in 1973, where she began to work with Graeme Segal on classifying space of categories which she considered as her second doctorate. From 1976 to 1978 she was a lecturer at the University of Warwick.

She was appointed assistant professor at the State University of New York at Stony Brook in 1978 and promoted to full professor in 1984. In the same year, she married famous mathematician John W. Milnor and they have one son.

Since 1998 she is Distinguished Professor of Mathematics at Stony Brook. McDuff was elected a Fellow of the Royal Society of London in 1994 and to the National Academy of Sciences in the United States in 1999.

Her hobbies include are chamber music, playing the cello, gardening and walking.[3]

Figure 182.   Margaret D. W. McDuff

Calyampudi Radhakrishna Rao

C.R. Rao was born on 10 September 1920 in Hadagali, Karnataka State, India. He received an M.S. in mathematics from Andhra University and an M.S. in Statistics from Calcutta University in 1943.

Rao received a Ph.D. at Cambridge University in 1948 with his thesis *Statistical problems in biological classification* under Ronald Fisher's supervision. In 1965 he also received a Doctor of Science degree at Cambridge. He was elected a Fellow of the Royal Society of London in 1967. Rao was Eberly Professor of Statistics at the Pennsylvania State University, 1988-2001. President George Bush awarded him the United State National Medal of Science in June 2002.[42]

Among his best known discoveries are the Cramér-Rao bound and the Rao-Blackwell theorem both related to the quality of estimators. He has also worked on multivariate analysis, estimation and differential geometry.

Shaun Wylie

Shaun Wylie was born on 17 January 1913 in Headington, Oxford, England. He graduated from New College, Oxford in 1934 and received his Ph.D. at Princeton University in 1937 for his thesis *Duality and intersection in general complexes.*[10] His formal adviser was Solomon Lefschetz but Albert Tucker actually helped him. Wylie met Alan Turing at Princeton.

Turing and Wylie worked at Bletchley Park on solving the Enigma machine used by the German Navy during World War II. Wylie solved "Tunny", a German teleprinter cipher.[110]

After the war, he was a fellow at Trinity Hall, Cambridge supervising five Ph.D. students in mathematics including Frank Adams, Crispin Nash-Williams, William Tutte, Christopher Zeeman and Gregory Kelly.[10]

Wylie became Chief Mathematician at Government Communications Headquarters in 1958 and retired from there in 1973. He taught at Cambridgeshire High school for Boys (later Hills Road Sixth Form College) for seven years.[110]

With Peter J. Hilton, he wrote *Homology Theory: An Introduction to Algebraic Topology* published by Cambridge University Press in 1960.

William Thomas Tutte

William Tutte was born on 14 May 1917 in Newmarket, Sulfolk, England. He graduated from Trinity College, Cambridge in 1938 taking chemistry as his major subjects. He worked at Bletchley Park from January 1941. His major achievement there was deciphering German military encryption codes known as FISH.[3] A computer called colossus was built to run Tutte's decoding algorithms at Bletchley Park.[3]

After the war ended, he returned to Cambridge and received his Ph.D. in 1948 for his thesis *An algebraic theory of graphs* supervised by Shaun Wylie.[10]

Donald Coxeter (1907-2003) invited Tutte for a faculty position at the University of Toronto in 1948. Tutte married Dorothea Mitchell in the following year.

Tutte moved to the University of Waterloo in 1962 where he created the Department of Combinatorics and Optimisation.

He was elected a Fellow of the Royal Society of Canada and a Fellow of the Royal Society of London. He received the Order of Canada in 2001.[3] Tutte wrote several books including *Connectivity in Graphs* (1966), *Introduction to the Theory of Matroids* (1971) and *Graph Theory* (1984). Tutte supervised eight Ph.D. students at Waterloo and Toronto.[10]

He died on 2 May 2002 in Waterloo, Canada.

Erik Christopher Zeeman

Chris Zeeman was born on 4 February 1925 in Japan. He served as a navigator with the Royal Air Force from 1943 to 1947. He received his B.A. from Cambridge in 1950 and his Ph.D. for his thesis *Dihomology* in 1953 supervised by Shaun Wylie.[3] Zeeman became a fellow of Gonville and Caius College, Cambridge in 1953 and also a Lecturer of Cambridge in 1955.

He married Rosemary Gledhill in 1960 and they had three sons and two daughters. He was appointed Foundation Professor of Mathematics at the University of Warwick in 1963 where his research flourished. He supervised 30 Ph.Ds, including 21 at Warwick.[10]

Zeeman became a Fellow of the Royal Society of London in 1975. He became Principal of Hertford College, Oxford in 1988 and retired from there in 1995. He was the 63rd President of the London Mathematical Society from 1986 until 1988. He was knighted in 1991 and received several honorary degrees.

Zeeman wrote *Catastrophe Theory* in 1977, *Geometry and Perspective* in 1987 and *Gyroseopes and Boomerangs* in 1989.[3]

Figure 183. J. Frank Adams

Figure 184. Harish-Chandra

John Frank Adams

Frank Adams was born on 5 November 1930 in Woolwich, London, England. He served in the Royal Engineers as a soldier in 1948 and 1949 before beginning his university education. He graduated from Trinity College, Cambridge in 1952 with a B.A. degree in mathematics. He married Grace Rhoda Carty in 1953 who later became a minister in the Congregational Church. They had one son and three daughters (one adopted).[3]

He received a Ph.D. at Cambridge in 1955 with his thesis *On special sequences of self-obstruction invariants*. He showed his gratitude to Gugenheim, Hilton, Moore, Wylie and Whitehead. From 1955 to 1958 he was a research fellow at Cambridge, but he spent the 1957-1958 academic year at the University of Chicago and Princeton as a Commonwealth Fellow. Adams was appointed a Reader in 1962 and Fielden Professor in 1964 at the University of Manchester.

Unfortunately, around that time he suffered the first of several psychiatric illnesses and was forced to take sick leave for some months.

In 1970 Adams was appointed as Lowndean Professor of Astronomy and Geometry at Cambridge succeeding William Hodge.

He was elected a Fellow of the Royal Society of London in 1964 and received the Sylvester Medal in 1982. He was elected to the U.S. National Academy of Sciences in 1985. Adams solved the famous conjecture about the existence of H-structures on spheres. Using the K-theory he solved another important conjecture about vector fields on spheres. His research was in the homotopy theory of classifying spaces of topological groups, finite H-spaces and equivalent homotopy theory.[3]

His health continued to cause him problems with another psychiatric illness in 1986. He went to London, despite feeling unwell, to the retirement of a friend. He was killed in a car accident near Brampton, Huntingtonshire, England on 7 January 1989.

Seven years after his death, his final book, *Lectures on Exceptional Lie Groups* was published in 1996.

Harish-Chandra

Harish-Chandra was born on 11 October 1923 in Kanpur, Uttar Pradesh, India. He received a B.Sc. in 1941 and a master's degree in 1943 at the University of Allahabad, northern India.

In 1945 Harish-Chandra went to Gonville and Caius College, Cambridge, where he studied for his doctorate in theoretical physics under Dirac's supervision. Colleges at Cambridge provide room and board to graduate students without research guidance. But College have substantial education activities for undergraduate students by tutoring. In 1945 Dirac was a fellow at St. John's College and Lucasian Professor at the University of Cambridge. Harish-Chandra attended Dirac's lecture course but he discovered that

Dirac was essentially reading from one of his books which Harish-Chandra was already familiar. So he attended the lecture courses of Littlewood and Philip Hall.

Harish-Chandra earned a Ph.D. in 1947 for his thesis *Infinite irreducible representation of the Lorentz group* supervised by Dirac. Harish-Chandra once complained to Dirac about the fact that his proofs were not rigorous and Dirac replied, "I am not interested in proof but only in what nature does." Harish-Chandra became more interested in mathematics. He spent his most productive years on the faculty at Columbia University, New York, from 1950 to 1963. In 1952 he married Lalitha Kale and they had two daughters.

Harish-Chandra worked intermittently at the Institute for Advanced Study in Princeton in 1947-48, 1955-56, 1961-63 (as a Sloan Fellow) and finally in 1968 he was appointed IBM-von Neumann Professor.

One day Harish-Chandra and physicist Freeman Dyson were walking and talking when Harish-Chandra said, "I am leaving physics for mathematics. I find physics messy, unrigorous, elusive." Dyson replied, "I am leaving mathematics for the exactly the same reasons."[33]

Harish-Chandra received the Cole Prize in 1954 for his papers on representations of semisimple Lie algebras and groups. He was elected a Fellow of the Royal Society in 1973 and to the National Academy of Sciences in the United States in 1981.

Harish-Chandra is credited with developing the theory of harmonic analysis on reductive p-adic groups. The Michel Plancherel's (1885-1967) formula in the p-adic context is one of his triumphs. Harish-Chandra was fond of the unifying principle that whatever is true for a real reductive group is also true for a p-adic reductive group. He called this the "Lefschetz principle."[33]

He died of a heart attack on 16 October 1983 in Princeton, New Jersey.

Figure 185. Francis Galton

Figure 186. Karl Pearson

Francis Galton

Francis Galton was born on 16 February 1822 in Sparkbrook near Birmingham, England. His maternal grandfather was Erasmus Darwin (1731–1802) and Charles Darwin was a grandson of Erasmus. Therefore Francis Galton and Charles Darwin are cousins.

Galton entered Trinity College, Cambridge in 1840, but he never graduated. He travelled Africa for many years and published *Tropical South Africa* in 1853. In 1856, he was elected a Fellow of the Royal Society of London. Galton devoted the latter part of his life to eugenics, which purports to improve the physical and mental make up of the human species by selected parenthood.[3] He wrote *Hereditary Genius* in 1869.

He defined an index of correlation as a measure of the degree to which the two different things were related. But he failed to understand the complexity of the mathematics involved. He wrote *Natural Inheritance* in 1889 which profoundly influenced Karl Pearson.[3]

Galton received the Royal Medal in 1876, the Copley Medal in 1910 and he was knighted in 1909. He died on 17 January 1911 in Greyshott House, Haslemere, Surrey, England.

Karl Pearson

Karl Pearson was born on 27 March 1857 in London, England. At King's College, Cambridge he was taught by Stokes, Maxwell, Cayley, Burnside and Routh. He graduated from King's College in 1879 as Third Wrangler in the Mathematical Tripos.

Pearson had a very broad range of interests, including physics, physiology, Roman Law, German Literature, and Socialism. He studied law at Lincoln's Inn and was called to the Bar in 1882, but he never practiced.

In 1885, he was appointed to the Chair of Applied Mathematics at University College, London.

From 1893 to 1912, he wrote 18 papers entitled *Mathematical Contributions to the Theory of Evolution.* These papers contain contributions to regression analysis, the correlation coefficient and the Chi-square test. Pearson also coined the term "Standard Deviation" in 1893.

Pearson was co-founder, with Walter Weldon and Francis Galton of the statistical journal *Biometrika* and he became editor until his death. He was elected a Fellow of the

Royal Society in 1896. From 1911 to 1933 he was the first Galton Professor of Eugenics at the University College.

He married Maria Sharpe in 1890, and they had three children. Maria, died in 1928 and one year later he remarried Margaret Victoria Child. When he retired in 1933, his chair was split into two. His son Egon Pearson became head of the Department of Statistics and Ronald Fisher became head of the Department of Eugenics. Pearson died on 27 April 1936 in Coldharbour, Surrey, England.

Figure 187.   Philip Hall

Philip Hall

Philip Hall was born on 11 April 1904 at his maternal grandfather's house in Hampstead, London, England. His parents were not married. His father left without making any provision for Philip and his mother. At Cambridge he was taught by William Hodge, Ernest Hobson, Henry Baker and Dudley Littlewood.

Hall graduated with a B.A. in 1925 from King's College, Cambridge. He was elected a fellow of King's College in March 1927 after submitting an essay *The isomorphisms of Abelian groups*.

Gaining Fellowship at Cambridge was almost equivalent to receiving a doctorate in Germany until 1930s. Hall's interest in group theory came from William Bumside's (1852-1927) book *The Theory of Groups of Finite Order*. Hall became a worthy successor to Burnside as the promoter of group theory in England.

In 1932 he wrote his most famous paper *A contribution to the theory of groups of prime power order*. In 1933 he was appointed as a lecturer at Cambridge which enabled him to offer courses at the University and supervise doctoral students.

During World War II he worked at the Code and Cypher School at Bletchley Park on Italian and Japanese ciphers. He learned about 1500 Japanese character's which are practically same to Chinese characters.

Hall returned to Cambridge in July 1945 and was promoted to Reader in 1949. He succeeded Mordell as the Sadlerian Chair in 1953. He retired in 1967 and in 1976 he was elected an honorary fellow at Jesus College. Hall was elected a Fellow of the Royal Society of London in 1942 and awarded the Sylvester Medal in 1961. He also received the De Morgan Medal in 1965.

Hall loved music, art, flowers, and country walks. He had a deep love of poetry, which he recited in English, Italian or Japanese. He was a wonderful person, gentle, amusing, kind, and a man of integrity.[3]

An investigation of the conditions under which finite groups are soluble led him in 1937 to postulate a general structure theory for finite soluble groups. In 1954 he published an examination of finitely generated soluble groups in which he demonstrated that they could be divided into two classes of unequal size.[5]

Hall died on 30 December 1982 in Cambridge, England.

John Charles Burkill

Charles Burkill was born on 1 February 1900 in Holt, Norfolk, England. He entered Trinity College, Cambridge in January 1919 and graduated in 1921. One year later, he was elected a Fellow of Trinity College after submitting a dissertation on surface areas. He won a Smith's Prize in 1923 for functions of intervals and the problem of area.

In 1924 he was appointed to the chair of pure mathematics at Liverpool University. On 1 August 1928 he married Margareta (Greta) Braun who had been born in Germany. They had three children. In 1929 Burkill returned to Cambridge to be appointed a University lecturer and fellow of Peterhouse, the oldest college in Cambridge founded in 1280.

After Hitler came to power in Germany in 1933, the Burkills helped to bring out of Germany and settle in England many hundreds of refugee children. The Burkills themselves took into their family and assumed responsibility for the education of two boys.[3]

He was elected a Fellow of the Royal Society in 1953 in recognition of his research. Burkill introduced the "Burkill integral". He introduced the notion of approximate differentiation extending and simplifying Besicovitch's work. He wrote several books on the Lebesgue integral and mathematical analysis. In 1961 he was promoted to Reader in Mathematical Analysis and retired in 1967. Burkill became Master of Peterhouse in 1968 and stepped down in 1973.

He also served as editor of the *Mathematical Proceedings of the Cambridge Philosophical Society*. Burkill died on 6 April 1993 in a Sheffield nursing home due to bronchopneumonia.

Figure 188.   William T. Gowers

Béla Bollobás

Béla Bollobás was born on 3 August 1943 in Hungary. He wrote his first paper with Erdös when Bollobás was seventeen. He received B.A. in 1966 and Dr.rer.nat in 1967 at the University of Budapest. He was a research scientist at Hungarian Academy of Sciences in 1966-1969. Bollobás earned a Ph.D. in 1972 at Cambridge supervised by Frank Adams. He also received Doctor of Science degree at Cambridge in 1984.

Since 1972 he has been a Fellow of Trinity College and Director of Studies in mathematics in 1972-96. During that time, he was a lecturer (1974-1985) and a Reader in pure mathematics (1985-1996). Since 1995, he has been Distinguished Professor of Excellence in Combinatorics at the University of Memphis, Tennessee.[42]

In 1990 he became a Foreign Member of the Hungarian Academy of Sciences. He married Gabriella Farkas, a sculptor, in 1969 and they have one son. He has written over 300 papers and many books on graph theory and combinatorics.

William Timothy Gowers

Timothy Gowers was born on 20 November 1963 in Wiltshire, in Southern England. He is the son of composer Patrick Gowers. He received a Ph.D. in 1990 at Cambridge for his thesis *Symmetric structures in Banach spaces* under the supervision of Béla Bollobás.

He was a research Fellow at Trinity (1989-1993), and Lecturer (1991-1994), Reader (1994-1995) at University College, London. He was a lecturer at Cambridge (1995-1998) and appointed Rouse Ball Professor of Mathematics in 1998. He was elected a Fellow of the Royal Society in 1999. In 1998 he was awarded a Fields Medal for his important results in Banach Space theory and combinatorics. The functional analysis and combinatorial theory apparently have little to do with each other, and a significant achievement of Gowers has been to combine these fruitfully.[44]

Gowers has been able to construct a Banach Space that has almost no symmetry. His contribution opened the way to the solution of "homogeneous space problem". He married Emily Joanna Thomas in 1988, and they have two sons and one daughter.

Alexander Victor Oppenheim

Alexander Oppenheim was born on 4 February 1903 in Salford, Lancashire, England and graduated from Oxford University in 1927. He received a Ph.D. at the University of Chicago with his thesis *Minima of indefinite quadratic quaternary forms* supervised by L.E. Dickson in 1930. He also received the Doctor of Science degree at Oxford University.

After teaching at Edinburgh University in 1930-1931, he went to Singapore and taught at Raffles College from 1931 to 1942. He was captured by the Japanese Army when they invaded Singapore during World War II. He and thousands of Allied prisoners of war were engaged as slave labor for the infamous railroad construction work in Burma.

After the war ended, he became Deputy Principal of Raffles College (1947-1949) and Dean of the Faculty of Arts at the University of Malaya. Oppenheim also served as

Vice-Chancellor of the Universities of Malaya in Singapore and Kuala Lumpur from 1957 to 1965. Later, he was a Visiting Professor of Mathematics at universities in England, Ghana and Nigeria.

In 1929 Oppenheim formulated the Oppenheim Conjecture.[90]

Let Q be an nondegenerate indefinite quadratic form in n variables. Let $L_Q = Q(Z^n)$ denote the set of values of Q at integral points. The Oppenheim Conjecture states that if $n \geq 3$, and Q is not proportional to a form with rational coefficients, then L_Q is dense.

The Oppenheim conjecture was proved affirmative by Gregori Margulis in 1987 using methods of ergodic theory. Oppenheim was knighted in 1961 and published *The Prisoner's Walk: An Exercise in Number Theory* in 1984.

He died on 13 December 1997.

Polish School

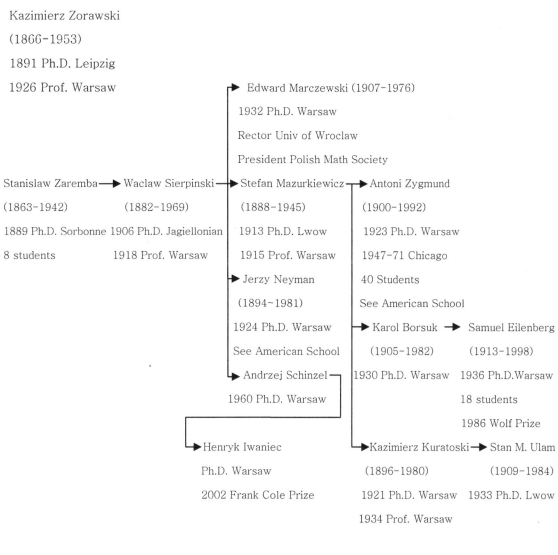

Kazimierz Zorawski
(1866-1953)
1891 Ph.D. Leipzig
1926 Prof. Warsaw

Edward Marczewski (1907-1976)
1932 Ph.D. Warsaw
Rector Univ of Wroclaw
President Polish Math Society

Stanislaw Zaremba → Waclaw Sierpinski → Stefan Mazurkiewicz → Antoni Zygmund
(1863-1942)   (1882-1969)   (1888-1945)   (1900-1992)
1889 Ph.D. Sorbonne 1906 Ph.D. Jagiellonian 1913 Ph.D. Lwow   1923 Ph.D. Warsaw
8 students   1918 Prof. Warsaw   1915 Prof. Warsaw   1947-71 Chicago

Jerzy Neyman   40 Students
(1894-1981)   See American School
1924 Ph.D. Warsaw
See American School   → Karol Borsuk → Samuel Eilenberg
                         (1905-1982)   (1913-1998)
Andrzej Schinzel   1930 Ph.D. Warsaw   1936 Ph.D.Warsaw
1960 Ph.D. Warsaw   18 students
                                           1986 Wolf Prize

Henryk Iwaniec   Kazimierz Kuratoski → Stan M. Ulam
Ph.D. Warsaw   (1896-1980)   (1909-1984)
2002 Frank Cole Prize 1921 Ph.D. Warsaw   1933 Ph.D. Lwow
                     1934 Prof. Warsaw

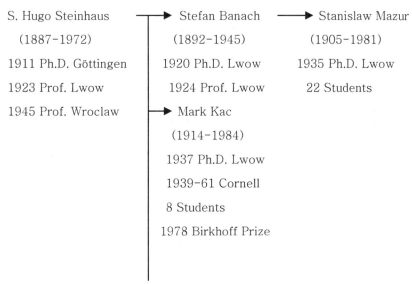

S. Hugo Steinhaus   → Stefan Banach   → Stanislaw Mazur
(1887-1972)   (1892-1945)   (1905-1981)
1911 Ph.D. Göttingen 1920 Ph.D. Lwow   1935 Ph.D. Lwow
1923 Prof. Lwow   1924 Prof. Lwow   22 Students
1945 Prof. Wroclaw → Mark Kac
                      (1914-1984)
                      1937 Ph.D. Lwow
                      1939-61 Cornell
                      8 Students
                      1978 Birkhoff Prize

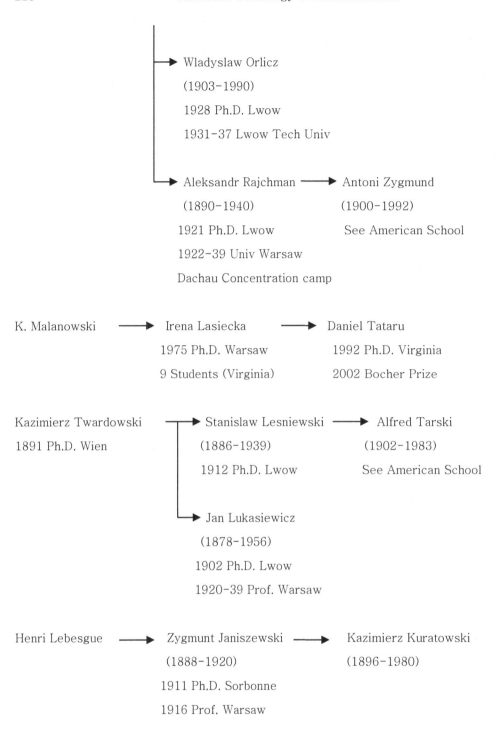

Wladyslaw Orlicz

(1903-1990)

1928 Ph.D. Lwow

1931-37 Lwow Tech Univ

Aleksandr Rajchman ⟶ Antoni Zygmund

(1890-1940)   (1900-1992)

1921 Ph.D. Lwow   See American School

1922-39 Univ Warsaw

Dachau Concentration camp

K. Malanowski ⟶ Irena Lasiecka ⟶ Daniel Tataru

1975 Ph.D. Warsaw   1992 Ph.D. Virginia

9 Students (Virginia)   2002 Bocher Prize

Kazimierz Twardowski   Stanislaw Lesniewski ⟶ Alfred Tarski

1891 Ph.D. Wien   (1886-1939)   (1902-1983)

1912 Ph.D. Lwow   See American School

Jan Lukasiewicz

(1878-1956)

1902 Ph.D. Lwow

1920-39 Prof. Warsaw

Henri Lebesgue ⟶ Zygmunt Janiszewski ⟶ Kazimierz Kuratowski

(1888-1920)   (1896-1980)

1911 Ph.D. Sorbonne

1916 Prof. Warsaw

Polish School

In 1815 the Congress of Vienna created the Kingdom of Poland consisting of about three-quarter of the territory of the former duchy of Warsaw, with the Russian Czar as its King. At the same time, it established Kraków as a city republic, and split the remainder of Poland among Russia, Austria and Prussia.[45]

The Poles expelled the Russian imperial authorities and in January 1831 proclaimed their independence. However, the Russians re-took Warsaw in September 1831 and their rule continued until 1864.

The University of Warszawski (Warsaw) founded in 1816 became a Russian university in 1869. The Russians restricted the use of the Polish language and the Russian language was introduced in schools.

The Poles in Prussian Poland were subjected to a policy of Germanization, but Poles in Austrian Poland were treated more liberally. The Austrian Poles became the country's second nationality, and they were allowed to participate in the government of Vienna and retained sovereignty over the Ruthenes in Galicia. Galicia itself became the haven of polish culture, with the Jagiellonian University in Karków (founded in 1364), the University of Lwów (founded in 1661 now the Ivan Franko National University of Lviv in Ukraine), and the Academy of Sciences attracting many Polish intellectuals.[108]

The University of Breslau (founded in 1702) is now University of Wroclaw in Poland.

Many Polish mathematicians moved to the United States around 1939 including Mark Kac, Jerzy Neyman, Antoni Zygmund, Alfred Tarski, Samuel Eilenberg and Stan Ulam.

Stanislaw Zaremba

Stanislaw Zaremba was born on 3 October 1863 in Romanowka, Ukraine. He graduated from a secondary school in St. Petersburg, Russia and received his engineering diploma in 1886 from St. Petersburg Institute of Technology.

Zaremba was awarded a doctorate at the Sorbonne in 1889 for his thesis, *Sur un probléme concernant l'état calorifique d'un corp homogéne indéfini*

He taught for 11 years at schools in France while publishing his research results in French mathematical journals.

Zaremba returned to Poland in 1900 to be appointed to a chair at the Jagiellonian University in Kraków.

When the Mathematical Society of Kraków was founded in 1919, Zaremba was elected its first president. The Society was renamed the Polish Mathematical Society in 1920.

With Kazimierz Zorawski, Zaremba created the golden age of Polish mathematics between the two World Wars.

When Zaremba received an honorary degree from the Jagiellonian University in 1930, Henri Lebesgue praised Zaremba's originality and his contribution to mathematical physics.

Hadamard said he could not forget Zaremba's splendid results in the domain of mixed boundary problems and of harmonic functions as well as of hyperbolic equations.[3]

Zaremba was elected to the Soviet Academy of Sciences in 1925.

He died on 23 November 1942 in Kraków, Poland.

Figure 189.   Stanislaw Zaremba

Kazimierz Zorawski

Kazimierz Zorawski was born on 22 June 1866 in Szczaczyn, Poland (now Belarus). In 1884 he entered the University of Warszawski (Warsaw) to study mathematics. He fell in love with Maria Sklodowska, an 18 year old girl, who was employed by his family as a governess. They wanted to marry. But his parents would not agree because Maria's family was so poor.

Maria went to Paris for her education in 1891 and married Pierre Curie in 1895. She is of course the famous Marie Curie who won two Nobel Prizes, one in physics, and the other in chemistry.

Zorawski married Leokadia Jewniewicz, a talented pianist, and they had three children.[3]

Zorawski graduated from the University of Warsaw in 1888 and went to the University of Leipzig where he received his doctorate in 1891 for his thesis on the application of Lie groups to differential geometry.

Zorawski and Friedrich Engel (1861–1941) published *Theorie der Transformationsgruppen* in three volumes between 1888 and 1893. Sophus Lie praised Zorawski's work on integral invariants.

Zorawski worked at the Lwów Technical University (not to be confused with Lwów University) from 1892 to 1894. He received his habilitation from the Jagiellonian University in Kraków in 1893. Kraków is about 300 *km* west of Lwów. Zorawskii was appointed extraordinary professor in 1895 and ordinary professor in 1898 at the Jagiellonian University. Zaremba was also appointed as professor at Jagiellonian and together the two of them built a major school of mathematics in Poland.

Zorawski's main topics of research were invariants of differential forms, integral invariants of Lie groups, differential geometry, and fluid mechanics.[3]

Zorawski moved to the Technical University of Warsaw in 1919. In 1926, he became a professor at the University of Warsaw from where he retired in 1935.

During World War II his house was destroyed, and among its contents was his nearly completed multi-volume work on analytical geometry.

He died on 23 January 1953 in Narutowicz, Poland.

Waclaw Sierpinski

Waclaw Sierpinski was born on 14 March 1882 in Warsaw, Russian Empire (now Poland). He entered the University of Warsaw (Czar's University at that time) in 1899.

The lectures at the University were all in Russian and the staff was entirely Russians.

He received Candidate of Science degree in 1904 for his thesis on a famous problem on lattice points. In 1913 Edmund Landau shortened Sierpinskis' proof and described the result as profound.

He decided to go to Kraków for his doctorate and enrolled at the Jagiellonian University where he attended lectures by Zaremba and received his doctorate in 1906. He was appointed extraordinary (associate) professor at the Lwów University in 1910. From 1919 to 1939 Lwów University was known as Jan Kazimierz University.[46]

During the years 1908 to 1914, when Sierpinski taught at the University of Lwów, he published three books in addition to many research papers.

During World War I, he worked with Nicolai Luzin at Moscow State University. He became a professor at the University of Warsaw in 1918 and retired from there in 1960.

In 1920 Sierpinski, together with his former student Stefan Mazurkiewicz, took over the editorship of the mathematics journal *Fundamenta Mathemticae.*[3]

He was eleted to the Polish Academy in 1921. During World War II the Nazis German Army burned his house, completely destroying his library. More than half of the mathematicians who lectured at Polish universities were killed.

Siepinski was the author of 724 papers and 50 books.

He received Poland's Scientific Prize of the first class in 1949. Sierpinski's former students include Alfred Tarski, and Jerzy Neyman. Antoni Zygmund, Samuel Eilenberg and Stan Ulam were students of Sierpinski's student.

Sierpinski died on 21 October 1969 in Warsaw, Poland.

Figure 190. Waclaw Sierpinski

Figure 191. Stefan Mazurkiewicz

Stefan Mazurkiewicz

Stefan Mazurkiewicz was born on 25 September 1888 in Warsaw, Russian Empire (now Poland). He studied at Kraków, München and Göttingen, he received a doctorate in 1913 at the University of Lwów supervised by Sierpinski.

The Russians withdrew from Warsaw in August 1915 and the University of Warsaw began operating as a Polish university in November 1915. At that point Mazurkiewicz was appointed professor there, and he remained for the rest of his life.

His main work was in topology and the theory of probability. He proved the strong law of large numbers in 1922.

The manuscript of his treatise on probability was destroyed during World War II.

He died on 19 June 1945 in Grodzisk Mazowiecki Hospital during an operation for a gastric ulcer.

Hugo Dyonizy Steinhaus

Hugo Steinhaus was born on 14 January 1887 in Jaslo, Galicia, Austrian Empire (now Poland) into a family of Jewish intellectuals. He was awarded a Ph.D. at the University of Göttingen in 1911 for his thesis *Neue Anwendugen des Dirichlet'schen Prinzips* supervised by David Hilbert.

On summer evening in 1916 while walking in the park that surrounds the old centre of Kraków, Steinhaus heard a couple of words *"Lebesgue integral"*. He found the speakers were Stefan Banach and Otto Nikodym talking about mathematics.[46]

Steinhaus stopped and told Banach, then a 24 year old student, about a problem he had been working for quite some time. A couple days later Banach came up with a solution. Banach's first paper was published jointly with Steinhaus in the Bulletin of the Kraków Academy[46] in 1918.

Steinhaus became a docent (lecturer) in 1917 and a professor in 1923 at the University of Lwów.

Banach followed Steinhaus to Lwów where he received his Ph.D. in 1920. Steinhaus published a joint paper with Banach in 1927 and in 1929 they started a new journal *Studia Mathematica*.

The mathematicians of the Lwów school did a great deal of mathematical research at the Scottish Café. They were Steinhaus, Banach, Andrej Turowicz, Kazimierz Kuratowski, Stan Ulam, and Stanislaw Mazur.

The Scottish Book consists of open questions posed by the mathematicians working there. Henri Lebesgue was awarded an honorary degree at Lwów in 1938 and the reception for Lebesgue was held at the Scottish Café.

During World War II Steinhaus was hiding from the Nazis, suffering great hardships, going hungry most of the time but always thinking about mathematics.[3] After the end of World War II, Steinhaus sent the Scottish Book to Stan Ulam in the United States which was translated into English by Ulam and published.

Steinhaus moved to the University of Wroclaw (formerly Breslau) in 1945 and stayed there until 1961. He taught at the University of Nortre Dame in Indiana from 1961 until 1962 and moved to the University of Sussex in 1966.

He authored over 170 works in mathematical analysis, probability theory and statistics.
He described mathematics as a "Science of nonexistent things".
Steinhaus died on 25 Febuary 1972 in Wroclaw, Poland.

Figure 192.   Hugo D. Steinhaus

Wladyslaw Roman Orlicz

Wladyslaw Orlicz was born on 24 May 1903 in Okocim, a village near Krakow, Austro-Hungarian Empire (now Poland). His father died when he was only four years old. In 1919 his family moved to Lwów, where he studied mathematics at the Jan Kazimierz University (later Lwów University). In 1928 he received a Ph.D. for his thesis *Some problems in the theory of orthogonal series* under the supervision of Eustachy Zylinski and Hugo Steinhaus.[3]

Eustachy Zylinski was born in 1889 in Winnica (in Ukraine) and became full professor of mathematics at Jan Kazimierz University in 1922. He died in 1954.

Orlicz worked at the Lwów Technical University from 1931 until 1937. He received his habilitation for his thesis *Investigations of orthogonal systems* in 1934. Orlicz participated in the mathematical discussions at the Scottish Café in Lwów. He became a professor at Poznan University (now Adam Mickiewicz University) in 1937 and at the Ivan Franko National University of Lviv (former Lwów University) from January 1940 to June 1941.

He returned to Poznan University in May 1945 where he was promoted to a full professor in 1948 and retired in 1974. Orlicz published 171 mathematical papers and supervised 39 doctoral dissertations.[3]

Orlicz was President of the Polish Mathematical Society from 1977 to 1979. He received many prizes including the Stefan Banach Prize of the Polish Mathematical Society in 1948.

Orlicz's name is associated with the Orlicz spaces, the Orlicz-Pettis theorem, Mazur-Orlicz bounded consistency theorem, Orlicz category theorem, Orlicz interpolation theorem, Orlicz norm, Orlicz function and many others.[3]

He died on 9 August 1990 in Poznan, Poland.

Figure 193.   Mark Kac

Figure 194.   Stefan Banach

Mark Kac

Mark Kac was born into a Jewish family on 3 August 1914 in Krzemieniec, Russian Empire (now in Poland). By the time he entered the Jan Kazimierz University (now the University of Lviv), he was able to speak Russian, French, Hebrew, Polish, Latin and Greek.

He was awarded his doctorate at the Jan Kazimierz University in 1937 supervised by Steinhaus. He received for a scholarship to go to Johns Hopkins University in 1938. After one year there, he joined the mathematics faculty at Cornell University, where he stayed for twenty-two years, promoted to full professor in 1947.

He was a professor at Rockefeller University from 1961 to 1981 then moved to the University of Southern California. He received the Birkhoff Prize in 1978 for his important contributions to statistical mechanics and to probability theory and its applications.

Kac published a classic text *Statistical Independence in Probability, Analysis and Number Theory* in 1959. When he was lecturing on complex function theory at a university, it became clear that one student was not understanding the central point. So Kac said to the student: "This function 1/z, what kind of singularity does it have at the origin?" The student still could not understand. Finally Kac said. "Look at me. What am I?" The student replied, "Ah, a simple Pole."[105]

Kac died on 26 October 1984 in California.

Stefan Banach

Stefan Banach was born on 30 March 1892 in Kraków, then part of Austro-Hungarian Galicia (now Poland). His father was Stefan Greczek, a young soldier in the Autro-Hungarian Army, and his mother was Katarzyna Banach who vanished when the baby was only four days old.

Stefan Greczek, arranged for his son to be brought up by Franciszka Plowa who lived with her daughter Maria in Kraków. Maria's guardian was French intellectual Juliusz Mien who taught Banach to speak French and gave him an appreciation for education.[3]

At Kraków Gymnasium No.4 Banach became good friends with Witold Wilkosz (1891-1941). Banach had to earn money to support his studies and graduated from gymnasium in 1910.

Banach went to Lwów which is 300*km* east of Kraków to study at Lwów Technical University. In 1914 he passed his half-diplom examinations.[47] This is probably equivalent to Vor-Diplom in German universities which a student receives after two-year study at a university.

He was exempt from military service during World War I due to his poor vision in the left eye. He worked building roads in Kraków and earned money by teaching in the local schools.

He also attended mathematics lectures at the Jagiellonian University in Kraków where Zaremba lectured.

On summer evening in 1916 Steinhaus was walking through the Planty gardens, he heard the words "Lebesgue measure". Steinhaus met two young men on a park bench discussing mathematics. They were Stefan Banach and Otto Nikodym. They said they had another friend by the name Witold Wilkosz, whom they highly praised.[46]

Steinhaus had been undertaking military service and was about to take up a post at Lwów University (Jan Kazimierz University from 1919 to 1939).

From then on Steinhaus and Banach would meet regularly in Kraków. Steinhaus told Banach a problem he had been working for some time, and a couple of days later Banach turned up with a solution. They wrote a joint paper, which they presented to Zaremba for publication. It was finally published in the Bulletin of the Kraków Academy in 1918.

On Steinhaus' initiative, the Mathematical Society of Kraków was founded in 1919 and Zaremba was elected as the first President. Banach lectured at the Society twice during 1919 and continued to produce top quality research papers.[3] Through Steinhaus Banach married Lucja Braus in 1920. Banach was offered an assistantship at Lwów Technical University in 1920 and submitted a dissertation for his doctorate under Lomnicki's supervision. Apparently Banach did not have a formal diploma in mathematics but an exception was made to allow him to award him a doctorate for his thesis *On operations on abstract sets and their applications to integral equations*, which is sometimes said to mark the birth of functional analysis. In his doctoral dissertation in 1920 he defined axiomatically what today is called a Banach Space. The name "Banach Space" was coined by Maurice Frechet (1878–1973).

A Banach Space is a real or complex normed vector space that is complete as a metric space.

Steinhaus was at Lwów University which is different from Lwów Technical University.

In 1922 Banach was awarded his habilitation for a thesis on measure theory at Lwów University (Jan Kazimierz University). Banach was appointed extraordinary professor on 22 July 1922. He was promoted to full professor in 1924.

A new series of mathematical monographs were launched under the editorship of Banach and Steinhaus from Lwów and Knaster, Kuratowski, Mazurkiewicz and Sierpinski from Warsaw. The first volume in the series *Théorie des Opérations linéaires* was written by Banach and appeared in 1932.[3]

Lwów mathematicians met regularly at the Scottish Café at 22 Akademicka Street called Ludwik Zalewskis' confectionery to discuss research problems. They were Banach, Steinhaus, Andrzej Turowicz, Kazimierz Kuratowski, Stan Ulam and Stanislaw Mazur.

In 1939 Banach was elected as President of the Polish Mathematical Society. When Soviet troops occupied Lwów the University was renamed the Ivan Franks University in September 1939.

Banach was allowed to continue to hold his chair and became the Dean of the Faculty of Science. Russian mathematicians Sobolev and Aleksandrov visited Banach in Lwów in 1940.[3]

When the German Army occupied Lwów in 1941, many Polish academics including his doctoral supervisor Lomnicki were murdered. Towards the end of 1941 Banach's body was used to breed lice in a German Institute to develop an anti-typhus serum.

When the Soviet troops retook Lwów in July 1944, Banach was released but by that time he was seriously ill. He died of lung cancer on 31 August 1945 in Lwów.

Banach is remembered by Banach algebra, the Hahn-Banach theorem, the Banach fixed point theorem, the Banach-Steinhaus theorem, the Banach-Alaoglu theorem and Banach-Tarski paradoxical decomposition of a ball.

Figure 195.   Stanislaw Mazur

Stanislaw Mazur

Stanislaw Mazur was born on 1 January 1905 in Lwów Poland. While he was a student at Lwów Technical University he submitted a paper to Steinhaus before it was to present it at a meeting. Steinhaus summoned Mazur to tell him that he had turned in four blank sheets of paper. It turns out that Mazur had diluted his ink with water to make it last longer, but he diluted it too much.

Mazur was awarded a doctorate in 1935 under Banach's supervision. They collaborated on several papers during the 1930s as members at the Scottish Café.

Mazur contributed 24 problems to the Scottish Book as the sole author and 19 more jointly with others.[3] Mazur was awarded a habilitation in 1936 and he taught at Lwów until 1946.

From 1948 he worked at the University of Warsaw. Since he was an active member of the Polish Communist Party in the 1930s, he became a high official in the science establishment in Poland after 1946.[3]

He died on 5 November 1981 in Warsaw, Poland.

Stanislaw Lesniewski

Stanislaw Lesniewski was born on 30 March 1886 in Serpukhov near Ivanovo-Vosniesiensk, Russia. His father was a Polish railway engineer. Lesniewski was awarded a doctorate in 1912 at Lwów University under Kazimierz Twardowski's supervision.

Jan Lukasiewicz (1878–1956) became extraordinary professor at Lwów University in 1911 who gave lectures on mathematical logic. Lesniewski was greatly influenced by his lectures and began to study formal logic and made strenuous attempts to understand Russell's paradox.

After World War I ended in 1918, Lesniewski joined Zygmunt Janiszewski and Stefan Mazurkiewicz in their school of mathematics in Warsaw. They set up the journal *Fundamenta Mathematicae* in 1919 and the first volume was published in 1920.[3]

Lesniewski died on 13 May 1939 in Warsaw, Poland.

Figure 196. Stanislaw Lesniewski

Figure 197. Jan Lukasiewicz

Jan Lukasiewicz

Jan Lukasiewicz was born on 21 December 1878 in Lwów, Austro-Hungarian Galicia (now Ukraine). He was awarded a doctorate in 1902 at Lwów University supervised by Kazimierz Twardowski. He also received his habilitation at Lwów in 1906. In 1911 he was appointed extraordinary professor at Lwów lecturing mathematical logic. In 1915 he moved to the University of Warsaw and he was appointed Polish Minister of

Education in 1919. He was a full professor at the University of Warsaw from 1920 to 1939. During this period he also served Rector of the University of Warsaw on two different occasions.[3]

He published *Elements of Mathematical Logic* in Polish in 1928.

Lukasiewicz married Regina Barwinska in 1928. They fled from Poland in 1946 opposing the communist regime to Belgium then to Ireland. He was offered a chair by the University of Dublin in Ireland.[3]

He introduced the "Poilsh notation" which allowed expressions to be written unambiguously without the use of brackets. A similar concepts underlies the reverse Polish notation (RPN) which was adopted by many Hewlett Packard calculators.

Lekasiewicz died on 13 February 1956 in Dublin, Ireland.

Zygmunt Janiszewski

Zygmunt Janiszewski was born on 12 June 1888 in Warsaw, Russian Empire (now Poland).

He studied at Zurich, München, Göttingen and Paris.

He was awarded a doctorate in 1911 at the Sorbonne under Lebesgue's supervision. His thesis entitled *Sur les continus irréductibles entre deux points* was on topology. The examining committee included Poincare, Lebesgue and Fréchet.[3]

He returned to Poland in 1913 to teach at the University of Lwów. At the outbreak of World War I, he became a soldier in the Polish Legions of Józef Pilsudski, believing he was fighting for Polish independence. When the soldiers were required to take an oath of loyalty to the Austrian government, he left the Legion and hid under the false name of Zygmunt Wickerkiewicz.[3]

He became a professor at the University of Warsaw in 1917 and along with Mazurkiewicz, became the main force in the creation of the strong school of mathematics in Warsaw.

The Warsaw School and the Lwów School were the most productive mathematical centers in Poland between the two World Wars.

Janiszewski, Sierpinski and Mazurkiewicz jointly set up the journal *Fundamenta Mathematicae* in 1919. The first volume was published in 1920.

Janiszewski inherited family property from his father and donated it to charity and educational institutes.

He died of influenza during an epidemic on 3 January 1920 in Lwów, Poland. He bequeathed his body to be used for medical research and his cranium for craniological study.

Figure 198. Zygmunt Janiszewski

Figure 199. Kazimierz Kuratowski

Kazimierz Kuratowski

Kazimierz Kuratowski was born on 2 February 1896 in Warsaw, Russian Empire (now Poland). He studied engineering for one year at the University of Glasgow in 1913 under the name of Casimir Kuratov.

World War I interrupted his studies in Glasgow, he transferred to the University of Warsaw. After graduating in 1919, Kuratowski began his doctoral studies under Mazurkiewicz and Janiszewski.

He was awarded his doctorate in 1921, although Janiszewski had died in 1920.[3]

Kuratowski was appointed professor at the Lwów Technical University in 1927 when Stan Ulam was a freshman.[48]

Kuratowski worked with Steinhaus and Banach but Kuratowski was not a regular member of the Scottish Café. He left Lwów in 1934 to become professor of mathematics at the University of Warsaw.

During World War II, Kuratowski risked his life to teach at an underground university in Warsaw.[3] He led the effort after the war to rebuild Poland's educational system. He was president of the Polish Mathematical Society for eight years.

He was appointed Director of the Mathematical Institute of the Polish Academy of Sciences in 1949, and he held that position for 19 years.

He succeeded Sierpinski as Editor-in-Chief of *Fundamenta Mathematicae* in 1952 and continued in that role for the rest of his life. His main work was in the area of topology and set theory.

He was elected to the USSR Academy of Sciences, the Hungarian Academy of Sciences, the Austrian Academy of Sciences and the Royal Society of Edinburgh.

Kuratowski died on 18 June 1980 in Warsaw, Poland.

Stanislaw Marcin Ulam

Stanislaw Ulam was born on 3 April 1909 in Lemberg, Autro-Hungarian Empire (now Lwów Uklaine). He entered Lwow Technical University in 1927 and received his Dr. Math.Sc. in 1933 under Kuratowski's supervision.

Between the two world wars, Lwów was a productive mathematical center with the University of Lwów and Lwów Technical University.

Many mathematicians met regularly at the Scottish Café at 22 Akademicka Street called Ludwik Żalewski's Confectionery. They were Banach, Steinhaus, Turowicz, Kuratowski, Mazur and Ulam.

Ulam stayed at the Institute for Advanced Study in Princeton at the invitation of von Neumann from 1935 to 1936.

G.D. Birkhoff then invited him to Harvard University where he was appointed Junior Fellow in the new Society of Fellows from 1936 to 1939 and he became a lecturer at Harvard in 1940.

Ulam was at the University of Wisconsin from 1941 to 1943. In 1943 he became an American citizen.

That same year he was asked by von Neumann to join the Los Alamos National Laboratory. Ulam's telegram to Los Alamos was delivered as "I was interested in principal"[48] instead of "in principle."

At Los Alamos, he developed the "Monte-Carlo method" which searched for solutions to mathematical problems using a statistical sampling method with random numbers.[3]

Ulam solved the problem of how to initiate fusion in the Hydrogen bomb jointly with Edward Teller by the method now known as the Teller-Ulam configuration.[3]

He became good friends with Pal Erdös. Ulam had a brain surgery to treat encephalitis in 1945. Erdös followed Ulam's home to stay.

As soon as they arrived at Ulam's home, Erdös proposed a game of chess. Ulam beat him twice.

Ulam spent one year at UCLA, then in 1946 he went back to Los Alamos where he stayed until 1965.

Ulam was a professor of mathematics at the University of Colorado from 1965 to 1976, and then he moved to the University of Florida in 1974.[94]

He died on 13 May 1984 in Santa Fe, New Mexico, USA.

Figure 200. Stanislaw M. Ulam

Figure 201. Karol Borsuk

Karol Borsuk

Karol Borsuk was born on 8 May 1905 in Warsaw, Russian Empire (now Poland). He received a master's degree in 1927 from the University of Warsaw. In 1930 he received his doctorate for his dissertation *Sur les rétractes* in which he invented the theory of retracts supervised by Mazurkiewicz.

He married Zofia Paczkowska in 1936 and they had two children.[3]

Borsuk introduced absolute neighborhood retracts (ANRs) and the cohomotopy groups, later called Borsuk-Spanier cohomotopy groups. His topological and geometric conjectures and themes stimulated research for more than half a century.

Borsuk was a member of the Scottish Café in Lwów before World War II. He was imprisoned by the Nazis but he escaped and was able to survive by hiding throughout the rest of the war.

He was promoted to professor at the University of Warsaw in 1946 and spent three years intermittently at the Institute for Advanced Study at Princeton, UC Berkeley and the University of Wisconsin in the United States.

He became Vice Director of the Mathematical Institute of the Polish Academy of Sciences in Warsaw in 1956.

He died on 24 January 1982 in Warsaw, Poland.

Hungarian School

Farkas Bolyai ——————→ Janos Bolyai
(1775–1856)   (1802–1860)

Leo Königsberger ——————→ Gyula König
(1837–1921)   (1849–1913)
1860 Ph.D. Berlin   1870 Ph.D. Heidelberg
                                      1874 Prof. TU Budapest

G.R. Kirchhoff ——————→ Lorand Eötvös
(1824–1887)   (1848–1919)
                                    1870 Ph.D. Heidelberg

Jozsef Kürschak ——→ Denes König ——→ Tibor Gallai ——→ Lászlo Lóvasz
(1864–1933)   (1884–1944)   (1912–1992)   (1948–   )
1890 Ph.D. TU Budapest   1907 Ph.D. TU Budapest   TU Budapest   1970 Ph.D.
1916–18 Rector   Eötvös Loránd
                                                                          1999 Wolf Prize
                                                                          7 students

Gyula Valyi—→ Frigyes Riesz ——→ Bela Szökefalvi-Nagy
                 (1880–1956)   (1913–1998)
       1902 Ph.D. Eötvös Loránd
       1920 Prof. Szeged

                              ——→ Tibor Rado (1895–1965)
                                      1923 Ph.D. Szeged
                                      21 students (Ohio State U)

                              └→ Alfred Rényi ——→ Domokos Szász (1941–   )
                                  (1921–1970)   1971 Ph.D. Moscow (B.V. Gnedenko)
                                  1947 Ph.D. Szeged   1981 Doctor of Math. Sciences
                                  Ph.D. Leningrad   Prof. TU Budapest
                                  (Yuri V. Linnik)

Lipót Fejér ——————▶ Marcel Riesz ——————▶ C. Einar Hille (1894–1980)

(1880–1959)   (1886–1969)   1918 Ph.D. Stockholm

1902 Ph.D. Eötvös   1907 Ph.D. Eötvös Loránd   27 students

Loránd Univ   1911 Stockholm   (25 Yale, 2 Princeton)

                            1926 Lund

                        ▶ Michael Fekete ——————▶ Menahem M. Schiffer

                            (1886–1957)   (1911–1997)

                            1909 Ph.D. Eötvös Loránd   1939 Ph.D. Hebrew U

                            1929 Prof. Hebrew U   20 students (Stanford)

                        ▶ Paul (Pál) Erdös (1913–1996)

                            1934 Ph.D. Eötvös Loránd U

                            1984 Wolf Prize

                        ▶ John von Neumann (1903–1957)

                            1926 Ph.D. Eötvös Loránd U

                            1938 Bócher Prize

                            1951–52 AMS President

                        ▶ Otto Szász (1884–1952)

                            1911 Ph.D. Wien

                            1921–33 Frankfurt

                            1936–52 Cincinnati

                            9 students

                        ▶ George Pólya (1887–1985)

                            1912 Ph.D. Eötvös Loránd U

                            1942–77 Stanford

                            30 students (21ETH, 9 Stanford)

                        ▶ Gabor Szegö (1895–1985)

                            1918 Ph.D. Wien

                            1926–35 Königsberg

                            1938–54 Stanford Dept Head

                            16 students

                        ▶ Pal Turan (1910–1976)

                            1935 Ph.D. Eötvös Loránd U

                            1949 Prof. Eötvös Loránd U

Hungarian School

János Bolyai

János Bolyai was born on 15 December 1802 in Kolozsvár, Hungary (now Cluj, Romania). His father was Farkas Bolyai (1775-1856) who studied at Göttingen with Gauss.

Farkas taught mathematics, physics and chemistry at the Marosvásárholy Calvinist College (secondary school) but he was not well paid. Marosvásárhely is now called Tirgu Mures and is located in what is now Romania, 60 *km* east of Cluj. János learned to play the violin and he was able to play difficult concert pieces.

He graduated from Marosvásárhely Calvinist College in June 1817. He then studied at the Royal Engineering College in Vienna from 1818 to 1822.[3]

In September 1823 he entered the army engineering corps as a sublieutenant and spent eleven years in military service. He had a reputation as the best swordsman and dancer in the Austro-Hungarian Imperial Army. He spoke ten languages.

He began to develop the basic idea of hyperbolic geometry around 1820. By 1824 he had developed most of a complete system of non-Euclidean geometry.

His work was published as 24 page appendix to his father's book, the Tentamen, in June 1831.

Farkas sent a copy of his son's appendix to Gauss who said "I regard this young geometer Bolyai as a genius of first order. Of the appendix, he said, "To praise it would amount to praising myself. For the entire content of the work··· coincides almost exactly with my own meditations which have occupied my mind for the past thirty or thirty-five years".[3]

Farkas resented Gauss remark very much and did not appreciate his taking credit for what he thought was his son's discovery. But Gauss had in fact written a letter to Franz Taurinus (1794-1874) on 8 November 1824 on non-Euclidean geometry and in 1848, Janos Bolyai discovered that Lobachevsky had published a similar piece of work on hyperbolic geometry in 1829.[3]

Janos Bolyai left more than 20,000 pages of mathematical writings when he died. They are now in the Bolyai-Teleki library in Tirgu-Mures (former Marosvásárhely) in Romania. Bolyai's eponymous Babes-Bolyai University in Cluj-Napoca is the largest university in Romania.

Bolyai died of pneumonia on 27 January 1860 in Marosvásárhely.

Figure 202. János Bolyai

(Courtesy of the Hungarian Academy of Sciences)

Figure 203. Lóránd von Eötvös

(Courtesy of the Hungarian Academy of Sciences)

Lóránd Baron von Eötvös

Lóránd (also known as Roland in Germany) Eötvös was born on 27 July 1848 in Pest (now part of Budapest), Hungary.

His father was Baron József Eötvös who was the Minister of Education when Lóránd was born.

Baron Eötvös was a friend of Franz Liszt, the famous pianist and composer.

Lóránd inherited his father's title and membership in the Upper House of the Hungarian Parliament after his father died in 1871.

His full title in Hungarian is Vásárosnaményi Báro Eötvös Lóránd.

He studied at the University of Heidelberg where his teachers were Kirchhoff, Helmholtz and Bunsen. He also studied at the University of Königsberg under Franz Neumann and Friedrich Richelot.

Eötvös received a doctorate in 1870 at Heidelberg for a thesis on problems of Armand Fizeau on the relative motion of light source.[3]

He returned to Hungary and became a professor of theoretical physics at the University of Budapest. He married Gizella Horváth in July 1876 and they had two daughters. He founded the Hungarian Mathematical and Physical Society in 1891 and became its first president.

In 1919 the Society was renamed the Eötvös Lóránd Mathematical and Physical Society. The Péter Pázmány University (founded in 1635) in Budapest was renamed the Lóránd Eötvös University in 1950 which is the most prestigious university in Hungary.

In 1889 he was elected President of the Hungarian Academy of Sciences, and he held that position until 1905. His hobbies included mountaineering and photography.

He died on 8 April 1919 in Budapest, Hungary.

Julius (Gyula) König

Gyula König was born on 16 December 1849 in Györ, Hungary. He studied medicine in Vienna then mathematics in Heidelberg. He was awarded his doctorate in 1870 for his thesis *Zur Theorie der Modulargleichungen der elliptischen Functionen* under Königsberger's supervision.

In 1874 he was appointed professor at the Technical University of Budapest (Budapesti Müszaki Egyetem, BME) which was founded in 1782 and has since produced three Nobel Laureates; Jenö Wigner, 1963 Physics, Dénes Gábor, 1971 Physics and György Oláh, 1994 Chemistry.

König raised the level of mathematics teaching at TU Budapest to a very high standard and he was Rector of the university three times.

In August 1904, at the International Congress of Mathematicians in Heidelberg, König announced that the continuum hypothesis was false. His proof was later found to be in error. The König theorem in set theory named after him. König retired from TU Budapest in 1905.

His son Dénes König (1884-1944) also became a famous mathematician in Hungary.

König died on 8 April 1913 in Budapest, Hungary.

Figure 204. Gyula König

(Courtesy of the Hungarian Academy of Sciences)

Figure 205. Dénes König

(Courtesy of the Hungarian Academy of Sciences)

József Kürschák

József Kürschák was born on 14 March 1864 in Buda (now part of Budapest), Hungary.

He entered the Technical University of Budapest in 1881, graduated in 1886 and then received his doctorate in 1890. He became a full professor in 1900.

Kürschák served as Rector of TU Budapest from 1916 until 1918.

The Loránd Eötvös Mathematics Competition, started in 1925, was renamed the József Kürschák Mathematics Competition in 1949.

He died on 26 March 1933 in Budapest, Hungary.

Dénes König

Dénes König was born on 21 September 1884 in Budapest, Hungary, a son of Gyula König. He received his doctorate in 1907 at the Technical University of Budapest where he became a full professor in 1935.

His book *Theorie der Endlichen und Unendlichen Graphen* (Theory of Finite and Infinite Graphs) was published in 1936 and an English translation was published in 1990.[3]

His classes were visited by Paul Erdös, as a freshman at the Eötvös Loránd University.

During the Nazi occupation of Hungary, König worked to help persecuted mathematicians. He committed suicide on 19 October 1944 in Budapest, Hungary, one day before he was to move to the Budapest ghetto.[49]

Figure 206.   László Lovász

(Courtesy of the Hungarian Academy of Sciences)

László Lovász

László Lovász was born on 9 March 1948 in Budapest, Hungary. He received his Dr.rer.nat in 1970 at Eötvös Loránd University supervised by Tibor Gallai. And he received his Doctor of Mathematical Science degree from the Hungarian Academy of Sciences in 1977.

Lovász was elected a regular member of the Hungarian Academy of Sciences in 1985 and Russian Academy of Sciences in 2006.

Lovász held the Chair of Geometry at József Attila University, in Szeged from 1978 until 1982 and the Chair of Computer Science at Eötvös Loránd University in Budapest from 1983 until 1993. He was then a professor of computer science at Yale University from 1993 until 2000.

Lovász was awarded the Tiber Szele Medal in 1992 by the Bolyai Society, the National Order of Merit of Hungary in 1998 and the Wolf Prize in 1999. He also received the Knuth Prize in 1999 by ACM-SIGACT and the Gödel Prize in 2001 by ACM-EATCS.

His research fields are combinatorial optimization, graph theory and theoretical computer science.

Since 2006 he has been the Director of Matematical Institute of Eötvös Loránd University. And he has been President of the International Mathematical Union since 2007.

Frigyes Riesz

Frigyes (Friedrich) Riesz was born on 22 January 1880 in Györ, Austro-Hungary (now Hungary). His younger brother, Marcel Riesz, was also a famous mathematician.

Frigyes Riesz received his doctorate in 1902 at the Eötvös Loránd University with his dissertation on geometry.

Eötvös Loránd University was called University of Budapest until 1921 when it was renamed Pázmány Péter University. It was renamed Eötvös Loránd University in 1950.

Riesz was a founder of functional analysis, which has found important applications to mathematical physics.

Many of Riesz's fundamental findings in functional analysis were incorporated with those of Stefan Banach. The Riesz-Fischer theorem in 1907 is fundamental in the Fourier analysis of Hilbert space. It was the mathematical basis for proving that matrix mechanics of Heisenberg and wave mechanics of Schrödinger are equivalent.

Ernst S. Fischer (1875-1954) was an Austrian. He was professor at Erlangen (1911-1920) and Köln (1920-1938).

In 1911 Riesz was appointed to a chair at the University of Kolozsvár in Hungary. Kolozsvár became a part of Romania and was renamed Cluj in 1920. The university was therefore moved to Szeged, Hungary in 1920.

In 1922 Riesz founded the János Bolyai Mathematical Institute in Szeged with Alfred Haar.

Kolozsvár was the birthplace of János Bolyai. Riesz became the editor of the newly founded journal *Acta Scientiarum Mathematicarum*.

In 1945 Riesz was appointed to the Chair of Mathematics at Eötvös Loránd University.

His book *Leçons d'analyse fonctionnelle* (Lessons of Functional Analysis) written with Bela Szökefalvi-Nagy in 1956 is one of the most readable books on functional analysis. The book was republished by Dover in 1990.

Riesz was elected to the Hungarian Academy of Sciences and the Paris Academy of Sciences.

He died on 28 February 1956 in Budapest, Hungary.

Figure 207. Frigyes Riesz

(Courtesy of the Hungarian Academy of Sciences)

Figure 208. Tibor Radó

(Courtesy of the Hungarian Academy of Sciences)

Tibor Radó

Tibor Radó was born on 2 June 1895 in Budapest, Hungary. During World War I, he was sent to the Russian front as a lieutenant in the Austro-Hungarian Army.

He was taken prisoner by the Russians in 1916 and spent four years in a Siberian prison camp near Tobolsk.

He studied at the University of Szeged after the war, and where he was awarded a doctorate in 1923 under Frigyes Riesz' supervivion. Rado wrote around 20 papers during the first five years of his formal mathematical education.[3]

In September 1924 he married Ida Barabás de Albis and they had two children.

In 1930 he published *The problem of least area and the problem of Plateau*

Joseph A.F. Plateau (1801-1883) was a Belgian mathematician and physicist. It is concerned with a minimal surface area of the soap film. Plateau's experiments suggested that for any bounding contour there is always a minimal surface bounded by that contour.[3]

Radó wrote books on Plateau Problems in 1933 and 1937.

He was appointed to the faculty of Ohio State University in 1930, and he retired from there in 1964.

In the last decade of his life, Radó was interested in theoretical computer science publishing *On Non-Computable Functions* in 1962 and *Computer Studies of Turing Machine Problems* in 1965.

He died on 12 December 1965 in New Smyrna Beach, Florida, USA.

Alfréd Rényi

Alfréd Rényi was born on 30 March 1921 in Budapest, Hungary.

During World War II, his parents were taken prisoners and were held in the Budapest ghetto. Alfréd wearing a soldier's uniform, walked into the ghetto and marched his parents out.[3]

Rényi received his Ph.D. in 1947 at the University of Szeged under Frigyes Riesz for work on Cauchy-Fourier series.

He went to the Soviet Union to work with Yuri V. Linnik (1915-1972) on the theory of numbers, in particular working on the Goldbach conjecture.

He published joint work with Erdös on random graphs and also studied random space filling curves. He is best remembered for proving that every even integer is the sum of a prime and an almost prime number (one with only two prime factors).[3]

He used to say "A mathematician is a machine for converting coffee into theorems".[3]

Rényi was the founder and director of the Mathematical Institute of the Hungarian Academy of Sciences.

He died on 1 February 1970 in Budapest, Hungary.

Figure 209. Alfréd Rényi

(Courtesy of the Hungarian Academy of Sciences)

Figure 210. Domokos Szász

(Courtesy of the Hungarian Academy of Sciences)

Domokos Szász

Domokos Szász was born on 18 August 1941 in Budapest, Hungary. He received a doctorate (Dr.rer.nat) in 1967 at Eötvös Loránd University supervised by Alfred Rényi and Candidate degree at Moscow State University in 1971 under Boris V. Gnedenko's supervision.

Szász was awarded a Doctor of Mathematical Sciences degree in 1981 from the Hungarian Academy of Sciences. He was elected a regular member of the Hungarian Academy of Sciences in 1995 and he has been a professor of mathematics at the Budapest University of Technology (Budapesti Müszaki és Gazdaságtudományi Egyetem) since 1999.

Szász is also Chairman of the Mathematics Section of the Hungarian Academy of Sciences.

His research fields are stochastic processes, dynamical processes and unequilibrium statistical physics. He has supervised 11 doctoral students.

Szász was awarded the Tiber Szele Prize in 1995 by the Bolyai Society and the Széchenyi Prize by the President of Hungary in 2005.

Lipót Fejér

Lipót Fejér was born Leopold Weiss on 9 February 1880 in Pécs, Hungary. Around 1900 he changed his name to Lipót Féjér. At the University of Berlin he was a student of Hermann Schwarz. After he changed his name from Weiss to Fejér in order to sound more Hungarian, Schwarz refused to talk to him.[3]

Fejér received a doctorate in 1902 from the University of Budapest for his thesis on a fundamental summation theorem for Fourier series. The University of Budapest was renamed to Pázmány Péter University in 1921 and again renamed to Eötvös Loránd University in 1950.

From 1905 to 1911 he taught at University of Kolozsvár (now Cluj in Romania). In 1911 he was appointed to the Chair of Mathematics at the University of Budapest and he had many notable students; Pál Erdös, John von Neumann, Pál Turán, George Pólya, Tibor Radó, Macel Riesz, Gábor Szegö and Michael Fekete.

Fejér's main work was in harmonic analysis. He collaborated to produce important papers, one with Carathéodory on entire functions in 1907 and another major work with Frigyes Riesz in 1922 on conformal mappings (a short proof of the Riemann theorem).

He loved music and was a good pianist.

George Pólya's wife used to take many pictures of mathematicians. One day she stopped Fejér while he was with several others, and she took photo of them posing on the street car tracks in front of the university. Fejér cried out, "What a good wife! She puts all these full professors on the tracks of the street car so that they may be run over and then her husband will get a job!"[50]

Fejér died on 15 October 1959 in Budapest, Hungary.

He is buried in Kerepesi Cemetery in Budapest.

Figure 211. Lipót Fejér

(Courtesy of the Hungarian Academy of Sciences)

Figure 212. Marcel Riesz

(Courtesy of the Hungarian Academy of Sciences)

Marcel Riesz

Marcel Riesz, a younger brother of Frigyes Riesz, was born on 16 November 1886 in Györ, Hungary.

He was awarded a doctorate in 1907 at the University of Budapest for his thesis *Summierbare trigonometrische Reihen und Potenzreihen* under Fejér's supervision.

He went to Sweden in 1908 on the invitation of Mittag-Leffler, and he spent most of his life there.

From 1911 to 1926 he was at the University of Stockholm where he had many outstanding doctoral students including Olof Thorin, Harald Cramér, Franz Berwald, and Einar Hille.[3]

From 1926 to 1952 he was a professor at Lund University where Otto Frostman and Lars Hörmander were among his students.

Riesz's interests include functional analysis, partial differential equations, mathematical physics, number theory, Clifford algebra and Spinors.

Riesz retired in 1952 from his Lund chair and spent ten years in several different universities. In the early 1950s he gave a lecture at Stanford University in a large crowded room. After Riesz filled up the board, Gábor Szegö, the Executive Head of the Mathematics Department at Standford, sprang up and washed the blackboard while Riesz stood by and observed the proceedings.[35]

Riesz and Szegö studied under Fejér at the University of Budapest, and Szegö was reverting to the role of a European-style assistant to Riesz.

George Pólya, only one year junior to Riesz, was in the audience and he was quite embarassed by the performance.

Riesz suffered a breakdown in 1962 and returned to Lund. He died on 4 September 1969 in Lund, Sweden.

Pál Erdös

Pál (Paul) Erdös was born on 26 March 1913 in Budapest, Hungary. His father was Lajos Engländer but he changed his family name to Erdös. Both of his parents were both teachers of mathematics, so Paul was introduced to mathematics at an early age.

Paul's parents were extremely protective of Paul because they had lost two young daughters due to the scarlet fever.

Paul was kept away from school and had a private tutor for many of his early years.

Paul's father was imprisoned by the Russian Army during World War I, and he came home from Siberia in 1920. He had taught himself English at the prison camp, but he did not know the correct pronunciation of the words.

He taught Paul to speak English, and the strange English accent remained in Paul throughout his life.

Paul entered Pázmány Páter University in Budapest (renamed Eötvös Leránd University in 1950) in 1930 and he was awarded a doctorate in 1934 under Fejér's supervision.

His dissertation was *Über die Primzahlen gewisser arithmetischer Reihen.*[10]

He took up a four-year post-doctoral fellowship at the University of Manchester working with Mordell. He met Hardy and Davenport in Cambridge in 1934 and he also met Stan Ulam in 1935. Erdös and Ulam became very close. Erdös then went to the Institute for Advanced Study in Princeton.

For most of his life Erdös did not have a job. He never wanted material possessions and based his lifestyle on an old Greek saying that the wise man has nothing he cannot carry in his hands. He never had a bank account or credit card and lived out of a suitcase, traveling from one mathematical center to another.[5]

His father died of a heart attack in 1942 and four of his uncles and aunts had been murdered by the Nazis during World War II.

Erdös accepted a temporary post at the University of Notre Dame in 1952 but he was unable to re-enter the United States during the McCarthy Era.

He spent much of the next ten years in Israel and he finally received a visa to the United States in November 1963.

When Stan Ulam was recovering from his brain surgery in 1945, Erdös went Ulam's home. No sooner had they arrived Ulam's home, Erdös proposed a game of chess. Ulam won twice.[48] Erdös always called a child "epsilon", a child's mother "capital epsilon", a man "slave", a woman "boss", the United States "Sam", and the USSR "Joe".[48]

He met Ronald Graham in 1963. Graham and his wife Fan Chung, also a mathematician, provided a room in their house where Erodös could live when he wanted.

The Grahams also stored Erdös papers at their home, managed his money and acted as his secretary.

Graham also popularized the concept of the Erdös number which is the minimum number of links away from Erdös. If someone wrote a paper jointly with Erdös, he or she would have an Erdös number of one.

Erdös had about 500 collaborators and published 1500 papers.

In 1983 Erdös won the Wolf Prize, which included a cash prize of 50,000 dollars. He gave most of the money to students or as prizes for solving problems he had posed. Erdös was a solver of problems, not a builder of theories.

His main areas of interest were combinatorics, graphs theory and number theory.

His mother died in 1971 in Calgary, Canada. Erdös died on 20 September 1996 at the age of 83 due to a heart attack in Warsaw, Poland.

Figure 213. Pál Erdös   **Figure 214. Garbor Szegö**

(Courtesy of the Hungarian Academy of Sciences)   (Courtesy of the Hungarian Academy of Sciences)

Garbor Szegö

Garbor Szegö was born into a Jewish family on 20 January 1895 in Kungheyes, Hungary. He won the Eötvös mathematics competition in 1912. Szegö studied in Budapest, Berlin and Göttingen but received his doctorate in 1918 at the University of Vienna.[10]

While he was in Budapest, he coached John von Neumann in mathematics. He spent some time in the Austro-Hungarian Air Force during World War I, where he met Richard von Mises, who was a pilot in the Air Force.

In 1921 he moved to Berlin where he became a friend of Issai Schur and worked with von Mises and Erhard Schmidt. Szegö was a privatdozent in Berlin and he received the Gyula König Prize in 1924.

Szegö and Pólya jointly published two volumes of *Aufgaben and Lehrsätze aus der Analysis* (Problems and Theorems in Analysis) in 1925. The book has had an enormous impact on later generations of mathematicians.

He was appointed to a chair (Zweter Lehrstahl) at the University of Königsberg in 1926 succeeding Konrad Knopp.

Szegö was a very popular professor there, but had to give up the chair in 1935 because of the Nazis law against Jews.

He went to the United States and was at Washington University in St. Louis, Missouri from 1935 to 1938. He moved to Stanford University in 1938.

In early 1950s the famous Hungarian mathematician Marcel Riesz visited Stanford University.

After Riesz had filled up the blackboard, Riesz motioned to Gabor Szegö, nine years junior to Riesz, to wash the board. While Szegö was clearing the board Riesz stood by and observed the proceedings. At that time Szegö was Executive Head of Mathematics Department.[35]

Szegö's name is honored by the Szegö kernel function, the Szegö limit theorem, Szegö polynomials orthogonal on the unit circle, the Szegö class of polynomials.

He wrote *Toeplitz Forms and their Applications* with Ulf Grenander in 1958.

He is Peter Lax's uncle by marriage.

Szegö died on 7 August 1985 in Palo Alto, California.

Figure 215. John von Neumann

(Courtesy of the Hungarian Academy of Sciences)

Figure 216. Neumann with his computer

John von Neumann

John von Neumann was born on 28 December 1903 in Budapest, Hungary. His original name was Margittai Neumann Janos Lajos. His father, Max Neumann, was a well-known banker who learnt languages from the German and French governesses. Although the family was Jewish, Max Neumann did not observe the Jewish religion. In 1913, he purchased a title and his son used the German form von Neumann.[3] His German name was Johann Ludwig von Neumann.

As a child John had an incredible memory and he could exchange jokes with his father in classical Greek.

He graduated from the Lutheran Gymnasium in Budapest in 1921. Eugene Wigner had graduated one year earlier.

During his gymnasium years, Gábor Szegö and Michael Fekete coached him in mathematics.

Von Neumann's first paper in transfinite diameter was written jointly with Fekete in 1922.

He entered the University of Budapest (renamed Pázmány Péter University in 1921) in 1921 but did not attend lectures. Instead he studied chemistry at the University of Berlin (1921-1923) and *Eidgenössiche Technische Hochschule Zürich* (1923-1926).

Von Neumann received his diploma in chemical engineering from ETH Zürich and a doctorate in mathematics at the Pázmány Péter University in Budapest in 1926.

While in Zürich he took courses from Hermann Weyl and George Pólya who said Johnny was the only student he feared. If Pólya stated an unsolved problem in his lecture, von Neumann would come to him as soon as the lecture was over, with the complete solution in a few scribbles on a slip of paper. Von Neumann took over one of Weyl's cources when he was absent from Zurich.[3]

He studied under Hilbert at Göttingen during 1926-27 and in 1927 von Neumann received his habilitation at the University of Berlin under Erhard Schmidt.

In 1930 he married Marietta Kovesi in Budapest who was studying economics in Budapest. Their daughter Marina born in 1935 became a member of the Council of Economic Advisers for President Nixon during 1973-74.

The von Neumanns divorced in 1937, and John remarried Klari Dan in Budapest in 1939.

Von Neumann became a professor at Princeton University in 1931. Two years later he became one of the original six mathematics professors (along with James W. Alexander, Albert Einstein, Marston Morse, Oswald Veblen, and Hermann Weyl) at the newly founded Institute for Advanced Study at Princeton, and he stayed for the remainder of his life.

He published *Mathematische Grundlagen der Quatenmechanik* (Mathematical Foundation of Quantum Mechanics) in 1932 which provided a solid framework for the quantum mechanics developed in 1925-28.

He introduced self-adjoint algebras of bounded linear operators on a Hilbert space in 1929. Such operator algebras were called "rings of operators", W∗-algebras or "von Neumann algebras."[3]

He also wrote a classic text *Theory of Games and Economic Behaviour* with Oskar Morgenstern in 1944.

In 1943 he joined the Manhattan Project to develop atomic bombs. He worked with three Hungarian physicists, Wigner, Szilard and Teller.

He turned his attention to electronic computers and decided to design one of his own. The trustees of the Institute for Advanced Study pledged $100,000 towards the total expense. Other major funding came from the Office of Naval Research, Army Ordnance, and the Radio Corporation of America (RCA). It took six years to complete the project. The final dedication of the computer took place on June 10, 1952.

Parts of the computer are now in the National Museum of American History in Washington, D.C.

Von Neumann received many honors. Just one year after becoming a US citizen he was elected to the National Academy of Sciences in 1932.

He was awarded the Bôcher Prize in 1938, and the Enrico Fermi Award in 1956 and he was President of the American Mathematical Society in 1951-52.

He also received two Presidential Awards, the Medal for Merit in 1947 and the Medal for Freedom in 1956.

When he received the Enrico Fermi Award, he knew that he was incurably ill with cancer.

He suffered more when his mind would no longer function and could not accept his fate. He knew how to live fully, did not know how to die.

He died on 8 February 1957 in Washington, D.C.

Klari von Neumann later remarried physicist Carl Eckert, but she committed suicide on 10 November 1963.

Otto Szász

Otto Szász was born on 11 December 1884 in Alsoszucs, Hungary. He received a doctorate in 1911 at the University of Budapest supervised by Lipót Fejér. From 1911 to 1914 he continued postdoctoral studies at München, Paris and Göttingen.

In 1914 he moved to the University of Frankfurt as a privatdozent and was promoted to extraordinary professor in 1921. At that time, Max Dehn and Carl Siegel were full professors, Szász and Paul Epstein were extraordinary (associate) professors at

Frankfurt. From 1921 to 1933 Frankfurt was one of the important centers for mathematical research.

Hilter came to power, and Szász left for the United States in 1933 and, Dehn in 1935. Carl Siegel went to Göttingen in 1937 and Epstein committed suicide in 1939.

In 1936 Szász was appointed through Norbert Wiener's effort to the faculty of the University of Cincinnati where he stayed until 1952.

In 1939 he was awarded the Gyula König Prize of the Hungarian Mathematical and Physical Society.

He contributed to the theory of Fourier series, summability theory, continued fractions and approximation theory.

Szász died of a heart attack on 19 December 1952 in Montreux, Switzerland while vacationing with his wife at her mother's estate.

Figure 217. Otto Szász

(Courtesy of the Hungarian Academy of Sciences)

Figure 218. George Pólya

(Courtesy of the Hungarian Academy of Sciences)

George Pólya

György (George) Pólya was born on 13 December 1887 into a Jewish family. His father's original name was Jakab Pollák but changed to Pólya in 1882 to sound more Hungarian. Jakab Pólya and his wife Anna converted to the Roman Catholic faith in 1886. He was a lawyer and became a privatdozent at the University of Budapest shortly before his death in 1897.

George enrolled at the University of Budapest in 1905 and studied at the University of Vienna in 1910-11. He was awarded a doctorate in 1912 at the University of Budapest.

He spent 1912 and 1913 in Göttingen and went to Paris early in 1914. He was offered the position privatdozent at *Eidgenössische Technische Hochschule Zürich* by Adolph Hurwitz and Pólya accepted it eagerly.[4]

In 1918 he married Stella Vera Weber who was the daughter of professor of physics at the University of Neuchâtel.

She used to take many photographs of mathematicians.

One day, she took a photo of Lipót Fejér in the company of three or four others posing on the street car tracks in front of the university.

Fejér cried out, "What a good wife! She puts all these full professors on the tracks of the street car so that they may be run over and then her husband will get a job!"[50]

Actually *The Pólya Picture Album, Encounters of A Mathematician* was published by Birkhäuser in 1987. Pólya was promoted to extraordinary professor at ETH Zürich in 1920 and to ordinary professor in 1928.

He received a Rockefeller Fellowship in 1924 to work at Oxford and Cambridge with Hardy and Littlewood. Hardy asked Pólya to take the Cambridge Mathematical Tripos to show that such a distinguished foreign mathematician would have failed to do well in the Tripos which will help to reform the Tripos system.

But Pólya did better than any of the normal candidate. Eventually the tripos system was reformed eliminating the word senior wrangler and the rank system. Since 1909 students have not been ranked individually by test score. Instead, they are grouped with other students in a tiered ranking system.

In 1928 Pólya became a Swiss citizen. Von Neumann took some of Pólya's course at ETH and Pólya said von Neumann was the only student he was ever afraid of. In the course of a lecture he stated an unsolved problem, the chances were he would come to Pólya as soon as the lecture was over, with the complete solution in a few scribbles on a slip of paper.[3] At that time von Neumann was studying chemical engineering at ETH.

In 1940 he took two-year appointment at Brown University and then in 1942 he was appointed associate professor at Stanford University. He became a US citizen in 1947.

Pólya retired from Stanford in 1953 but continued teaching and conducting research as a Professor Emeritus. In 1978 at the age of ninety he taught a course on Combinatorics in the Computer Science Department at Stanford.

He wrote *Aufgaben und Lehrsätze aus der Analysis* with Gábor Szegö in 1925.

Szegö became a collegue at Stanford for many years.

Pólya coined the term "Central limit theorem" in 1920. In 1920 he proved his famous theorem on random walks on an integer lattice.[3] In 1923 he showed that the Fourier transform of a probability density function is a characteristic function.

Pólya's book *How to Solve It* published in 1945 by Princeton University Press has been translated in 17 languages and sold over one million copies.

He was elected to the National Academy of Sciences of the United States in 1976.

Pólya died on 7 September 1985 in Palo Alto, California at the age of 97. In his 97th year, he published his 250th paper.[94]

Finnish School

Ernst L. Lindelöf ⟶ Rolf Herman Nevanlinna ⟶ Lars Valerian Ahlfors

(1870–1946)   (1895–1980)   (1907–1996)

1893 Helsingin Yliopisto   1919 Helsingin Yliopisto   1930 Ph.D. Helsingin Yliopisto

  14 students   22 students   1946–77 Harvard

                                            1936 Fields Medal

                                            1981 Wolf Prize

                                            25 students (Harvard)

                                            See American School

Finnish School

Ernst Leonard Lindelöf

Ernst Lindelöf was born on 7 March 1870 in Helsingfors, Russian Empire (now Helsinki, Finland). Lindelöf received his doctorate in 1893 at Helsingfors (Helsinki) University, and then taught there as a docent. He visited Göttingen in 1901 and, he became an extraordinary professor in 1902 and a full professor in 1903 at Helsinki.[3]

Lindelöf retired from professorship in Helsinki in 1938.

He published a book *Le calcul des résidus et ses applications á la théorie des fonctions* in 1905 and *Differential and Integral Calculus and Their Applications* in four volumes between 1920 and 1946.[3] Lindelöf space, Lindelöf hypothesis, Lindelöf lemma, Lindelöf theorem and Picard-Lindelöf theorem are named after him.

He died on 4 June 1946 in Helsinki, Finland.

Figure 219. Ernst L. Lindelöf

Figure 220. Rolf H. Nevanlinna

Rolf Herman Nevanlinna

Rolf Nevanlinna was born on 22 October 1895 in Joensuu, Russian Empire (now Finland).

He was awarded a doctorate in 1919 at the University of Helsinki under Lindelöf's supervision. Lindelöf was a cousin of Nevanlinna's father.

Nevanlinna was appointed professor at Helsinki in 1926.

He was offered Herman Weyl's chair at ETH Zürich in 1930 when he left Zürich for Göttingen but Nevanlinna refused. He became Rector of Helsinki University in 1941.

In November 1939 Andre Weil was arrested by the Finnish Police. He was rescued by Nevanlinna before Weil was executed as a spy.[20]

Nevanlinna invented harmonic measure in 1936 and developed the theory of value distribution, which is now named after him.

Since 1982 the Rolf Nevanlinna Prize has been presented at the International Congress of Mathematicians for young mathematicians dealing with the mathematical aspects of information science.[3]

Nevanlinna died on 28 May 1980 in Helsinki, Finland.

Figure 221. Lars V. Ahlfors

Lars Valerian Ahlfors

Lars Ahlfors was born on 18 April 1907 in Helsingfors, Russian Empire (now Helsinki, Finland). He followed Nevanlinna to ETH Zürich in 1928 when Nevanlinna replaced Weyl for the academic year 1928/29 while Weyl was on leave. Ahlfors was able to prove Denjoy's conjecture on the number of asymptotic values of an entire function with the help from Nevanlinna and Pólya[3] in 1929.

Ahlfors was awarded a doctorate in 1930 at the University of Helsinki. In 1936 he was one of the first two recipients of the Fields Medal at the International Congress of Mathematicians in Oslo. He was cited for methods he had developed to analyze Riemann surfaces of inverse functions in terms of covering surfaces.

In 1938 he was appointed to a chair of mathematics at the University of Helsinki. He went to Zürich during World War II.

He was offered a professorship at Harvard in 1946 and he stayed at Harvard for 31 years.

The Ahlfors principal theorem, the Ahlfors five-disk theorem, the Ahlfors principal theorem are named after him.

He published many books including *Complex Analysis* (1953), *Riemann Surfaces* with Leo Sario (1960), *Lectures on Quasi-Conformal Mappings* (1966) and *Conformal Invariants* (1973).

Ahlfors died on 11 October 1996 in Pittsfield, Massachusetts, USA.

Swedish School

Carl Björling ⟶ Anders Wiman ⟶ Arne Beurling ⟶ Lennart Carleson (1928-)
1863 Ph.D. Uppsala   1892 Ph.D. Lund   (1905-1986)   1950 Ph.D. Uppsala
                                            1933 Ph.D. Uppsala   1992 Wolf Prize
                                            1954 IAS Princeton   2006 Abel Prize
                                                                  26 students

Göran Dillner ⟶ Gösta Mittag-Leffler ⟶ Erik Ivar Fredholm
                (1846-1927)   (1866-1927)
                1872 Ph.D. Uppsala   1893 Ph.D. Uppsala
                3 students   1898 D.Sc. Uppsala

Marcel Riesz ⟶ Carl Harald Cramer ⟶ Ulf Grenander (1923-   )
(1886-1969)   (1893-1975)   1950 Ph.D. Stockholm
               Ph.D. 1917 Stockholm   21 Students (Brown)
               10 students   NAS member

⟶ Lars Garding ⟶ Lars Hörmander (1931-   )
                  1955 Ph.D. Lund
                  1962 Fields Medal
                  1988 Wolf Prize

Swedish School

Arne Carl-August Beurling

Anne Beurling was born on 3 February 1905 in Gothenburg, Sweden. He received a doctorate in 1933 at Uppsala University and taught there from 1932 to 1952.

In 1954 he became a professor at the Institute for Advanced Study in Princeton.

During World War II he decoded the German G-Schreiber message code in two weeks. As a result, the Swedish Command was aware of all German troop movements.[3]

Beurling died on 20 November 1986 in the United States.

Figure 222. Arne C. A. Beurling   Figure 223. Lennart A. E. Carleson

Lennart Axel Edvard Carleson

Lennart Carleson was born on 18 March 1928 in Stockholm, Sweden. He was awarded a doctorate in 1950 at Uppsala University for his thesis *On a class of meromorphic functions and its exceptional sets* under Beurling's supervision.

In 1954 he was appointed professor at the University of Stockholm but he returned to Uppsala one year later. He retired from his chair of mathematics at Uppsala in 1993.

Carleson was Director of the Mittag-Leffler Institute, in Stockholm from 1968 to 1984. He was also editor of *Acta Mathematica* from 1956 to 1979 and President of the International Mathematical Union from 1978 to 1982.[3]

In 1962 he solved the famous "Corona Problem". The "Carleson measures" became a fundamental tool in complex analysis and harmonic analysis.

In 1966 he proved Luzin's conjecture of 1913 in his paper *On convergence and growth of partial sums of fourier series*.

He was elected to the Russian Academy of Sciences, the Royal Society of London and the French Academy of Sciences.

Carleson received the Wolf Prize in 1992, the Lomonosov Gold Medal in 2002, the Sylvester Medal in 2003 and the Abel Prize in 2006.

Magnus Gösta Mittag-Leffler

Gösta Mittag-Leffler was born on 16 March 1846 in Stockholm, Sweden. His father was Johan Olof Leffler and his mother was Gustava Wilhelmina Mittag. Gösta took Mittag-Leffler as his last name when he was 20 years old. Both sides of the family were of German Origin.

He entered Uppsala University in 1865 and received a doctorate in 1872. He spent one and half years in Paris meeting many mathematicians including Hermite, Liouville, Darboux and Chasles.

In the spring of 1875 he went to Berlin where he attended Weierstrass' lectures.

Weierstrass arranged for Mittage-Leffler to be appointed extraordinary professor at the University of Berlin but instead he took a chair at Helsinki in 1876.

Five years later, he was appointed to a chair at the University of Stockholm.[3]

In 1882 he founded *Acta Mathematica* and served as the chief editor of the journal for 45 years.

In the same year he married Signe at Lindfors whom he met in Helsinki.

In 1884 Sofya Kovalevskaya arrived in Stockholm at his invitation where she died in 1891.

His best known work concerned the analytic representation of a one-valued function, which culminated in what is known as the Mittag-Leffler theorem.

In his home in the suburbs of Stockholm, he had the finest mathematical library in the world. The library was donated to the Swedish Academy of Sciences.

The Mittag-Leffler Institute was set up based on his house and it is a major mathematical research centre today.

He was elected a Fellow of the Royal Society of London in 1896.

There is a rumor that Mittag-Leffler ran off with Alfred Nobel's wife and that is why there is no Nobel Prize for mathematics.

This is completely false as Nobel never married. Nobel once considered donating part of his fortune to the University of Stockholm, but eventually formed the Nobel Foundation.

Mittag-Leffler died on 7 July 1927 in Stockholm, Sweden.

Figure 224. M. Gösta Mittag-Leffler

Figure 225. E. Ivar Fredholm

Erik Ivar Fredholm

Ivar Fredholm was born on 7 April 1866 in Stockholm, Sweden. After studying one year at the Royal Institute of Technology in Stockholm (Kungl Techniska Högskolan) he transferred to the University of Uppsala in 1886. He wanted to study under Mittag-Leefler who was at the newly founded University of Stockholm (Stockholms Högskola) but it did not award doctorates.

So Fredholm registered for his doctorate at Uppsala but also studied at Stockholm under Mittag-Leffler. In 1893 he was awarded his Ph.D. and in 1898 he received the degree of Doctor of Science at Uppsala.[3]

Fredholm is best remembered for his work on integral equations and spectral theory. Hilbert extended Fredholm's work to include a complete eigenvalue theory for the Fredholm integral equation. This work led directly to the theory of Hilbert space.[3]

In 1906 Fredholm was appointed to a chair in mechanics and mathematical physics at Stockholm. He was an actuary in the Skandia Insurace Company proposing an elegant mathematical formula to determine the surrender value of a life insurance policy.[3]

He played the flute and the violin. He even built his own violin from half of a coconut. He received the Poncelet Prize from the French Academy of Sciences in 1908.

He died on 17 August 1927 in Danderyd, Stockholm, Sweden.

Carl Harald Cramér

Harland Cramér was born on 25 September 1893 in Stockholm, Sweden. He was awarded a Ph.D. in 1917 at the University of Stockholm for his thesis *On a class of Dirichlet series* under Marcel Riesz's supervision.

His main concerns were number theory and probability theory. He worked as an actuary at Svenska Life Assurance Company. In 1929 he was appointed to a newly created Chair of Actuarial Mathematics and Mathematical Statistics at the University of Stockholm.

He published *Random Variables and Probability Distributions* in 1937.

During World War II he gave shelters to Willi Feller (1906-1970) who had been forced out of Germany by Nazis anti-Jewish policies in 1934.

He also published *Mathematical Methods of Statistics* in 1945. In 1950 he became the President of Stockholm University and he retired in 1961.[3]

One of his theorem states that of the sum of two independent random variables is normal then all are normal.

He died on 5 October 1985 in Stockholm, Sweden.

Figure 226. Carl H. Cramér

Figure 227. Lars Hörmander

Lars Hörmander

Lars Hörmander was born on 24 January 1931 in Mjällby, Blekinge, Sweden. He received a doctorate in 1955 at the University of Lund.

He was appointed professor at the University of Stockholm in 1957 and continued to spent time at Stanford University and the Institute for Advanced Study in Princeton.

In 1962 at the International Congress of Mathematicians he was awarded a Fields Medal for his work on partial differential equations. The work is contained in his book *Linear Partial Differential Operators* (1963).

From 1964 to 1968 he was at the Institute for Advanced Study at Princeton. In 1968 he took up the chair of mathematics at the University of Lund and he retired from there in 1996.

From 1984 until 1986 he was director of the Mittag-Leffler Institute in Stockholm.

He is one of the ten mathematicians that has received both the Fields Medal and the Wolf Prize.

Norwegian School

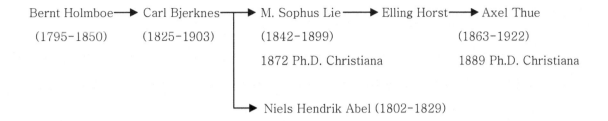

Bernt Holmboe ⟶ Carl Bjerknes ⟶ M. Sophus Lie ⟶ Elling Horst ⟶ Axel Thue

(1795–1850)   (1825–1903)   (1842–1899)   (1863–1922)

                               1872 Ph.D. Christiana   1889 Ph.D. Christiana

                  ⟶ Niels Hendrik Abel (1802–1829)

?   ⟶   Atle Selberg (1917–2007)

              1943 Ph.D. Oslo

              1950 Fields Medal

              1986 Wolf Prize

Norwegian School

Bernt Michael Holmboe

Bernt Holmboe was born on 23 March 1795 in Vang, Norway. He became a teacher at the Cathedral School of Christiana in 1818 where he taught Niels Abel. He even helped to pay for Abel's university education.

Holmboe was appointed to the chair of pure mathematics at the University of Christiana in 1834. After Abel's death, Holmboe edited Abel's complete works in 1839.[3] He died on 28 March 1850 in Christiana, Norway.

Figure 228. Bernt M. Holmboe

Figure 229. Niels H. Abel

Niels Henrik Abel

Niels Abel was born on 5 August 1802 in Frindoe, Norway.

His father Soren Georg Abel was a Lutheran minister and his mother Ane Marie Simonson was the daughter of a merchant and ship owner.

Soren Abel was a Norwegian nationalist who was active in the movement to make Norway independent, but his political career ended in disaster and he became an alcoholic. When Soren Abel died in 1820, his wife was left heavily in debt, and she became an alcoholic as well.

With Holmboe's help, Abel was able to enter the University of Christiana in 1821 where the wife of professor Christopher Hansteen cared for Abel as if he was her own son.

In 1824 he proved the impossibility of solving the general equation of the fifth degree in radicals. In 1825, he was given a scholarship from the Norwegian government to travel abroad.

Abel went to Berlin and met August Leopold Crelle who had just founded the *Journal für die reine und angewandte Mathematik* (Journal for Pure and Applied Mathematics), commonly known as Crelle's Journal. Abel's paper on the insolubility of the quintic equation as well as six of his other papers were published in the first volume of *Crelle's Journal* in 1827. Crelle's Journal also published early works of Abel, Dirichlet, Eisenstein, Grassmann, Hesse, Jacobi, Kummer, Lobachevski, Möbius, Plünker, von Stadt, Steiner and Weierstrass.[4]

In July 1826 Abel went to Paris where he met Cauchy, Legendre, Poisson, Lacroix and Liouville. Legendre was courteous but Cauchy received Abel with characteristic discourtesy. Cauchy was a devout Catholic and Abel was anti-Catholic. Liouville regretted later that he could not appreciate Abel's genius at that time. The memoir on transcendental function by Abel was presented to the Paris Academy was lost. It was found by Cauchy in 1830 after Abel's death and printed in 1841.

Abel met Christine (Crelly) Kemp in 1823 and they became engaged in 1824.[109]

When Abel returned to Christiana in May 1827, he found no prospect of a suitable job, but plenty of debts.

In early 1828 he found a paper by Carl Jacobi on transformations of elliptic integrals. Abel quickly showed that Jacobi's results were an extension of his own and tired to compete with Jacobi in the theory of elliptic functions.

Legendre saw the new ideas in the papers by Abel and Jacobi and said, "through these works you two will be placed in the class of the foremost analysts of our times".[3]

In 1830 the Paris Academy awarded Abel and Jacobi the Grand Prix for their outstanding work.

Abel's health deteriorated and he was diagnosed with tuberculosis. He died on 6 April 1829 in Froland, Norway. Crelle wrote to Abel on 8 April 1829 to tell him of his appointment as a professor at the Berlin Gewerbeschule (now Technische Universität Berlin). The letter arrived in Christiana a few days after Abel's death.

Carl Anton Bjerknes

Carl Bjerknes was born on 24 October 1825 in Christiana (now Oslo), Norway. He graduated from the University of Christiana in 1848 with a degree in mining engineering, and he worked at the Kongsberg silver mines until 1852.

He studied mathematics in Göttingen and Paris in 1856 and 1857. He attended Dirichlet's lectures on hydrodynamics and spent the rest of his life researching in that area.

In 1859 he married Aletta Koren and they had three sons.

Bjerknes was appointed lecturer in applied mathematics at the University of Christiana in 1861. He was promoted to a reader in 1863 and professor in 1866. In 1869 he accepted the chair of pure mathematics at the same university.[3]

His two volume book *Hydrodynamic action at a distance* was published in 1900 and 1902.

Bjerknes died on 20 March 1903 in Oslo, Norway.

Figure 230. M. Sophus Lie

Figure 231. Axel Thue

Marius Sophus Lie

Sophus Lie was born on 17 December 1842 in Nordfjordeide, Norway. He graduated from the University of Christiana in 1865. His paper, *Über eine Darstellung des Imaginären in der Geometrie*, published in 1869 in *Crelle's Journal* proved vital.[53]

He was awarded a scholarship to travel to Germany and France.

In Berlin, Lie became good friends with Felix Klein who had been a student of Plücker. Lie studied Plücker's papers and he felt as if he were also Plücker's student.

When Lie was in France in 1871 the Franco-Prussian war began. Klein quickly returned to Berlin from France. But Lie was arrested as a German spy because his mathematics notebook was assumed to contain top secret coded messages.

Lie spent one month in prison and was released only after the intervention of Darboux.

Lie was awarded his doctorate in 1872 at the University of Christiana for a dissertation *On a class of geometric transformations* (written in Norwegian).

He was appointed associate professor in the same year at his alma mater. In 1874 he married Anna Birch and they had two sons and one daughter.

Anna Birch's mother's grandfather was Niels Abel's maternal grandfather.[53]

In 1873 Lie began to develop what became his theory of continuous transformation groups, later called Lie groups.

Klein sent Friedrich Engel (1861–1941) to Christiana to work with Lie. Engel and Lie worked together for nine months. Engel was appointed as a privatdozent at Leipzig in 1885.

Klein left his chair at Leipzig in 1886 to go to Göttingen, and Lie was appointed to succeed him.

The collaboration between Engel and Lie nevertheles continued for nine years culminating with their publication *Theorie der Transformationsgruppen* in three volumes between 1888 and 1893.

Lie was troubled by constant homesickness in Leipzig. He missed the forests and mountains of Norway. He began to suffer from insomnia, and in the autumn of 1889 he experienced a nervous breakdown. He stayed in a psychiatric clinic near Hannover for seven months.

He recovered fully two years later. Lie's relationship with Engel broke down at the end of the 1880s. His long friendship with Klein also broke down and Lie publicly attacked Klein in 1893 saying, "I am no pupil of Klein, nor is the opposite the case, although this might be closer to the truth".[53]

Lie returned to Christiana in 1898 to his chair. He received the Lobachevsky Prize in 1897 on Klein's recommendation, despite their strained relationship.

During twelve years at Leipzig, Lie supervised 26 doctoral candidates.[53]

He died of pernicious anaemia on 18 February 1899 in Christiana (Oslo), Norway.

Axel Thue

Azel Thue was born on 19 February 1863 in Tönsberg, Norway. He graduated from the University of Oslo in 1889. He also studied in Leipzig under Lie and in Berlin as well.

In 1894 he married Lucie Collett Lund.

Between 1894 and 1903 he worked at Trondheim Technical College. He was appointed professor of applied mathematics at the University of Oslo in 1903 where he remained until his death in 1922.

He wrote 35 papers on number theory, mostly on the theory of Diophantine equations.

His work was extended by Carl Siegel in 1920 and by Klaus Roth in 1958.

Thue died on 7 March 1922 in Oslo, Norway.

Figure 232. Atle Selberg

Atle Selberg

Atle Selberg was born on 14 June 1917 in Langesund, Norway.

He was influenced by Rámanujuan and Erich Hecke. He was awarded a Ph.D. (Dr. Philos) in 1943 at the University of Oslo.

In 1947 he married Hedvig Liebermann and moved to the United States. They had one son and one daughter. Hedvig died in 1995.

Selberg spent one year at the Institute for Advanced Study in Princeton at the invitation of Carl Siegel in 1947-48.

He spent one year at Syracuse University, then returned to the Institute in 1949. Two years later he was appointed professor in the Institute's School of Mathematics where he retired in 1987.[3]

In 1950 he was awarded a Fields Medal for his work on generalizations of the sieve methods of Viggo Brun, and for his major work on the zeros of the Riemann Zeta function.

Selberg was one of the four editors of Axel Thue's *Selected Mathematical Papers* published in 1977.

He also received the Wolf Prize in Mathematics in 1986. Selberg is one of ten people who have received both the Fields Medal and the Wolf Prize.

Many mathematical terms bear his name: Selberg Trace Formula, Selberg Sieve, Selberg Integral, Selberg Class, Rankin-Selberg L-Function, Selberg Eigenvalue Conjecture and Selberg Zeta Function.

The collected papers of Selberg were published in two volumes in 1989 and 1991.

Selberg died on 6 August 2007 in Princeton, New Jersey at the age of ninety.

Italian School

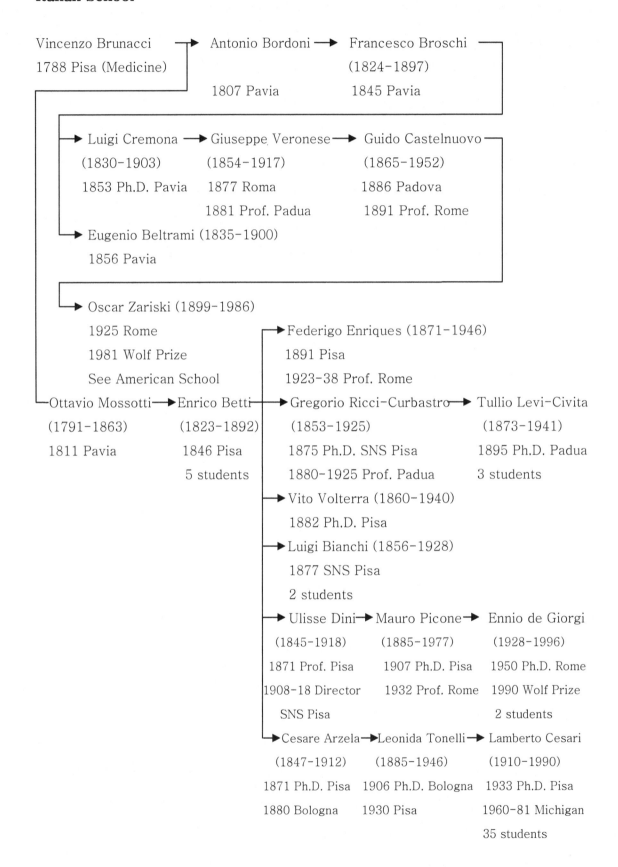

Vincenzo Brunacci
1788 Pisa (Medicine)

Antonio Bordoni ➔ Francesco Broschi
1807 Pavia

Francesco Broschi
(1824-1897)
1845 Pavia

Luigi Cremona ➔ Giuseppe Veronese ➔ Guido Castelnuovo
(1830-1903)
1853 Ph.D. Pavia

Giuseppe Veronese
(1854-1917)
1877 Roma
1881 Prof. Padua

Guido Castelnuovo
(1865-1952)
1886 Padova
1891 Prof. Rome

Eugenio Beltrami (1835-1900)
1856 Pavia

Oscar Zariski (1899-1986)
1925 Rome
1981 Wolf Prize
See American School

Federigo Enriques (1871-1946)
1891 Pisa
1923-38 Prof. Rome

Ottavio Mossotti ➔ Enrico Betti ➔ Gregorio Ricci-Curbastro ➔ Tullio Levi-Civita
(1791-1863)
1811 Pavia

Enrico Betti
(1823-1892)
1846 Pisa
5 students

Gregorio Ricci-Curbastro
(1853-1925)
1875 Ph.D. SNS Pisa
1880-1925 Prof. Padua

Tullio Levi-Civita
(1873-1941)
1895 Ph.D. Padua
3 students

Vito Volterra (1860-1940)
1882 Ph.D. Pisa

Luigi Bianchi (1856-1928)
1877 SNS Pisa
2 students

Ulisse Dini ➔ Mauro Picone ➔ Ennio de Giorgi
(1845-1918)
1871 Prof. Pisa
1908-18 Director
SNS Pisa

Mauro Picone
(1885-1977)
1907 Ph.D. Pisa
1932 Prof. Rome

Ennio de Giorgi
(1928-1996)
1950 Ph.D. Rome
1990 Wolf Prize
2 students

Cesare Arzela ➔ Leonida Tonelli ➔ Lamberto Cesari
(1847-1912)
1871 Ph.D. Pisa
1880 Bologna

Leonida Tonelli
(1885-1946)
1906 Ph.D. Bologna
1930 Pisa

Lamberto Cesari
(1910-1990)
1933 Ph.D. Pisa
1960-81 Michigan
35 students

Leonida Tonelli   ⟶   Emilio Baiada   ⟶   Calogero Vinti

 (1885–1946)   (1914–1984)   (1926–1997)

                    1937 Ph.D. SNS Pisa   1949 Ph.D. Palermo

                    1952–1961 Palermo   1962–1970 Modena

                    1961–1984 Modena   1970–1997 Perugia

Giovanni Ricci ⟶ Enrico Bombieri (1940–   )

 (1904–1973)   Ph.D. Milano

1925 SNS Pisa   1974 Fields Medal

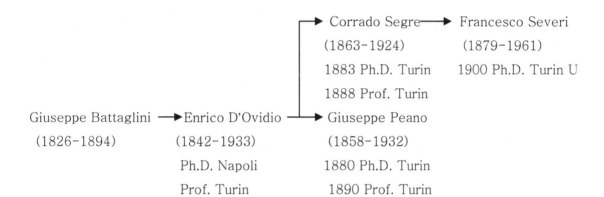

Corrado Segre ⟶ Francesco Severi

                              (1863–1924)   (1879–1961)

                              1883 Ph.D. Turin   1900 Ph.D. Turin U

                              1888 Prof. Turin

Giuseppe Battaglini ⟶ Enrico D'Ovidio ⟶ Giuseppe Peano

 (1826–1894)   (1842–1933)   (1858–1932)

                          Ph.D. Napoli   1880 Ph.D. Turin

                          Prof. Turin   1890 Prof. Turin

Italian School

Until the introduction of the *dottorato di ricerca* in the mid-1980s, the *laurea* was the most advanced academic degree and allowed the holder access to the highest academic careers. *Laureati* are customarily addressed as *dottore* (for a man) or *dottoressa* (for a woman), i.e. "doctor". To earn a *laurea*, the student had to complete 4 to 5 years of university cources. It was customary to describe progress in terms of the number of exams passed, rather than years. In most cases, a student was required to conduct experimental work and write a thesis to earn a *laurea*. Italian students graduate from high school (*scuola secondaria superiore*) at the age of nineteen.

Consequently, the old Italian degree system was very similar to that of German system before 1940s. The first and highest degree in German Universities before 1940s was doctorate. There were no bachelor and master degrees.

Many authors misunderstood the Italian degree *laurea* as an undergraduate degree.

Famous Nobel Laurate physicists Enrico Fermi and Carlo Rubbia held laurea degrees in physics as their highest degree. They both studied at the *Scuola Normale Superiore di Pisa* and received their laurea degrees at the University of Pisa.

The *Scuola Normale Superiore di Pisa* was founded in 1810 by Napoleon as a part of *Ecole Normale Superieure* in Paris. Following the French tradition, the *Scuola Normale Superiore* do not award degrees which are awarded by the nearby University of Pisa.

The *dottorato di ricerca* consisted in three years of Ph.D. level courses and experimental work with research all paid for by the state. But the Italian government has never made the *dottorato di ricera* a requirement to become a professor in Italian universities.

In 1999 the *laurea* was split into a three year *laurea triennale* degree (equivalent to a bachelor's degree) and a two-year *laurea magistrale* (equivalent to a master's degree).[59]

A student can undertake a *dottorato di ricerca* only after achieving a *laurea magistrale*.

Through this reform, the Italian degree system is now be compatible to the French, German and British systems.

Antonio Luigi Gaudenzio Giuseppe Cremona

Luigi Cremona was born on 7 December 1830 in Pavia, Lombardy (now Italy). He fought against the Austrians for Italian independence before he entered the University of Pavia in 1849. In 1853 he was awarded a *laurea* in Civil Engineering at the University of Pavia.

After Lombardy was liberated from Austrian rule, and on 10 June 1860 Cremona was appointed by Royal decree to a position as an ordinary professor at the University of Bologna. He stayed there for seven years. While in Bologna he was awarded the Steiner Prize in 1866 and he developed the theory of birational transformations, later known as Cremona transformation.[3]

In October 1867 he moved to the Polytechnic Institute of Milan then in October 1873 he was appointed Director of the newly established Polytechnic School of Engineering in Rome. In November 1877 he became professor of higher mathematics at the University of Rome.

He became Minister of Education in 1879, and his mathematical work ended at that point. He went on to become Vice-President of the Senate.

Cremona died on 10 June 1903 in Rome, Italy.

Figure 233.   A. Luigi G.G. Cremona

Giuseppe Veronese

Giuseppe Veronese was born on 7 May 1854 in Chioggia, Italy. He studied at Zurich Polytechnic (now ETH Zürich), then moved to the University of Rome on Cremona's advice and graduated in 1877.

He did research under Klein at Leipzig in 1880 and 1881. Veronese was appointed to the chair of algebraic geometry at the University of Padua in 1881, and he held that chair for the rest of his life.

Veronese invented non-Archimedean geometries around 1890. He served as a member of Parliament from 1897 to 1900 and as a Senator from 1904 to 1917.

He died on 17 July 1917 in Padua, Italy.

Guido Castelnuovo

Guido Castelnuovo was born into a Jewish family on 14 August 1865 in Venice, Italy.

He attended the University of Padua, where he took Veronese's classes and consequently acquired an interest in geometry. He graduated in 1886 and became an assistant to Enrico D'Ovidio at the University of Turin where he was influenced by Corrado Segre (1863-1924).

In 1891 Castelnuovo was appointed to the Chair of Analytic and Projective Geometry at the University of Rome.

He published a book on algebraic geometry *Geometria Anralitica e proiettiva* in 1903 and a book on probability *Calcolo della probabilitá* in 1919.

He collaborated with Federigo Enrique (1871-1946) over a period of 20 years. The Academia dei Lincei awarded the Royal Prize in Mathematics to Castelnuovo in 1905. He was elected to the *Académie des Sciences* of Paris.

Castelnuovo had to retire from teaching at the University of Rome in 1935 due to the Mussolini government's anti-semitic policies. After Rome was liberated he became president of the Academia dei Lincei. In 1949 he was named Senator of the Italian Republic.

His name is honored in the Castelnuovo-Severi inequality and Kronecker-Castelnuovo theorem.

He died on 27 April 1952 in Rome, Italy.

Oscar Zariski

Oscar Zariski was born into a Jewish familiy on 24 April 1899 in Kobrin, Belarus, Russian Empire. His name was originally Ascher Zaritsky, but he changed it to the Italian sounding Oscar Zariski at Federigo Enriques's suggestion in Rome.

He was awarded a *laurea* in 1924 at the University of Rome for a thesis on a topic related to Galois theory supervised by Castelnuovo. He married Yole Calgi in Kobrin on 11 September 1924. The Fascist policies against Jews made life in Italy very difficult for Zariski. With the help of Solomon Lefschetz, he went to the United States in 1927 to join the faculty of Johns Hopkins University. He became a full professor at Johns Hopkins in 1937.

He was in Sao Paolo in 1945 and 1946 and at the University of Illinois in 1946 and 1947. He was then appointed to a chair at Harvard University, and he retired in 1969.

In the United States he supervised 16 doctoral candidates, most of them at Harvard University.[10]

Two of his students, Heisuke Hironaka and David Mumford received the Fields Medal.

Zariski published *Algebraic Surfaces* in 1935 and *Commutative Algebra* jointly with P. Samuel, in two volumes in 1958 and 1960.

Zariski used to say "geometry is the real life".

He was awarded the Cole Prize in Algebra in 1944, the National Medal of Science in 1965, the Steele Prize and the Wolf Prize in 1981.

He was president of the American Mathematical Society in 1969 and 1970. Zariski was elected to the United States National Academy of Sciences in 1944 and the Academia dei Lincei in 1958.

He died of Alzheimer's disease on 4 July 1986 in Brookline, Massachusetts, USA.[4]

Figure 234. Guido Castelnuovo

Figure 235. Oscar Zariski

(Courtesy of the American Mathematical Society)

Enrico Betti

Enrico Betti was born on 21 October 1823 in Pistoia, Tuscany (now Italy). He graduated from the University of Pisa in 1846 with a *laurea* degree in mathematics. He fought in two battles for Italian independence.

After teaching at secondary schools for a few years, he was appointed professor of higher algebra at the University of Pisa in 1857.

He visited Göttingen and became friendly with Riemann. Riemann visited Italy in 1863, and the two renewed their friendship.[3] Riemann directed Betti's mind to problems of potential theory and elasticity.

He served a term as Rector of the University of Pisa, and then in 1864 he became director of the *Scuola Normale Superiore* holding that post until his death.[3]

The *Scuola Normale Superiore* had been founded by Napoleon in 1810 as a branch of the *Ecole Normale Superieure de Paris*.

Students at the *Scuola Normale Superiore* are awarded degrees from the University of Pisa. Betti was the first to give a proof that the Galois group is closed under multiplication. In 1854 Betti showed that the quintic equation could be solved in terms of integrals resulting in elliptic functions.[3]

Henri Poincaré gave the name *Betti numbers* to certain nambers that characterize the connectivity of a manifold (the higher-dimensional analog of a surface).

Betti's lectures at the University of Pisa inspired a generation of Italian mathematicians, the most famous of them being Vito Volterra and Luigi Bianchi.

Betti died on 11 August 1892 in Soiana, Pisa, Italy.

Figure 236.   Enrico Betti

Cesare Arzela

Cesare Arzela was born on 6 March 1847 in Santo Stefano di Magra, La Spezia, Italy. He received a *laurea* degree at the University of Pisa in 1871 under the supervision of Enrico Betti. After teaching at high Schools from 1870 in Macerata, Siena, Savona,

Como, Florence and the University of Palermo (1878-1880), he was appointed in 1880 to the Chair of Higher Analysis at the University of Bologna. He is known to every mathematician for the Theorem of Ascoli-Arzela (1889) and for the notion of convergence today known as "almost uniform convergence".[115]

Arzela was the first mathematician to bring Bologna's mathematical teaching up to international standards. Arzela-Ascoli theorem can be stated as follows[113]:

Consider a sequence of continuous functions $(f_n)_{n \in N}$ *defined on a closed and bounded internal* $[a, \ b]$ *of the real line with real values. If this sequence is uniformly bounded and uniformly equicontinuous, then there exists a subsequence* (f_{nk}) *that converges uniformly.*

Arzela died on 15 March 1912 in Santo Stefano di Magra, Italy.

Leonida Tonelli

Leonida Tonelli was born on 19 April 1885 in Gallipoli, Italy. He received a *laurea* degree in 1907 at Bologna University under the supervision of Cesare Arzela and Salvatore Pincherle. After teaching at the universities of Cagliari (1913), Parma (1914), and Bologna (1922), he finally settled at the University of Pisa in 1930. He fully reorganized in 1932 the "Annali della Scuola Normale Superiore di Pisa", founded in 1871, to a prestigious international journal.

His fundamental contributions to mathematics include calculus of variations, length and area functionals, Fourier series, differential and integral equations, integration theory. He was author of some 165 publications including articles and books. His well-known research books are *Fondamenti del Calcolo delle Variazioni*, Vol 1 and 2 (1921-1923) and *Serie Trigonometriche* published in 1928.[115]

His prominent students included Lamberto Cesari, Alessandro Faedo, Guido Stampacchia, Emilio Baiada, Silvio Cinquini, Adolfo del Chiaro and Basilio Mani.

Tonelli died on 12 March 1946 in Pisa, Italy.

Lamberto Cesari

Lamberto Cesari was born on 23 September 1910 in Bologna and he received a *laurea* in 1933 at the *Scuola Normale Superiore* at Pisa under the direction of Leonida Tonelli.

Cesari worked with Constantin Carathéodory in Germany from 1934 to 1935, at the *Scuola Normale Superiore* and the Istituo Nazionale per le Applicazioni del Calcolo in Rome. He became a professor of mathematical analysis at the University of Bologna in 1947. He went to the United States as a visiting professor at the Institute for Advanced Study at Princeton in 1948. Cesari also worked at Purdue University, the University of California, Berkeley and the University of Wisconsin-Madison.

In 1960 he was appointed as a professor of mathematical analysis at the University of Michigan at Ann Arbor where he remained until his retirement in 1981. He became a naturalized US citizen in 1976.[114]

He is remembered for his achievements on the Plateau's problem and on the Lebesque measure of continuous parametric minimal surfaces and related variational problems. He also worked in the field of optimal control and studied periodic solutions of systems nonlinear ordinary differential equations by using methods of nonlinear functional analysis.[114] He wrote about 250 scientific works about nonlinear functional analysis including *Optimization-Theory and Applications: Problems with Ordinary Differential Equations* published by Springer Verlag in 1983. Cesari died on 12 March 1990 in Ann Arbor, Michigan.

Emilio Baiada

Emilio Baiada was born on 12 January 1914 in Tunis, Tunisia. He was a student of Leonida Tonelli at the Scuola Normale Superiore in Pisa. His *laurea* thesis in 1937 resulted in the award of the Michel Prize. After teaching mathematical analysis at Pisa he went to the United States in 1949. Baiada collaborated with Otto Szasz and Charles N. Moore at the University of Cincinnati, then with Marston Morse at the Institute of Advanced Study at Princeton.

He returned to Italy in 1952, working as full professor at the University of Palermo until 1961 when he finally settled at the University of Modena. He was a member of the *Academia Modenese di Scienze, Lettere ed Arti* and was honored with Gold Medal of Science, Arts and Culture by the President of the Italian Republic. He was the author of 63 scientific papers. Baiada had 18 students mostly at Modena. He died on 14 May 1984 in Modena, Italy.

Calogero Vinti

Calogero Vinti was born on 12 July 1926 in Agrigento, Sicily. He received a *laurea* in mathematical sciences in 1949 from the University of Palermo officially under Michele Cipolla and Benedetto Pettineo. However, he was a true scientific disciple of Emilio Baiada with whom he wrote two important papers on perimeter of sets. He was appointed to the Chair of Mathematical Analysis in Modena in 1966. In 1970 he moved to the University of Perugia, where he remained until his death. He founded the Faculty of Engineering at Perugia in 1986, and was its Dean until 1995. He was Editor-in-Chief of the Journal *Atti del Seminario Mathematico e Fisico dell Universita degli Studi di Modena* since 1984. Vinti had 16 *laurea* students. In 1996 he received the Gold Medal of Science, Arts and Culture from the President of the Italian Republic. The Italian Mathematical Union founded the Premio Calogero Vinti Prize.

Vinti died on 25 August 1997 in Perugia, Italy.

Figure 237. Gregorio Ricci-Curbastro

Figure 238. Vito Volterra

Gregorio Ricci-Curbastro

Gregorio Ricci-Curbastro was born on 12 January 1853 in Lungo, Papal State (now Italy).

He studied at the University of Rome, University of Bologna and the *Scuola Normale Superiore* where he was awarded a *laurea* for his thesis *On Fuch's research concerning linear differential equations.*

Technically his *laurea* degree was from the University of Pisa because degrees from *Scuola Normale Superiore* were conferred in the name of the University of Pisa. His habilitation thesis was *On a generalization of Riemann's problem concerning hypergeometric functions.*[3]

In 1877 and 1878 he attended lectures by Felix Klein and Alexander W. von Brill at the Technische Hochschule München. From 1880 until his death, Ricci-Curbastro was a professor of mathematical physics at the University of Padua.

In 1884 he married Bianca Bianchi Azzarani and they had three children. Between 1884 and 1894 he invented the absolute differential calculus which became the foundation of tensor analysis used by Einstein in his theory of general relativity.

Much of Ricci-Curbastro's work after 1900 was done jointly with his student Tullio Levi-Civita.

In a joint paper in 1900 *Méthod de calcul differentiel absolu et leurs applications* he used the name Ricci instead of his full name.

Therefore the term Ricci tensor is used today in books on general relativity. This work of 1900 contains references to the use of intrinsic geometry as an instrument of computation dealing with normal congruencies, geodetic laws and isothermal families of surfaces.[54]

He was honored with membership to several academies. Ricci-Curbastro died on 6 August 1925 in Bologna, Italy.

Vito Volterra

Vito Volterra was born on 3 May 1860 in Ancona, Papal Sates (now Italy). His father died when he was just two years old, and his family was extremely poor. At the age of eleven he began to study Legendre's geometry.

He was awarded a *laurea* degree in physics in 1882 at the University of Pisa with a thesis on hydrodynamics. Volterra became professor of mechanics at Pisa in 1883, professor of mechanics at Turin in 1892. He was appointed to the Chair of Mathematical Physics at the University of Rome in 1900, succeeding Eugenio Beltrami (1835-1899). Volterra conceived the idea of a theory of functions which depend on a continuous set of values of another function in 1883.

Hadamard later introduced the word "functional" to replace Volterra's original terminology.[3]

Volterra's papers on partial differential equations of the early 1890s included the solutions of equations for cylindrical waves

$$\frac{\partial^2 u}{\partial t^2} = \frac{\partial^2 u}{\partial x^2} + \frac{\partial^2 u}{\partial y^2}.$$

Volterra constructed a model for population change, in which the predator population at time t by $x(t)$, the prey population at time t by $y(t)$, interact in a continuous manner expressed in the following nonlinear simultaneous differential equations;

$$\frac{dx}{dt} = -ax + bxy$$

$$\frac{dy}{dt} = cy - kxy$$

where a, b, c, k are positive constants. The linear terms $-ax$ and cy model the natural decay and growth respectively. Since A.J. Lotka, an American biologist, independently derived the equations, they are known as Lotka–Volterra predator–prey model.

Volterra refused to take an oath of allegiance to the fascist government in 1931, and he was forced to leave his chair at Rome.

He was elected a Fellow of the Royal Society of London in 1910. Volterra died on 11 October 1940 in Rome, Italy.

Luigi Bianchi

Luigi Bianchi was born on 18 January 1856 in Parma, Italy.

He was awarded a *laurea* in 1877 at the University of Pisa where he studied under Enrico Betti and Ulisse Dini.

After studying at Göttingen under Klein, he was appointed professor at the *Scuola Normale Superiore di Pisa* in 1882. He became a full professor of analytic geometry in 1890.

His work on non-euclidean geometries was used by Einstein in his general theory of relativity. The Bianchi identity is essential in the general theory of relativity. He wrote many books on differential geometry, theory of groups of substitutions, theory of continuous groups and theory of algebraic numbers. He was editor of *Annali di Mathematica Pura ed Applicata*.

Bianchi died on 6 June 1928 in Pisa, Italy.

Eugenio Beltrami

Eugenio Beltrami was born on 16 November 1835 in Cremona, Lombardy, Autro-Hungarian Empire (now Italy). He studied at the University of Pavia from 1853 to 1856. He transferred to the University of Milano, where he continued his mathematical work again and in 1862 he published his first paper.

In the same year he was appointed visiting professor of algebra and analytic geometry at the University of Bologna.

From 1864 to 1866 he was at the University of Pisa, then back at Bologna from 1866 to 1873. In 1873 he was appointed to the Chair of Rational Machanics at the University of Rome.

He moved to Pavia in 1876, then returned to Rome in 1891 and remained there for the rest of his life.[3]

He was President of the *Academia dei Lincei* in 1898. His four-volume work, *Opere Matematiche* published posthumously contains his comments on a broad range of mathematical physics including thermodynamics, elasticity, optics, electricity and magnetism.

He gave a generalized form of the Laplace operator and a generalization of Green's theorem. He indirectly influenced the development of tensor analysis by providing a basis for the ideas of Ricci-Curbstro and Levi-Civita.[3]

Beltrami died on 18 February 1900 in Rome, Italy, and Volterra succeeded his chair at the University of Rome in 1900.

Figure 239.   Luigi Bianchi

Figure 240.   Eugenio Beltrami

Ulisse Dini

Ulisse Dini was born on 14 November 1845 in Pisa, Italy.

He attended the *Scuola Normale Superiore* in Pisa and studied for one year in Paris producing seven papers.

Dini returned to Pisa in 1866 and he was promoted to professor of analysis and higher geometry at the University of Pisa in 1871.

In 1888 he became Rector of the University of Pisa. In 1908 he was appointed director of the *Scuola Normale Superiore* and held this position until his death.

He was involved in politics and served on the Pisa City Council, in the Italian Parliament and as senator.

He solved a problem posed by Beltrami (1835–1900) of representing one surface on a second surface in such a way that geodesic lines in the first correspond to geodesic lines in the second.[3]

He published many books on Fourier series, the theory of functions of a real variable and infinitesimal analysis.

Luigi Bianchi was one of his famous students.

Dini died on 28 October 1918 in Pisa, Italy.

Tullio Levi-Civita

Tullio Levi-Civita was born into a Jewish family on 29 March 1873 in Padua, Veneto, Italy.

He received a *laurea* degree in 1895 at the University of Padua supervised by Ricci-Curbastro.

He was appointed an instructor at Padua in 1898 and then was promoted to Professor of Rational Mechanics in 1902.

He married Libera Trevisani, one of his students in 1914.

In 1918 he was appointed to the Chair of Higher Analysis at the University of Rome.[55]

He was one of the leading figures in the creation of the International Congress of Applied Mechanics in 1922.

In 1938 Levi-Civita was dismissed from his Chair in Rome and forced to leave the editorial board of *Zentralblatt für Mathematik* due to the policies the fascist government.

He is best known for his work on the absolute differential calculus and its application to the theory of general relativity. With Ricci-Curbastro, Levi-Civita wrote the pioneering work on the calculus of tensors, *Méthodes de calcul différentiel absolu et leurs applications* (Methods of the Absolute Differential Calculus and Their Applications) in 1900.

He also made contributions to differential geometry, hydrodynamics and three-body problem which involves the motion of three bodies as they revolve around each other.

When asked what he liked best about Italy, Einstein responded "Spaghetti and Levi-Civita".

Levi-Civitta was awarded the Sylvester Medal in 1922 and elected a Fellow of the Royal Society of London in 1930. He died isolated from the rest of the scientific world in his apartment in Rome on 29 December 1941.

Figure 241.   Ulisse Dini   Figure 242.   Tullio Levi-Civita

Federigo Enriques

Federigo Enriques was born into a Jewish family on 5 January 1871 in Leghorn (now Livorno), Tuscany, Italy. He was awarded a *laurea* degree in 1891 at the University of Pisa after studying under Betti at the *Scuola Normale Superiore* in Pisa.

He also studied with Castelnuovo in Rome, and Corrado Segre in Turin.

Enrique was appointed a professor of geometry at the University of Bologana in 1896 and professor of higher geometry at the University of Rome in 1923. While in Rome, Enrique suggested that Ascher Zaritsky change his name to the more Italian-sounding Oscar Zariski.

Because of the Racial Law, Enriques had to resign his chair in 1938.

His work on algebraic surfaces gained world-wide recognition at the International Congress of Mathematicians in Cambridge in 1912.

Enriques and Francesco Severi received the Prix Bordin in 1907 for their work on hyperelliptic surfaces.[3]

He died on 14 June 1946 in Rome, Italy.

Enrico D'Ovidio

Enrico D'Ovidio was born on 11 August 1842 in Campobasso, Italy. He did not have any university degree but he published some articles on determinants and conics.

He was granted an honorary degree in mathematics by the University of Naples due to his reputation as a mathematician.

Beltrami persuaded him to apply for the Chair of Algebra and Analytic Geometry at the University of Turin. He was successful and remained at Turin for 46 years. He also served as the Rector of the University of Turin from 1880 until 1885.

His most important work is *The fundamental metric functions in spaces of arbitrary many dimensions with constant curvature* published in 1877.

D'Ovidio died on 21 March 1933 in Turin, Italy.

Figure 243. Enrico D'Ovidio

Figure 244. Federigo Enriques

Giuseppe Peano

Giuseppe Peano was born on 27 August 1858 in Cuneo, 75 km south of Torino, Italy. In 1880 he graduated from the University of Turin (Torino) with a *laurea* degree in mathematics.

He worked as an assistant to D'Ovidio and later to Angelo Genocchi.

Peano received his qualification to be a university professor (habilitation) in December 1884. He became a lecturer of infinitesimal calculus at the University of Turin in 1884 and then a professor in 1890. He also taught at the Academia Militare in Turin from 1886 to 1901.

He made several important discoveries, including a continuous mapping of a line onto every point of a square.

His *Formulario Mathematico* (Mathematical Formularity), written with collaborators and published from 1894 to 1908, was intended to develop mathematics in its entirety from its fundamental postulates, using his logic notation and his simplified international language called Latino Sine Flexione.

This artificial language (Interlingua) was based on a synthesis of Latin, French, German and English vocabularies with greatly simplified grammar.

It was hard to read, though, and after World War I Peano's influence declined markedly.

Part of Peano's logic notation, however, was adopted by Russell and Whitehead in their *Principia Mathematica.*[56]

Peano taught at the University of Turin until the day before he died of a heart attack on 20 April 1932 in Turin, Italy.

Corrado Segre

Corrado Segre was born on 20 August 1863 in Saluzzo, 50 km south of Torino (Turin), Italy. In 1883 he was awarded a *laurea* degree for a thesis on quadrics in higher dimensional spaces.

In 1828 he was appointed to the chair of higher geometry in Turin succeeding D'Ovidio. He held that post until his death in 1924.

Segre worked on geometric properties invariant under linear transformations, algebraic curves and ruled surfaces.[3]

He introduced bicomplex points into geometry.

Segre is considered one of the most illustrious members of the Italian school of geometry second only to Luigi Cremona.

Segre died on 18 May 1924 in Turin, Italy.

Francesco Severi

Francesco Severi was born on 13 April 1879 in Arezzo, 60 km southeast of Firenze, Italy.

In 1900 he was awarded a *laurea* degree at the University of Turin supervised by Corrado Segre.

He worked as an assistant to D'Ovido, Enriques and Eugenio Bertini. In 1904 he was appointed to the chair of Projective and Descriptive Geometry at the University of Parma then in 1905 at the University of Padua. During World War I he served with distinction as artillery officer.[3]

In 1922 he moved to the University of Rome. In 1907 Enrique and Severi won the Prix Bordin from the *Académie des Sciences* in Paris for their work on hyperelliptic surfaces.

He also worked in the solution of the Dirichlet problem and the development of the theory of rational equivalence.

Severi published 400 papers, but some of his work has subsequently been shown to be inadequate by Oscar Zariski and David Mumford.[57] He died on 8 December 1961 in Italy, Rome.

Figure 245. Giuseppe Peano

Figure 246. **Corrado Segre**

Figure 247. **Francesco Severi**

Ennio De Giorgi

Ennio De Giorgi was born on 8 February 1928 in Lecce, Southern Italy (the heel of the boot). He was awarded a *laurea* degree in 1950 at the University of Rome supervised by Mauro Picone.

In 1956 De Giorgi proved what has become known as "De Giorgi's Theorem" concerning the Hölder continuity of solutions of elliptic partial differential equations of second order.

In 1958 he was appointed to the Chair of Mathematical Analysis at the University of Messina. He stayed there less than one year and in 1959, he moved to the *Scuola Normale Superiore* at Pisa as a professor of analysis. He stayed there for the rest of his life.

He had no life outside mathematics, no family of his own or other close relationship and literally lived in his office. Despite occupying the most prestigious mathematical chair in Italy, he lived a life of ascetic poverty, completely devoted to his research and teaching. He attempted to prove the existence of God through mathematics.[12]

De Giorgi made contributions to geometric measure theory, the solution of Hilbert's 19th problem in any dimension, the solution of the n-dimensional Plateau problem, the solution of the n-dimensional Bernstein problem, convergence problems for functional and operators, free boundary problems, semicontinuity and relaxation problems, minimum problems with free discontinuity set and motion by mean curvature.[3]

Figure 248. Ennio De Giorgi

Figure 249. Enrico Bombieri

He received the National Prize of *Academia dei Lincei* in 1973, and the Wolf Prize in 1990.

He was elected to the *Académié des Sciences* in Paris and the National Academy of Sciences of the United States.

He died on 25 October 1996 in Pisa, Italy following an operation.

Enrico Bombieri

Enrico Bombieri was born on 26 November 1940 in Milan, Italy.

He was awarded a *laurea* degree in 1963 at the University of Milan and then he studied with Harold Davenport at Trinity College, Cambridge.

In 1966 Bombieri was appointed to a chair of mathematics at the University of Pisa.

In 1974 he moved to the *Scuola Normale Superiore* in Pisa. Then he held the IBM von Neumann Chair at the Institute for Advanced Study at Princeton, New Jersey.

Bombieri was elected to the National Academy of Sciences of the United States in 1966 and to the French Academy of Sciences in 1984.

He was awarded a Fields Medal at the International Congress of Mathematicians in Vancouver in 1974.

The award was made for his major contributions in primes, in univalent functions and the local Bieberbach conjecture, in theory of functions of several complex variables and in theory of partial differential equations and minimal surfaces in particular, to the solution of Bernstein's problem in higher dimensions.

The Bombieri-Vinogradov theorem is one of the major applications of the large sieve method. It improves Dirichlet's theorem on prime numbers in arithmetic progressions by showing that by averaging over the modulus over a range, the mean error is much less than can be proved in a given case.[58]

Between 1979 and 1982 Bombieri served on the executive committee of the International Mathematical Union.

He explored for wild orchids and other plants as a hobby in the Alps when he was young.

Dutch School

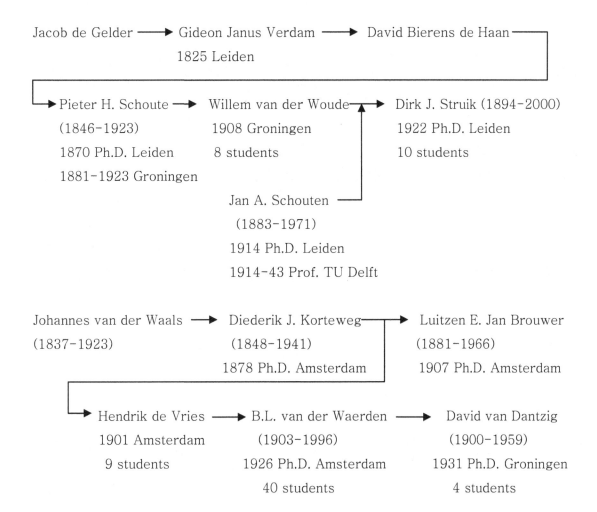

Jacob de Gelder ⟶ Gideon Janus Verdam ⟶ David Bierens de Haan
1825 Leiden

Pieter H. Schoute ⟶ Willem van der Woude ⟶ Dirk J. Struik (1894–2000)
(1846–1923)   1908 Groningen   1922 Ph.D. Leiden
1870 Ph.D. Leiden   8 students   10 students
1881–1923 Groningen

Jan A. Schouten
(1883–1971)
1914 Ph.D. Leiden
1914–43 Prof. TU Delft

Johannes van der Waals ⟶ Diederik J. Korteweg ⟶ Luitzen E. Jan Brouwer
(1837–1923)   (1848–1941)   (1881–1966)
                        1878 Ph.D. Amsterdam   1907 Ph.D. Amsterdam

Hendrik de Vries ⟶ B.L. van der Waerden ⟶ David van Dantzig
1901 Amsterdam   (1903–1996)   (1900–1959)
9 students   1926 Ph.D. Amsterdam   1931 Ph.D. Groningen
                   40 students   4 students

Thomas J. Stietjes (1856–1894)
1884 Hon Doctorate Leiden
1889 Prof. Toulouse

Dutch School

The degree system in the Netherlands has been different from other European countries. Greek and Latin languages were required for entry into universities. Max Dresden describes in: "H.A. Kramers Between Tradition and Revolution"[60]:

At universities there was no well-defined time schedule which the students were compelled to follow. They could take classes in the subjects of their choice and attended classes as they saw fit; the whole structure was very loose. It was required to take examinations in a number of subjects, but the examinations could be taken at any time. The examinations were all oral and consisted of a private discussion between student and professor. After a number of these exams called tentamina, there was more formal oral exam with several professors present. Passing this exam made the student officially a "candidate". It would typically take 3 or 4 years to pass the examination for a candidate. Therefore, the candidate is equivalent to a bachelor's degree.

It takes another 2 or 3 years to get a doctorandus degree (Latin: he who should become a doctor, abbreviation Drs.) *with very little or no research is required. The doctorandus degree in the Netherlands is equivalent to a master degree. Some excellent students can get a doctorandus degree only four years after matriculation.*

A fair number of students do not go beyond doctorandus degree and many high school teachers have a doctorandus degree.

To obtain a Ph.D. degree, a considerable amount of research has to be done, the writing of a thesis can take many years.

Pieter Hendrik Schoute

Pieter Hendrik Schoute was born on 21 January 1846 in Wormerveer, Netherlands. He graduated from the Polytechnic School in Delft (now Delft University of Technology) in 1867 with a degree in Civil Engineering. He received a doctorate in 1870 at the University of Leiden for his dissertation *Homography Applied to the Theory of Quadric Surfaces*. He taught at secondary schools from 1871 to 1881.[3]

From 1881 until his death in 1923 he was a professor of mathematics at the University of Groningen. He studied various topics in geometry including quadrics and algebraic curves. Schoute wrote two papers jointly with Alicia Boole Stott (1860–1940) on regular polytopes which generalize the concept of regular polyhedra. Alicia Boole Scott was a daughter of George Boole the inventor of Boolean Algebra. In 1890 she married Walter Stott, an actuary. She learned of Schoute's work on central sections of the regular polytopes and sent him photographs of her cardboard models which led to their collaboration.

Schoute came to England and worked with Alicia Stott and published two papers in Amsterdam in 1900 and 1910.

Schoute was an editor of the journal *Nieuw Archief Wiskunde* from 1898 to 1923.

He died on 18 April 1923 in Groningen, Netherlands.

Figure 250.   Jan A. Schouten   **Figure 251.   Dirk J. Struik**

Jan Arnoldus Schouten

Jan Schouten was born on 28 August 1883 in Nieuweramstel (now part of Amsterdam), Netherlands. He studied electrical engineering at the *Technische Hogeschool Delft* (now Delft University of Technology) and worked for several years as an electrical engineer.

He received a Ph.D. in 1914 at the University of Leiden with a thesis on tensor analysis. He was professor of mathematics at Delft from 1914 to 1943, and at the University of Amsterdam from 1948 to 1953.

He was an excellent lecturer and a meticulous author. He wrote 180 papers and six books on tensor analysis.[3] He applied it to Lie groups, relativity, unified field theory and systems of differential equations. Schouten was president of the 1954 International Congress of Mathematicians at Amsterdam.

He died on 20 January 1971 in Epe, Netherlands.

Dirk Jan Struik

Dirk Struik was born on 30 September 1894 in Rotterdam, Netherlands. He entered the University of Leiden in 1912 commuting 40km each day by train from Rotterdam.

At Leiden he studied mathematics and physics with Lorentz, de Sitter and Ehrenfest.

He ran out of money in 1917 and he taught mathematics at a school in Alkmaar, north of Amsterdam. Then he became an assistant to Jan Schouten at Delft studying tensor analysis.

He was awarded a Ph.D. in 1922 at the University of Leiden for his dissertation on the application of tensor methods to Riemanian manifolds. His formal supervisor was the geometer Willem van der Woude.

In 1923 he married Ruth Ramler who had received a doctorate in mathematics in 1919 at Charles University in Prague. Her field was the axiomatics of affine geometry. In 1924, funded by a Rockefeller fellowship, Struik worked with Tullio Levi-Civita in Rome and with Richard Courant in Göttingen.

In 1926 Struik received an offer from Norbert Wiener to become a visiting professor at Massachusetts Institute of Technology, where he spent the rest of his academic career. He was made full professor at MIT in 1940.

Struik joined the Communist Party of the Netherlands in 1919 and remained a Party member his entire life.

He was asked on his 100th birthday, how he managed to write peer-reviewed journal articles at such an advanced age. Struik replied that he had the "3Ms": Marriage, Mathematics and Marxism.

During the McCarthy era he was accused of being a Soviet spy, a charge he vehemently denied. He refused to answer any of the 200 questions asked of him at the House Committee on Un-American Activities (HUAC) hearing. MIT suspended him from teaching for five years (with full salary). He was re-instated in 1956, and retired in 1960.[3]

He wrote *Lectures on Classical Differential Geometry, A Concise History of Mathematics* and *A Source Book in Mathematics, 1200-1800.*[3]

His wife Ruth Ramler Struik died in 1993.

He died on 21 October 2000 at his home in Belmont, Massachusetts. He was 106 years old.

Figure 252.   Diederik J. Korteweg

Figure 253.   Luitzen E. J. Brouwer

Diederik Johannes Korteweg

Diederik Korteweg was born on 31 March 1848 in 's-Hertogenbosch, Netherlands.

He studied at the Polytechnic School in Delft (now Delft University of Technology) which qualified him to become a high school teacher. He taught in high schools for several years.

Korteweg was awarded a doctorate in 1878 at the University of Amsterdam for his dissertation *On the propagation of waves in elastic tubes* supervised by Johannes Van der Waals.

Korteweg's Ph.D. was the first one awarded by the University of Amsterdam.[3] Johannes van der Waals (1837-1923) received a Nobel Prize for Physics in 1910.

The "Korteweg Stress" is named after him.

He is honored by the Korteweg-de Vries equation on solitary waves. Many mathematicians, including George Stokes, were convinced such waves could not exist.

Gustav de Vries was a student of Korteweg. They found explicit, closed-form, travelling-wave solutions to the Korteweg-de Vries equation that decay rapidly. Korteweg also edited Huygens complete works from 1911 to 1927.

He also worked in pure mathematics on algebraic equations and properties of surfaces in the neighborhood of singular points.

Korteweg died on 10 May 1941 at the age of 93 in Amsterdam, Netherlands.

Luitzen Egbertus Jan Brouwer

Luitzen Egbertus Jan Brouwer was born on 27 February 1881 in Overschie (now a suburb of Rotterdam), Netherlands. He entered the University of Amsterdam in 1897 and received his doctorandus degree in 1904.

His first paper was on continuous motions in four dimensional space which was published by the Royal Dutch Academy of Science in Amsterdam in 1904.

In 1904 he married Lize de Holl who was eleven years his senior and had a daughter from a previous marriage. She later became a pharmacist.

Brouwer was awarded a Ph.D. in 1907 for a thesis on the logical foundations of mathematics. He continued to develop the ideas from his thesis in the book *The Unreliability of the Logical Principles* published in 1908.

He was elected to the Royal Dutch Academy of Science in 1912 and appointed extraordinary professor at the University of Amsterdam on Hilbert's recommendation. Brouwer was appointed ordinary professor in 1913.

Brouwer did not want any questions by students in his class. He was always looking at the blackboard, never towards the students. Although he was a major contributor to the theory of topology, he was no longer convinced of his results in topology because they were not correct from the point of view of intuitionism and he judged everything he had done before false according to his philosophy.[3]

Brouwer founded the mathematical philosophy of Intuitionism as an opponent to the then prevailing formalism of Hilbert. The intuitionism is simplistically characterized by saying that its adherents refuse to use the law of excluded middle in mathematical reasoning.

In late 1921s Brouwer came to Göttingen to deliver a talk on his ideas to the Mathematics Club. "You say that we can't know whether in the decimal representation of π ten 9's occur in succession," someone objected after Brouwer finished "Maybe we can't know – but God knows!" To this Brouwer replied dryly, "I do not have a pipeline to God."

After a lively discussion Hilbert finally stood up.

"With your methods," he said to Brouwer, "Most of the results of modern mathematics would have to be abandoned and to me the important thing is not to get fewer results but to get more results." He sat down to enthusiastic applause.[9]

Brouwer actively helped the Dutch resistance during World War II, and he supported Jewish students. In 1943, however, the Germans insisted that students sign a declaration of loyalty to Germany and Brouwer encouraged his students to do so.

After Amsterdam was liberated he was suspended from his post for a few months because of his actions on signing a declation of loyalty to Germany.[3]

After retiring in 1951 he lectured in South Africa, the United States and Canada.

He was elected a Fellow of the Royal Society of London in 1948 and many other academies.

He was made a Knight in the Order of the Dutch Lion in 1932.

Brouwer died on 2 December 1966 in Blaricum, Netherlands after being struck by a vehicle while crossing the street in front of his house.

Figure 254.   Bartel L. van der Waerden

Bartel Leendert van der Waerden

Bartel Leendert van der Waerden was born on 2 February 1903 in Amsterdam, Netherlands.

He was awarded a Ph.D. in 1926 for a thesis on the foundations of algebraic geometry supervised by Hendrik de Vries at the University of Amsterdam. He was heavily influenced by Emmy Noether at Göttingen where he received his habilitation in 1928.

When he was just 27 years old, he published his two-volume work *Algera* which synthesized research done by Emmy Noether, David Hilbert, Richard Dedekind and Emil Artin.

In 1931 he was appointed professor of mathematics at the University of Leipzig where he became a colleague of Werner Heisenberg.

He became a professor of mathematics at the University of Amsterdam in 1948. Then in 1951 moved to the University of Zurich where he spent rest of his career, supervising more than 40 Ph.D. students. He did a great deal of work in algebraic geometry, defining the notions of dimension of an algebraic variety, using the ideal theory in polynomial rings created by Artin, Hilbert and Emmy Noether. He also edited *Sources of Quantum Mechanics* in 1967.

Van der Waerden died on 12 January 1996 in Zurich, Switzerland.

David van Dantzig

David van Dantzig was born on 23 September 1900 in Rotterdam, Netherlands. After receiving a doctorandus degree at the University of Amsterdam he became an assistant to Schouten in 1927 at Delft University of Technology.

He was awarded a Ph.D. in 1931 at Groningen for his thesis *Studiën over topologische Algebra*. He became an extraordinary professor at Delft in 1938 and then an ordinary professor in 1940. In 1946, he was appointed professor at the University of Amsterdam, where he stayed for the rest of his life.[3]

Dantzig and Schouten founded the research and service institution, *Mathematisch Centrum* in Amsterdam in 1946. His most important work was in topological algebra. He also studied metrisation of groups, rings and fields.

Dantzig died on 22 July 1959 in Amsterdam, Netherlands.

Figure 255.   David van Dantzig

Figure 256.   Thomas J. Stieltjes

Thomas Jan Stieltjes

Thomas Stieltjes was born on 29 December 1856 in Zwolle, Overijssel, Netherlands. His father also named Thomas Stieltjes was a well-known civil engineer who constructed harbors in Rotterdam. A memorial statue of him stands on the Noordereiland at the Burgemeester Hoffman Plein in Rotterdam.

Stieltjes studied at the Polytechnic School in Delft in 1873. He did not attend lectures and, instead, he spent his time reading books by Gauss and Jacobi. He failed his examinations and as a result he was unable to graduate. Stieltjes began correspondence with Charles Hermite exchanging 432 letters over twelve years. He became assistant at Leiden Observatory in April 1877.

The director of Leiden Observatory, H. G. van der Sande-Bakhuyzen allowed Stieltjes to work more on mathematical issues than observational work.

In 1883 Stieltjes married Elizabeth Intveld who encouraged him to work in mathematics.

From September to December 1883 Stieltjes lectured on analytical geometry and descriptive geometry at Delft, substituting for F. J. van Berg who had become ill.

Stieltjes applied for a chair at Groningen in 1884, but he was rejected because he did not have a degree. In May 1884 Hermite attended the tricentennial anniversary of the founding of the University of Edinburgh where he convinced Professor Bierens de Haan from the Netherlands to award an honorary degree to Stieltjes.

As a result, Stieltjes received an honorary degree from the University of Leiden in June 1884. He moved to Paris in April 1885 and he eventually received his doctorate of science in 1886 for a thesis on asymptotic series.

Later that year he was appointed to a position the University of Toulouse, where he took a chair of differential and integral calculus in 1889. He became a naturalized French citizen.

His work on continued fractions earned him the Ormoy Prize of the *Académie des Sciences* in 1893. His name is honored in the Lebesgue-Stieltjes integral, Laplace-Stieltjes transform, Riemann-Stieltjes integral, Stieltjes moment problem and Chebyshev-Markov-Stieltjes inequalities.

He died on 31 December 1894 at the age of 38 in Toulouse, France.

Belgian School

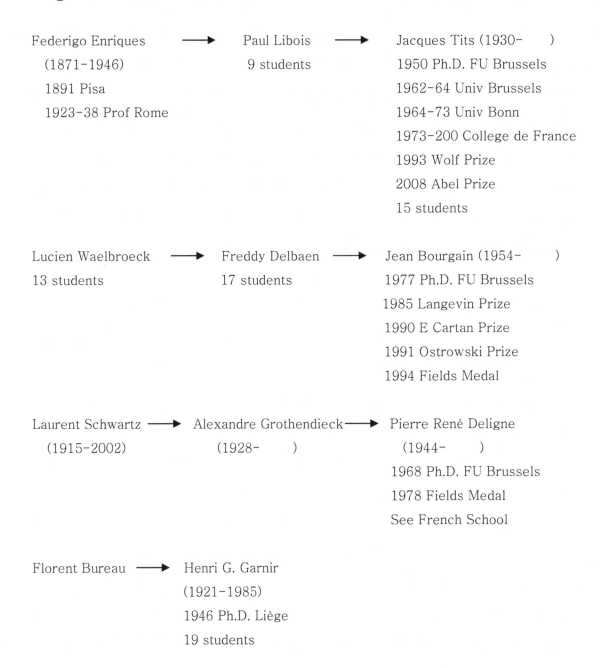

Federigo Enriques   ⟶   Paul Libois   ⟶   Jacques Tits (1930-   )
  (1871-1946)   9 students   1950 Ph.D. FU Brussels
  1891 Pisa   1962-64 Univ Brussels
  1923-38 Prof Rome   1964-73 Univ Bonn
                                     1973-200 College de France
                                     1993 Wolf Prize
                                     2008 Abel Prize
                                     15 students

Lucien Waelbroeck   ⟶   Freddy Delbaen   ⟶   Jean Bourgain (1954-   )
13 students   17 students   1977 Ph.D. FU Brussels
                                     1985 Langevin Prize
                                     1990 E Cartan Prize
                                     1991 Ostrowski Prize
                                     1994 Fields Medal

Laurent Schwartz   ⟶   Alexandre Grothendieck   ⟶   Pierre René Deligne
  (1915-2002)   (1928-   )   (1944-   )
                                     1968 Ph.D. FU Brussels
                                     1978 Fields Medal
                                     See French School

Florent Bureau   ⟶   Henri G. Garnir
                           (1921-1985)
                           1946 Ph.D. Liège
                           19 students

Jacques Tits

Jacques Tits was born on 12 August 1930 in Uccle, Belgium. He received his doctorate in 1950 at the Free University of Brussels for his dissertation *Généralisation des groupes projectifs bases sur la notion de transitivité* under the supervision of Paul Libois.[3]

Tits married historian Marie-Jeanne Dieuaide in September 1956. He was a professor at the University of Brussels from 1962 to 1964, and a professor at the University of Bonn from 1964 to 1973. Tits was appointed the Chair of Group Theory at the Collège de France in 1973 and retired in 2000. He became a naturalized French citizen in 1974.[3]

After retiring in 2000, he became the first holder of the Vallée-Poussin Chair from the University of Louvain. Tits received many honors including the Grand Prix of the French Academy of Sciences in 1976, the Wolf Prize in Mathematics in 1993 and the Cantor Medal from the German Manthematical Society in 1996. He was elected to many academies including the Royal Belgium Academy of Science in 1991 and the National Academy of Sciences of the United States in 1992. Tits also received the Abel Prize in 2008.

Figure 257. Jean Bourgain

Jean Bourgain

Jean Bourgain was born on 28 February 1954 in Ostende, Belgiam. He received a Ph.D. in 1977 and his habilitation in 1979 from the Free University of Brussels. Bourgain received the highest science honor from Belgium, the Damry-Deleeuw-Bourlart Prize in 1985. He received the Langevin Prize in 1985 and the Elie Cartan Prize in 1990 from the French Academy of Sciences.[3] He also received the Ostrowski Prize in 1991 from the Ostrowski Foundation in Switzerland. At the International Congress of Mathematicians in Zurich in 1994, Bourgain received a Fields Medal. Caffarelli described Bourgain's work.[3]

Bourgain's work touches on several central topics of mathematical analysis: the geometry of Banach Spaces, convexity in high dimensions, harmonic analysis, ergodic theory, and finally, nonlinear partial differential equations from mathematical physics.

Since 1994 he has been at the Institute for Advanced Study at Princeton.

Austrian School

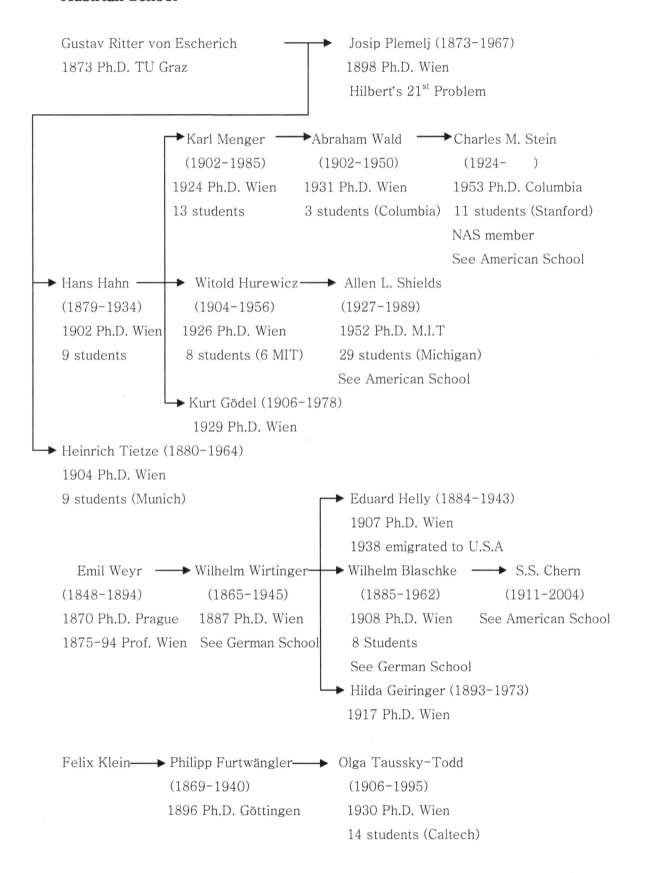

Gustav Ritter von Escherich
1873 Ph.D. TU Graz

Josip Plemelj (1873-1967)
1898 Ph.D. Wien
Hilbert's 21st Problem

Karl Menger
(1902-1985)
1924 Ph.D. Wien
13 students

Abraham Wald
(1902-1950)
1931 Ph.D. Wien
3 students (Columbia)

Charles M. Stein
(1924-   )
1953 Ph.D. Columbia
11 students (Stanford)
NAS member
See American School

Hans Hahn
(1879-1934)
1902 Ph.D. Wien
9 students

Witold Hurewicz
(1904-1956)
1926 Ph.D. Wien
8 students (6 MIT)

Allen L. Shields
(1927-1989)
1952 Ph.D. M.I.T
29 students (Michigan)
See American School

Kurt Gödel (1906-1978)
1929 Ph.D. Wien

Heinrich Tietze (1880-1964)
1904 Ph.D. Wien
9 students (Munich)

Eduard Helly (1884-1943)
1907 Ph.D. Wien
1938 emigrated to U.S.A

Emil Weyr
(1848-1894)
1870 Ph.D. Prague
1875-94 Prof. Wien

Wilhelm Wirtinger
(1865-1945)
1887 Ph.D. Wien
See German School

Wilhelm Blaschke
(1885-1962)
1908 Ph.D. Wien
8 Students
See German School

S.S. Chern
(1911-2004)
See American School

Hilda Geiringer (1893-1973)
1917 Ph.D. Wien

Felix Klein

Philipp Furtwängler
(1869-1940)
1896 Ph.D. Göttingen

Olga Taussky-Todd
(1906-1995)
1930 Ph.D. Wien
14 students (Caltech)

Austrian School

Austrian degree system has been identical to Germany. Before 1940s, the first degree at universities was doctorate. The Bachelor and Master degrees did not exist. Habilitation thesis was required to teach at universities.

However, at the Techniche Hochschule, the degree of Diplom was awarded after five years study.

Hans Hahn

Hans Hahn was born on 27 September 1879 in Vienna, Austria. He was awarded a Ph.D. in 1902 at the University of Vienna for his thesis *Zur Theorie der Zweiten Variation einfacher Integrale* supervised by Gustav Ritter von Escherich. While in Vienna Hahn was one of the "inseparable four", along with Paul Ehrenfest, Heinrich Tietze and Gustav Herglotz.

Hahn became a privatdozent in 1905 at Vienna after submitting his habilitation thesis. He was an extraordinary professor at Czernowitz (now Cernauti in Romania) from 1909 to 1914.

He was severely wounded in World War I while serving in the Austro-Hungarian Army.

Hahn was an extraordinary professor at Bonn from 1917 to 1920. He was appointed to a chair at the University of Vienna in 1921.

His three most famous students at Vienna were Karl Menger, Witold Hurewicz and Kurt Gödel, all of whom moved to the United States.

Hahn was a pioneer in set theory and functional analysis. The Hahn sequence space, Hahn-Banach theorem and Hahn-Mazurkiewicz theorem are named after him.

Hahn was also a mathematical philosopher. During the 1920s Hahn, together with Philipp Frank, Otto Neurath, and Moritz Schlick, founded the Vienna Circle of Logical Positivists a discussion group of gifted scientists and philosophers.[3]

He was elected to the *Kaiserliche Akademie der Wissenschaften in Vienna.*

Hahn died on 24 1934 at the age of 54 in Vienna, Austria.

Karl Menger

Karl Menger was born on 13 January 1902 in Vienna, Austria. He was awarded a Ph.D. in 1924 at the University of Vienna for his thesis *Über die Dimensionalität von Punktmengen* under Hahn's supervision. Menger completed his thesis in a sanatorium due to lung disease.

He worked with Brouwer in Amsterdam in 1925 and 1926. In 1927 Menger was appointed to the chair of geometry at the University of Vienna succeeding Reidemeister.

In 1938 Menger moved to the University of Notre Dame in the United States, then in 1948 he went to the Illinois Institute of Technology.

At Notre Dame Menger organized a mathematical colloquium modelled after the Vienna circle, but it failed to become as influential as the Vienna Circle had been.[3]

Menger died on 5 October 1985 in Chicago, Illinois.

Figure 258. Hans Hahn   **Figure 259. Karl Menger**

Kurt Gödel

Kurt Gödel was born on 28 April 1906 in Brünn, Austro-Hungarian Empire (now Brno, Czech Republic). He had mastered university mathematics and was proficient in Latin by his final Gymnasium years.

Gödel was awarded a Ph.D. at the University of Vienna in 1929 for his thesis proving the completeness of the first order functional calculus supervised by Hahn.

When he was 24 years old, Gödel solved the second Hilbert problem.

His habilitation thesis was accepted by Hahn, and Gödel became a privatdozent at the University of Vienna in March 1933. In 1936 Moritz Schlick, whose seminar had aroused Gödel's interest in logic, was murdered by a Nazis student, and Gödel had a nervous breakdown.

In 1938 he married Adele Porkert whom he had met in a night club in 1927. She was six years older than Gödel and had been previously married.

Although he was not Jewish, he was once attacked by a gang of Vienna youths who believed that he was Jewish. Gödel and his wife moved to the United States in 1940 and he became a US citizen in 1948. In the naturalization examination he had found an inconsistency in the United States Constitution so that the United States could be transformed into a dictatorship.[12]

When Gödel was trying to argue with the judge, Oskar Morgenstern who went to the court with him, kicked Gödel under the table.[35] Einstein also went together with them, and the judge enjoyed talking with Einstein.[12]

Gödel was an ordinary member of the Institute for Advanced Study at Princeton from 1940 to 1946, meaning his membership status was reviewed every year. He became a permanent member in 1946, then he became a professor in 1953.

His most important work was *Consistency of the Axiom of Choice and of the Generalized Continuum-Hypothesis with the Axioms of Set Theory* published in 1940.

He received the Einstein Award in 1951 and the US National Medal of Science in 1974. He was elected to the National Academy of Sciences, the Royal Society of London and the Institute of France.

However, he refused membership in the Academy of Sciences in Vienna and he refused to accept the highest National Medal for scientific and artistic achievement that Austria offered him because he resented his difficult life in Austria.[3]

When Paul Cohen proved the independence of the continuum hypothesis from the other axioms of set theory, he took his manuscript to Gödel in Princeton. Gödel cracked the door open and grabbed the manuscript. Several days later, Gödel pronounced the result correct and invited Cohen for tea.[35]

Gödel suffered from a duodenal ulcer, so he maintained an extremely strict diet consisting of butter, baby food and laxatives. Towards the end of his life, he became convinced that he was being poisoned. He refused to eat as a result, essentially starved to death.

Gödel died sitting in a chair in his Princeton hospital room on 14 January 1978.

Witold Hurewicz

Witold Hurewicz was born into a Jewish family on 29 June 1904 in Lodz, Russian Empire (now Poland). Instead of going to University of Warsaw, he studied at the University of Vienna. He was awarded a Ph.D. in 1926 for his thesis *Über eine Verallgemeinerung des Borelschen Theorems* supervised by Hans Hahn.[10]

Hurewicz was an assistant to Brouwer in Amsterdam from 1928 until 1936. He took a one year leave to study in the United States. Because of the impending war in Europe he decided to remain in the United States.

He worked on servomechanism at the University of North Carolina at Chapel Hill during World War II. From 1945 until his death in 1956 he worked at the Massachusettes Institute of Technology. Hurewicz is best remembered for two remarkable contributions, his discovery of the higher homotopy groups in 1935-1936 and his discovery of exact sequences in 1941.

He was very absentminded, and this probably contributed to his death. He attended the International Symposium on Algebraic Topology in Mexico, and during an outing on 6 September 1956, he fell off of a Mexican pyramid (ziggurat) and died.[5]

Figure 260. Kurt Gödel

Figure 261. Witold Hurewicz

Josip Plemelj

Josip Plemelj was born on 11 December 1873 in Bled, Slovenia. After graduating from high school in Ljubljana he entered the University of Vienna in 1894. He was awarded a Ph.D. in 1898 for his thesis *Über lineare homogene Differentialgleichungen mit eindeutigen periodischen Koeffizienten* under von Escherich's supervision. He studied with Frobenius and Fuchs in Berlin in 1899 and 1900, and then he went to Göttingen to work with Klein and Hilbert. Plemelj made a major contribution to integral equations and potential theory. For this work he was awarded the Prince Jablonowski Prize.

From 1902 to 1906 he was a privatdozent at the University of Vienna. After one year at the Technical University of Vienna he moved to the University of Chernivtsi in south-western Ukraine in 1907. In 1908 he was appointed a full professor there.

In 1919 the University of Ljubljana reopended as a Slovene University and Plemelj was appointed Professor of Mathematics and its first Rector. He retired in 1957 when he was 83 years old. The University of Ljubljana awarded him an honorary doctorate in 1963 on the occasion of his 90th birthday.[3]

Plemelj published a book in English in 1964, *Problems in the Sense of Riemann and Klein*, that is in part about his solution of Hilbert's 21st problem.

He died on 22 May 1967 in Ljubljana, Slovenia thinking he had solved the problem.[12]

Andrei Bolibruch (1950-2003) discovered a counterexample to Plemelj's result in 1989.[12]

Figure 262.   Josip Plemelj

Figure 263.   Abraham Wald

Abraham Wald

Abraham Wald was born into a Jewish family on 31 October 1902 in Kolozsvár, Hungary (now Cluj, Romania). He entered the University of Vienna in 1927 and was awarded his doctorate on geometry in 1931 under Karl Menger's supervision.

Vienna was a difficult place for even an extremely talented Jewish man to obtain an academic position, so he became a mathematics tutor to Karl Schlesinger, a leading Austrian banker and economist. He consequently developed an interest in mathematical applications to economics and econometrics. Wald published 10 papers on economics and econometrics as well as an important monograph on seasonal movements in time series in 1936.

In the summer of 1938 he moved to the United States and was appointed to the faculty of Columbia University in 1941. He invented the topic of sequential analysis in response to the demand for more efficient methods of industrial quality control during World War II.

He was the first to solve the general problem of sequential tests of statistical hypothesis.[3]

In 1950 Wald received an invitation from the Indian government to lecture on statistics.

He and his wife were tragically killed in a plane crash on 13 December 1950 in Travancore, India.

Figure 264.   Emil Weyr

Emil Weyr

Emil Weyr was born on 1 July 1848 in Prague, Bohemia (now Czech Republic). He studied at the Prague Polytechnic and also at the Charles University in Prague where he received his doctorate in 1870. Charles University (Universitas Carolina in Latin) was founded in 1347, the second oldest university after Heidelberg in the German speaking countries.

In 1875 he was appointed professor of mathematics at the University of Vienna, and he stayed there for the rest of his life. His main field was geometry and he wrote *Die Element der projectivischen Geometrie* and *über die Geometrie der alten Aegypter.*[3]

Weyr was one of the first 19 founding members of the Royal Czech Academy of Sciences in 1891.

Weyr died on 25 January 1894 in Vienna, Austria.

Hilda Geiringer von Mises

Hilda Geiringer was born into a Jewish family on 28 September 1893 in Vienna, Austria.

She received a doctorate in 1917 at the University of Vienna for a thesis on Fourier series in two variables under Wirtinger's supervision.

In 1921 Geiringer moved to Berlin to become an assistant to Richard von Mises in the Institute of Applied Mathematics at the University of Berlin. In the same year she married Felix Pollaczek who obtainded his Ph.D in 1922 under Issai Schur in Berlin.

Geiringer and Pollaczek had a daughter but their marriage broke up in 1922. After that Geiringer continued working for von Mises.

After Hitler came to power in 1933 von Mises left Germany for Istanbul and Geiringer followed him in 1934. She was appointed professor of mathematics at the University of Istanbul. In 1939 von Mises went to the United States, and Geiringer again followed him.

She became a lecturer at Bryn Mawr College, and von Mises was appointed a full professor at Harvard.

Around this time there was a rumor in Cambridge, Massachusetts that von Mises had a mistress and Norbert Wiener considered to call the FBI.[35]

Geiringer and von Mises married in 1943. She accepted a post as professor and chairman of the mathematics department at Wheaton College in Norton, Massachusetts.

There were only two members in the department at the time, and Geiringer was anxious to carry out research.

Von Mises died in 1953 and Geiringer began to edit von Mises works. In 1956 Humboldt University (former University of Berlin) elected her as professor emeritus. She had been discriminated against in Germany because of her Jewish background and had difficulties in the United States because she was a woman.

Geiringer died on 22 March 1973 in Santa Barbara, California.

Eduard Helly

Eduard Helly was born into a Jewish family on 1 June 1884 in Vienna, Austria. He received a Ph.D. in 1907 at the University of Vienna for a thesis on Fredholm equations under the direction of Wirtinger and Mertens. He also studied at Göttingen under Hilbert, Klein, Minkowski and Runge in 1907 and 1908.

Returning to Vienna in 1908, he taught in a Gymnasium and gave private lessons. He proved the Hahn–Banach theorem in 1912, fifteen years before Hahn published essentially the same proof.[3]

While serving as a lieutenant in September 1915, he was shot. The bullet went through his lung and he was captured by the Russians. He was in prisoner of war camps in Siberia and came home in 1920.[3]

Helly married to Elise Bloch in 1921 who also had a doctorate in mathematics from Vienna. In the same year he received his habilitation and he was appointed to an unpaid post in Vienna.

Helly and his family emigrated to the United States in 1938. He taught at Junior Colleges in New Jersey from 1939 to 1942. Helly and his wife were employed by the US Army Signal Corps as mathematicians in 1942. He died on 28 November 1943 in Chicago, Illinois after a heart attack.

He is remembered for Helly's theorem, Helly's selection principle, and the uniform boundedness principle for linear functionals.

Figure 265. Hilda Geiringer von Mises

Figure 266. Heinrich F. F. Tietze

Heinrich Franz Friedrich Tietze

Heinrich Tietze was born on 31 August 1880 in Schleinz, Austria. He entered *Technische Hochschule* in Vienna in 1898. At Vienna he was one of the 'inseparable four', the others being Paul Ehrenfest, Hans Hahn and Gustav Herglotz.

Tietze was awarded a Ph.D. in 1904 at the University of Vienna for his thesis *Funktional Gleichungen, deren Lösungen keiner algebraischen Differentialgleichung genügen* under von Esherich's supervision. Tietze received his habilitation in 1908 for his thesis on a topological topic considering topological invariants. In 1910 he was an extraordinary professor at TH Brünn (now Brno), and three years later he was promoted to ordinary professor.

In 1919 he was appointed to a chair of mathematics (Zweiter Lehrstuhl) at the University of Erlangen succeeding Max Noether.

Then in 1925 he was appointed to a chair of mathematics (Zweiter Lehrstuhl) at the University of Munich. He retired from there in 1950.

Tietze contributed to the foundations of general topology and developed important work on subdivision of cell complexes. He wrote six books and 104 papers, most of them in Munich.[3]

He died on 17 February 1964 in Munich, Germany.

Philipp Furtwängler

Philipp Furtwängler was born on 21 April 1869 in Elze, Germany. He received a Ph.D. in 1896 at the University of Göttingen for his thesis *Zur Theorie der in Linearfaktoren zerlegbaren ganzzahligen ternären Kubischen Formen* under Felix Klein's supervision.

From 1904 to 1907 and again from 1910 to 1912 he was a professor of mathematics at the *Landwirtschaftlichen Akademie* in Bonn. From 1907 to 1910 he held a chair of mathematics (Zweiter Lehrstuhl) at the *Technische Hochschule Aachen*.

In 1912 he was appointed to a chair of mathematics at the University of Vienna, and he stayed there for the rest of his life.

He was paralyzed, and taught from a wheelchair while an assistant wrote on blackboard.

Furtwängler died on 19 May 1940 in Vienna, Austria.

Olga Taussky-Todd

Olga Taussky-Todd was born into a Jewish family on 30 August 1906 in Olmütz, Austro-Hungarian Empire (now Olomouc, Czech Republic).

In 1930 she was awarded a Ph.D. at the University of Vienna for her thesis *Über eine Verschärfung des Hauptidealsatzes* supervised by Furtwängler.

She worked as an assistant at Göttingen editing the first volume of Hilbert's complete works on number theory. She also edited Artin's lectures in class field theory, and assisted Emmy Noether and Richard Courant.

Figure 267. Olga Taussky-Todd

She received a three-year research fellowship with a generous stipend of 300 pounds a year at Girton College, Cambridge in 1935.[94] A research fellow has a great freedom for research with no teaching obligation.

In London she met an Irishman John (Jack) Todd (1911-2007) and they were married on 29 September 1938. There was another John Todd (1908 -1994), a mathematician at Cambridge who was the teacher of Roger Penrose.

Todd and his wife Olga wrote six mathematical papers during World War II while sitting in London bomb shelters.

After the war Jack Todd went to Germany, where he examined computers made by Konrad Zuse (1910-1995). He also went to the mathematics research centre at Oberwolfach in the Black Forest run by Wilhelm Süss. Todd persuaded Moroccan troops who arrived shortly after he did, not to destroy the building and the books. For this, Jack Todd became known as "The Savior of Oberwolfach".[3]

Olga and Jack accepted appointments at the California Institute of Technology in 1957.

Olga was a research associate and then was promoted to full professor in 1971. Jack developed the first undergraduate courses in numerical analysis and numerical algebra, prerequisites to learning computing.

In 1977 she became Professor Emeritus. Her most influential work was in the field of matrix theory and number theory.

In 1963 she was given the "Woman of the Year" award by the Los Angeles Times.[94]

She received the Cross of Honor in Science and Arts, First Class in 1978 from the Austrian government. She was Vice-President of the American Mathematical Society in 1985.

Jack published the classic two volume work *Basic Numerical Mathematics* in 1977.

Olga Taussky-Todd died on 7 October 1995 in Pasadena, California and Jack Todd died on 21 June 2007 in Pasadena.

Japanese School

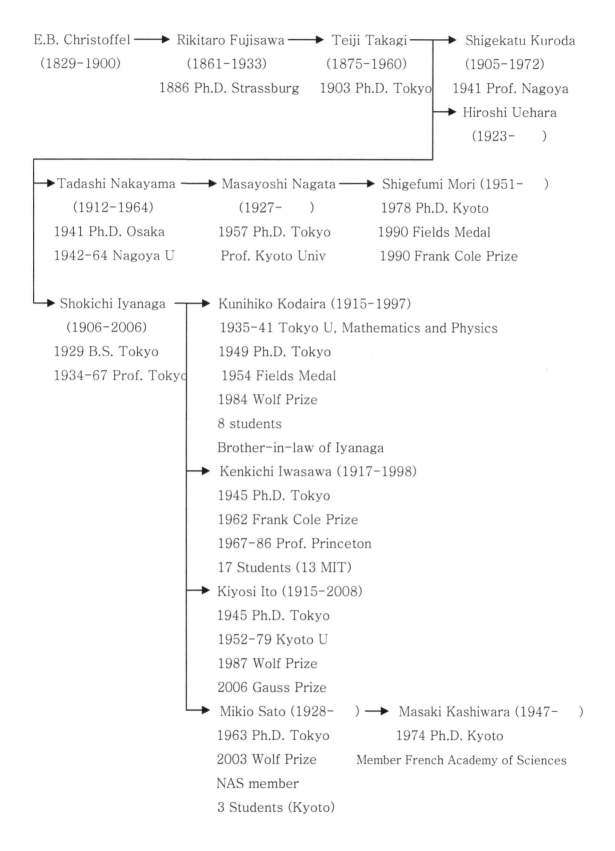

E.B. Christoffel ⟶ Rikitaro Fujisawa ⟶ Teiji Takagi ⟶ Shigekatu Kuroda
(1829-1900)   (1861-1933)   (1875-1960)   (1905-1972)
                 1886 Ph.D. Strassburg 1903 Ph.D. Tokyo 1941 Prof. Nagoya

⟶ Hiroshi Uehara
   (1923-   )

⟶ Tadashi Nakayama ⟶ Masayoshi Nagata ⟶ Shigefumi Mori (1951-   )
   (1912-1964)   (1927-   )   1978 Ph.D. Kyoto
   1941 Ph.D. Osaka   1957 Ph.D. Tokyo   1990 Fields Medal
   1942-64 Nagoya U   Prof. Kyoto Univ   1990 Frank Cole Prize

⟶ Shokichi Iyanaga ⟶ Kunihiko Kodaira (1915-1997)
   (1906-2006)   1935-41 Tokyo U, Mathematics and Physics
   1929 B.S. Tokyo   1949 Ph.D. Tokyo
   1934-67 Prof. Tokyo 1954 Fields Medal
                        1984 Wolf Prize
                        8 students
                        Brother-in-law of Iyanaga

                     ⟶ Kenkichi Iwasawa (1917-1998)
                        1945 Ph.D. Tokyo
                        1962 Frank Cole Prize
                        1967-86 Prof. Princeton
                        17 Students (13 MIT)

                     ⟶ Kiyosi Ito (1915-2008)
                        1945 Ph.D. Tokyo
                        1952-79 Kyoto U
                        1987 Wolf Prize
                        2006 Gauss Prize

                     ⟶ Mikio Sato (1928-   ) ⟶ Masaki Kashiwara (1947-   )
                        1963 Ph.D. Tokyo   1974 Ph.D. Kyoto
                        2003 Wolf Prize   Member French Academy of Sciences
                        NAS member
                        3 Students (Kyoto)

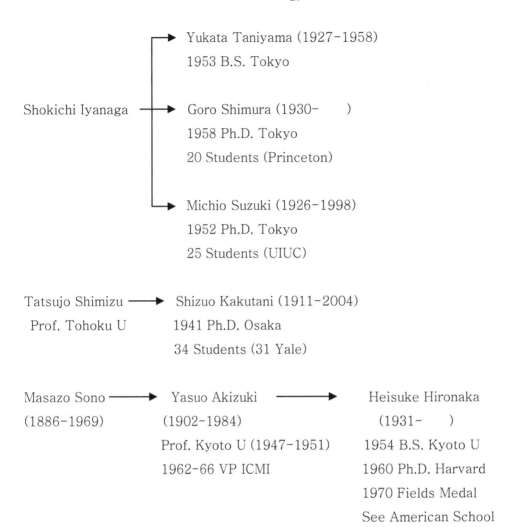

Shokichi Iyanaga ⟶ Yukata Taniyama (1927–1958)
1953 B.S. Tokyo

Goro Shimura (1930–   )
1958 Ph.D. Tokyo
20 Students (Princeton)

Michio Suzuki (1926–1998)
1952 Ph.D. Tokyo
25 Students (UIUC)

Tatsujo Shimizu ⟶ Shizuo Kakutani (1911–2004)
Prof. Tohoku U   1941 Ph.D. Osaka
34 Students (31 Yale)

Masazo Sono ⟶ Yasuo Akizuki   ⟶ Heisuke Hironaka
(1886–1969)   (1902–1984)   (1931–   )
Prof. Kyoto U (1947–1951)   1954 B.S. Kyoto U
1962–66 VP ICMI   1960 Ph.D. Harvard
1970 Fields Medal
See American School

Japanese School

The Japanese degree system before 1945 was different from those in western countries. Graduates of prestigious three-year high school could enter universities through a highly competitive entrance examinations. The Japanese high schools before 1945 were comparable to the final three years of French Lycée and the first two years of college in the United States.[12]

After three years of study at a university, students were awarded bachelor's degrees.

Graduate of four-year middle school could enter three-year colleges. Therefore three-year college graduates were roughly equivalent to high school graduates, but the latter had advantages for further studies at universities.

Doctoral degrees were awarded to scholars with many years of research experiences. Doctoral candidates were not required to register at universities and there were no official advisers or Doktor-Vaters. Even those without bachelor's degrees were permitted to submit doctoral dissertations to a university.

There was no Ph.D. degrees. Scientists received Doctor of Science, physicians Doctor of Medicine, lawyers Doctor of Law, and engineers Doctor of Engineering.

Consequently it was very rare to see young persons in their twenties with doctorates in Japan. Even many professors did not have doctorates.

Before 1945 there were only seven imperial universities in Japan, and the most prestigious universities were Tokyo and Kyoto. The others were Osaka, Nagoya, Tohoku, Kyushu and Hokkaido.

After World War II, Japan adopted a new system similar to that in the United States. However, it retained the "Ronbun Hakushi" (doctorate by dissertation) system, which allows the awarding of doctorates by submission of dissertation without registering or taking courses at universities.

The professorial ranks are, from lowest to highest, instructor - assistant professor - professor - modeled after the German privatdozent - extraordinary professor - ordinary professor.

Teiji Takagi (髙木貞治)

Teiji Takagi was born on 21 April 1875 in Kazuya Village, near Gifu, Japan. He entered the Third High School in Kyoto in 1891 where many bright pupils studied to prepare for university education. The mathematics teacher there was Jittaro Kawai who studied under Klein and Weber in Germany. Takagi graduated in 1894 and entered the Tokyo Imperial University after passing an extremely competitive entrance examination. He read Serret's *Algébre Supérieure* and Heinrich Weber's *Algebra* during his undergraduate years.

Takagi graduated from Tokyo Imperial University in 1897. In the following year he was chosen by the Government as one of the twelve students to study abroad. Professors Rikitaro Fujisawa (1861–1933) and physicist Hantaro Nagaoka (1865–1950) saw him off as he departed from a Yokohama wharf on the French ship Messageries Maritimes to Europe.[67]

He studied for three semesters at the University of Berlin, attending lectures by Fuchs, Frobenius and Schwarz. Then he went to Göttingen and told Hilbert that he wanted to study algebraic number theory. Takagi stayed in a house at 15 Kreuzbergweg (now called Kreubergring) where Hilbert had been living before he moved to 29 Wilhelm-Weber Strasse.

Takagi returned to Japan in December 1901 and was appointed an assistant professor in algebra at the Tokyo Imperial University. Takagi married Tochi Tani in April 1902 and they had three sons and five daughters.[3]

He was awarded Doctor of Science (Rigakuhakushi) degree in December 1903 at Tokyo with his thesis he had worked in Göttingen.

He was promoted to a professor in May 1904.

Rikitaro Fujisawa, who had been his teacher at Tokyo Imperial University, received a Ph.D. in 1886 at Strassburg under Christoffel's supervision, probably the first Japanese Ph.D. in mathematics. Fujisawa became a professor at Tokyo Imperial University and he was a pioneer of life insurance and pension theory in Japan.

Kikuchi Dairoku (1855–1917) was Takagi's another teacher. He graduated from St. John's College, Cambridge and London University. He taught at the Tokyo Imperial University from 1877 to 1898. Kikuchi served as President of Tokyo Imperial University (1898–1901) and Minister of Education (1901–1903).

In 1903 Takagi proved the Kronecker conjecture on abelian extension of imaginary number fields. He wrote his most important paper in 1920, introducing the Takagi class-field theory generalizing Hilbert's class field. Hilbert published that paper in *Mathematische Annalen* in 1925.

Takagi's class field theory was central to Artin's solution of Hilbert's 9th problem and is part of the context of Hasse's solution of the Hilbert's 11th problem.

Takagi's class field theory is also fundamental to progress made on the twelfth problem.[12]

Takagi was Vice-President of the International Congress of Mathematicians at Zurich in 1932. He also served on the committee to award the first Fields Medals for the 1936 Congress. He retired from Tokyo Imperial University on 31 March 1936.

He wrote 13 texts and became a father of modern Japanese mathematics. His wife died of lung cancer in 1952. He died of a stroke on 28 February 1960 at the hospital of Tokyo Imperial University.

Figure 268.   Teiji Takagi

Figure 269.   Shokichi Iyanaga

Shokichi Iyanaga (彌永昌吉)

Shokichi Iyanaga was born on 2 April 1906 in Tokyo, Japan. He entered Tokyo Imperial University in 1926 and studied under Takagi and Yosiye. Iyanaga published three papers while he was an undergraduate and he graduated in 1929.

He studied for a total of three years in Paris and Hamburg where he studied with Artin. Iyanaga was appointed assistant professor in 1934 at Tokyo Imperial University and promoted to professor in 1940.

He wrote a paper solving a question of Artin on generalizing the principal ideal theorem which was published in 1939 to honor Philipp Furtwängler.

After retiring from his Chair at Tokyo Imperial University in 1967, he worked at Gakushuin Uiversity for ten more years.[3] It is customary in Japan for a famous professor from a national university to continue to work at a private university after retiring.

Iyanaga had many excellent students who became famous mathematicians; including Kodaira, Iwasawa, Ito and Sato. Iyanaga was elected to the Japan Academy in 1978 and received the Order of *Legion d'Honneur* from France in 1980.

He died on 1 June 2006 at Seibo Hospital in Tokyo at the age of 100.

Tadashi Nakayama (中山正)

Tadashi Nakayama was born on 26 July 1912 in Tokyo, Japan.[99]

He was a student of Takagi at Tokyo Imperial University where he graduated in 1935 receiving his bachelor's degree (Rigakushi). Later that year, he was appointed assistant to K. Shoda at Osaka Imperial University and was then promoted to assistant professor in 1937.

He was at the Institute for Advanced Study at Princeton from 1937 to 1939, greatly influenced by Hermann Weyl and Richard Brauer.

In 1914 he was awarded a Doctor of Science (Rigakuhakushi) at Osaka Imperial University for his thesis On Frobenius Algebras, I and II. He moved to Nagoya Imperial University in 1942 and became a full professor there in 1944.

In 1953 he received a Japan Academy Prize in recognition of his research on the theory of rings and representations.

He was elected a member of the Japan Academy in 1963. Nakayama was one of the founders and the Editor-in-Chief of the Nagoya Mathematical Journal.

He wrote six mathematical books in Japanese including *Local Class Field Theory* (1935), *Sets, Topology and Algebraic System* (1949) and *Homological Algebra* (1957).

He died on 5 June 1964 in Nagoya after a long and fatal illness.

Sigekatu Kuroda (黒田成勝)

Sigekatu Kuroda was born on 11 November 1905. He graduated from Tokyo Imperial University in 1928 receiving a bachelor degree and received his Doctor of Science degree in 1945 at the same university.

He was appointed a professor in 1942 at Nagoya Imperial University, and he remained there until 1962. He then moved to the University of Maryland, College Park to a special chair in the Department of Mathematics.

In 1955 and 1956 he was at the Institute for Advanced Study at Princeton where Gödel took an interest in Kuroda's work. Beginning in the early sixties he became interested in the use of computers in number theory. He died on 3 November 1972. At the time of his death he was Professor Emeritus at both Maryland and Nagoya.[100]

Masayoshi Nagata (永田雅宜)

Masayoshi Nagata was born on 9 February 1927 in Obu, a town south of Nagoya, Japan. He graduated from Nagoya University in 1950 where Tadashi Nakayama (1912-1964) was his teacher. Nagata was appointed lecturer at Kyoto University in 1953 and received his Doctor of Science degree at Kyoto University in 1957.

In 1953 Artin came to Kyoto and asked Nagata what he was working on. On hearing Nagata's reply, Artin asked him if he knew of Hilbert's 14th problem. Nagata realized what he was working on was very similar to the 14th problem. He was a research fellow at Harvard in 1959 when he found his specific counterexample of the 14th problem.[12]

One of his students at Kyoto University, Shigefumi Mori, won the 1990 Fields medal.

Nagata's conjecture on curves concerns the minimum degree of a plane curve specified to have given multiplicities at given points. Nagata's conjecture on automorphisms concerns the existence of wild automorphisms of polynomial algebras. Recent work by I.P. Shestakov and U.U. Umirbaev in 2004 apparently solved this latter problem.[61]

Shigefumi Mori (森重文)

Shigefumi Mori was born on 23 February 1951. He received B.A in 1973. M.A in 1975 and Ph.D. in 1978 at Kyoto University. His Ph.D. dissertation was *The Endomorphism Rings of Some Abelian Varieties* under Masayoshi Nagata's supervision.

Mori was a faculty member at the University of Nagoya from 1980 to 1990. But he spent much time in the United States from 1977 until 1989 at Harvard, Columbia, Institute for Advanced Study at Princeton and the University of Utah.

In 1990 Mori was appointed to a chair at Kyoto University. Mori was awarded the Fields Medal at the 1990 International Congress of Mathematicians at Kyoto for his work in algebraic geometry developing the Minimal Model Program or Mori Program in connection with the classification problems of algebraic varieties of dimension three.

That same year Mori was awarded the Cole Prize in Algebra from the American Mathematical Society.

Figure 270. Shigefumi Mori   **Figure 271. Kenkichi Iwasawa**

Kenkichi Iwasawa (岩澤健吉)

Kenkichi Iwasawa was born on 11 September 1917 in Shinshuku-mura, Gumma Prefecture, Japan. He graduated from Tokyo Imperial University in 1940 and received his Doctor of Science (Rigakuhakushi) degree in 1945. As mentioned already, there was no formal doctoral adviser in Japan. Iyanaga was most likely in the committee that evaluated Iwasawa's dissertation and the candidate need not to defend his or her dissertation.

Iwasawa was seriously ill with pleurisy and this prevented him returning to Tokyo University until April 1947. From 1949 to 1955 he was an assistant professor at Tokyo University.

He spent two years (1950–1952) at the Institute for Advanced Study at Princeton, where he changed his research direction to algebraic number theory. From 1952 to 1967 he was a professor at Massachusetts Institute of Technology.

And from 1967 to 1986 he held the Henry Burchard Fine Chair of Mathematics at Princeton University.

He received both the Prize of the Academy of Japan and the Cole Prize from the American Mathematical Society in 1962. Iwasawa is best known for introducing what is now known as Iwasawa theory, whose central goal is to seek analogies between algebraic number fields and algebraic function fields in one variable over a finite field. The Iwasawa conjecture became known as "the main conjecture on cyclotomic fields" and it remained one of the most outstanding conjectures in algebraic number theory until it was solved by Barry Mazur and Andrew Wiles in 1984.[3]

Iwasawa returned to Japan after retiring from Princeton in 1986. He died on 26 October 1998 in Tokyo, Japan.

Figure 272. Kiyoshi Ito   **Figure 273. Kunihiko Kodaira**

Kiyoshi Ito (伊藤清)

Kiyoshi Ito was born on 7 September 1915 in Hokusei-cho, Mie Prefecture, Japan. After graduating from Tokyo Imperial University in 1938 he worked for five years at the Cabinet Statistics Bureau.

It was during this period that he made his most outstanding contributions.

Introducing the concept of regularization, developed by Joseph Leo Doob (1910-2004), Ito finally devised stochastic differential equations. In 1940 Ito published *On the Probability Distribution on a Compact Group* jointly with Yukiyosi Kawada and in 1942 *On Stochastic Processes* in the Japanese Journal of Mathematics.[3]

He was appointed assistant professor at Nagoya Imperial University in 1943 and he was awarded Doctor of Science degree in 1945 at Tokyo University. Iyanaga was Ito's undergraduate teacher but not a doctoral adviser.

In 1952 Ito was appointed to a chair at Kyoto University where he stayed until 1979. He also held professorship at Stanford University from 1961 to 1964, University of Aarhus in Denmark from 1967 to 1969 and Cornell University from 1969 to 1975.

After retiring from Kyoto University in 1979 he was appointed at Gakushuin University. It is customary when a renowned professor at a National University retires, he or she is appointed at a private university working for a few more years in Japan.

He is best known for Ito calculus, Ito integral and Ito Formula. Ito received the Japanese Imperial Prize, the Kyoto Prize in Basic Sciences and the Wolf Prize in 2002.

He was President of Japanese Mathematical Society from 1979 until 1981. Ito was awarded the inaugural Carl Friedrich Gauss Prize in 2006 from the King of Spain in Madrid. His daughter Junko Ito collected the Prize on behalf of her father.

Ito died on 10 November 2008.

Kunihiko Kodaira (小平邦彦)

Kunihiko Kodaira was born on 16 March 1915 in Tokyo, Japan.

He graduated from Tokyo Imperial University in 1938 with B.S. in mathematics in the same class as Kiyoshi Ito. He received a second bachelor's degree in physics, but it took another three years because the Japanese system did not permit double majors.

If they allowed double major, he could have obtained another bachelor's degree probably in one more year. Kodaira was able to read papers by Weyl, Stone, von Neumann, Hodge, Weil and Zariski during his undergraduate days.[3]

Kodaira married Seiko Iyanaga, a sister of his teacher Shokichi Iyanaga, on 30 May 1943. So Kodaira and Iyanaga became brothers-in-law.

Seiko was a violinist and Kodaira played the piano.

Kodaira received Doctor of Science (Rigakuhakushi) degree in 1949 at Tokyo University with his thesis *Harmonic Fields in Riemannian Manifolds (generalized potential theory)*.

From 1944 to 1949 he was assistant professor of mathematics at Tokyo University.

Invited by Hermann Weyl, he arrived San Fransisco on 24 August 1949 with physicist Tomonaga. Kadaira was at Johns Hopkins University from 1950 to 1951 for an annual salary of $6,000 (strictly speaking for nine months). After working at the Institute for Advanced Study at Princeton in 1951, he was appointed associate professor at Princeton University in 1952 and promoted to full professor in 1955.

He moved to Johns Hopkins University in 1962 at an annual salary of $18,000. Then to Stanford University in 1965 for $24,000.[62]

He went back to the University of Tokyo in 1967 and retired in 1975. At that time the retirement age at Japanese national university was 60. He spent another 10 years at Gakushuin, a private university.

Kodaira was awarded a Fields Medal in 1954 for achieving results in the theory of harmonic integrals and numerous applications to Kählerian manifolds and more specifically to algebraic varieties. He demonstrated by sheaf cohomology, that such varieties are Hodge manifolds.

He also received the Wolf Prize in 1984 and became one of only twelve mathematicians to receive both the Fields Medal and the Wolf Prize.

Kodaira died of prostate cancer on 26 July 1997 in Kofu, Japan.

Figure 274. Mikio Sato

Mikio Sato (佐藤幹夫)

Mikio Sato was born on 18 April 1928 in Japan. He graduated from Tokyo Imperial University with B.S. in mathematics and another B.S. in physics and then did graduate study in physics as a student of Tomonaga (1906-1979). Sato was awarded a Ph.D. in 1963 from Tokyo University. He then taught ten years at a high school.

From 1958 to 1960 he was an assistant to Kosaku Yosida (1909-1990) at Tokyo University and in 1960 he became a lecturer at Tokyo Kyoiku Daigaku (later to become Tsukuba University).

From 1963 to 1966 he was a professor at Osaka University, but he spent the academic years 1964-1966 at Columbia University.

In 1969 he became a professor at the Komaba Campus (Faculty of General Education) of Tokyo University. Then in June 1970 he moved to the Research Institute for Mathematical Research (RIMS) at Kyoto University.

He is known for his innovative work in a number of fields such as prehomogeneous vector spaces (PHVS), Bernstein-Sato polynomials and particularly for his hyperfunction theory.

It led to the theory of microfunctions, interest in microlocal aspects of linear partial differential equations and Fourier theory such as wave fronts and ultimately to the current developments in D-module theory. He also contributed basic work to non-linear soliton theory with the use of Grassmannians of infinite dimension. In number theory he is known for the Sato-Tate conjecture on L-functions of elliptic curves over the rational number fields.[63]

Sato received the Schock Prize in 1997 by the Royal Swedish Academy of Sciences and the Wolf Prize in 2003.

Figure 275.   Yutaka Taniyama

Yutaka Taniyama (谷山豊)

Yutaka Taniyama was born on 12 November 1927 in Kisai, north of Tokyo, Japan. He graduated from the University of Tokyo in 1953. He remained there as a research student then as an assistant professor.

Taniyama developed a friendship with Goro Shimura and they wrote *Modern Number Theory* in 1957 in Japanese. At the symposium on Algebraic Number Theory held in Tokyo and Nikko in 1955 he met Andre Weil who influenced Taniyama's later work. At the symposium, Taniyama presented some problems that dealt with the relationship between modular forms and elliptic curves. He had noticed some extremely peculiar similarities between the two types of entities. His observation led him to believe that every elliptic curve is somehow matched up with some modular form. Shimura later discussed with Taniyama on this idea that modular forms and elliptic curves are linked and this form the basics of the Taniyama-Shimura conjecture: *Every elliptic curve defined over the rational field is a factor of the Jacobian of a modular function field* In 1986 Kenneth Ribet proved that if the Taniyama-Shimura conjecture held, then so would Fermat's last theorem, which inspired Andrew Wiles to work for a number of years on it and eventually to prove Fermat's last Theorem.

Due to the pioneering contributions of Wiles and the efforts of a number of other mathematicians the Taniyama-Shimura conjecture was finally proved in 1999. The original Taniyama conjecture for elliptic curves over arbitrary number fields remain open.[64]

Taniyama suffered from depression and he committed suicide on 17 November 1958.

About a month later his fiancé, Misako Suzuki, also committed suicide.

Goro Shimura (志村五郎)

Goro Shimura was born on 1930 in Hamamatsu, Japan.

At the University of Tokyo, he became a friend of Taniyama who was three years senior to him. They wrote a book *"Modern number theory"* in 1957. Shimura received a Ph.D. at the University of Tokyo in 1958.

He is best known for the Taniyama-Shimura conjecture. Before he went to the United States, he worked at Tokyo Kyoiku Daigaku (now Tsukuba University) and Osaka University. Since 1962 he has been a professor at Princeton University, and later he held the Michael Henry Starter Chair. Shimura wrote a long series of major papers, extending the phenomena found in the theory of complex multiplication and modular forms to higher dimensions. This work provided some of the "raw data" later incorporated into the Langlands program. It also introduced the concept of *Shimura Variety*, which is the higher-dimensional equivalent of modular curves.[65]

Shimura received the Cole Prize for number theory in 1976 and the Steele Prize in 1996, both from the American Mathematical Society.

Michio Suzuki (鈴木通夫)

Michio Suzuki was born on 2 October 1926 in Chiba, Japan. After graduating from the third High School in Kyoto, he entered the University of Tokyo in 1945. He graduated in 1948 and continued his graduate studies there receiving his Ph.D. in 1952.

Suzuki went to the University of Illinois at Urbana-Champaign in September 1952 and in November of that year he married Naoko Akizuki and they had one daughter. Naoko

Akizuki is a daughter of Yasuo Akizuki (1902–1984), professor of mathematics at Kyoto University. Suzuki was promoted to full professor in 1959. In 1960 he discovered a new class of finite simple groups, which stunned the world of mathematics. In 1967 he discovered another new finite simple group. In 1974 Suzuki was awarded the Academy Prize from the Japan Academy.

He died of liver cancer on 31 May 1998 in Mitaka, Japan.

Figure 276. Michio Suzuki

Figure 277. Shizuo Kakutani

Shizuo Kakutani (角谷靜夫)

Shizuo Kakutani was born on 28 August 1911 in Osaka, Japan. Since his father wanted Shizuo to become a lawyer, his high school grades were not good enough to enter either Tokyo or Kyoto Imperial University. He entered Tohoku Imperial University in Sendai where he studied mathematics. After graduating from Tohoku he became a teaching assistant at Osaka Imperial University in 1934 where he collaborated with Kosaku Yosida (1909–1990) on a paper on Nevanlinna theory.[3]

Kakutani was invited in 1940 by Weyl to spend two years at the Institute for Advanced Study at Princeton. During his stay in Princeton the war broke out between the United States and Japan. He returned to Japan in the summer of 1942.

Upon his return Kakutani became an assistant professor at the Osaka Imperial University. In 1948 he was invited back to the Institute for Advanced Study at Princeton. He accepted an appointment at Yale University in 1949 and he retired from there in 1982.

Kakutani married Kay Uchida in 1952 and they had one daughter.[3]

His fixed-point theorem is a generalization of Brouwer's fixed-point theorem, holding for generalized correspondences instead of functions. Its important use is in proving the existence of Nash equilibria in game theory.

Kakutani's other well-known mathematical contributions include the Kakutani skyscraper, a concept in ergodic theory and his solution of the Poisson equation using the methods of stochastic analysis.[66]

He received the Academy Award and the Imperial Award of the Academy of Japan in 1982.

He died on 17 August 2004 in New Haven, Connecticut, USA.

Kiyoshi Oka (岡潔)

Kiyoshi Oka was born on 19 April 1901 in Osaka, Japan. He entered Kyoto Imperial University in 1919 and graduated in 1925 with a bachelor's degree. Upon graduation, he became a lecturer in mathematics at Kyoto Imperial University.

He married Michi Koyama in 1925.

Oka studied in France from 1929 to 1932 and became acquainted with Gaston Julia. He chose the theory of analytic functions of several variable as the theme of his research.

In May 1932 he accepted a position at Hiroshima University. Oka received his Doctor of Science (Rigakuhakushi) degree at Kyoto Imperial University in October 1940.

He asked Yukawa to hand his paper *Sur quelques notions arithmetiques* to Henri Cartan in Paris on his way to receive the Nobel Prize in 1948. This paper was published in *Bulletin de la Societe Mathematique de France* in 1950.

From 1949 to 1964 he was a professor at Nara Women's University. He then moved to Kyoto Sangyo University when he stayed until his death on 1 March 1978. He received the Asahi Culture Prize in 1954 and the Cultural Medal in November 1960.

In 1963, Oka went to a lecture by Heisuke Hironaka who gave a talk in Japan on the resolution of singularities. At that time Hironaka was an assistant professor at Brandeis University.

Hironaka explained how he would first do it under such and such conditions to solve the problem.

After the lecture, Oka said to Hironaka, "A difficult problem of mathematics should not be solved in that way. You are trying to specialize the problem. On the contrary, you should generalize it. If you make the problem abstract to convert it the most ideal way, it will be solved easily."[107]

Hironaka resented Oka's comments at that time, but later found them to be quite helpful.[107]

.

Korean School

Richard Brauer ⟶ Stephen A. Jennings ⟶ Rimhak Ree
(1901-1977)   1939 Ph.D. Toronto   (1922-2005)
1926 Ph.D. Berlin   1944 Keijo Imperial U
                                        1955 Ph.D. UBC
                                        FRS Canada

Mark Ingraham ⟶ George Whaples ⟶ Dock Sang Rim
1924 Ph.D. Chicago   1939 Ph.D. Wisconsin   (1928-1982)
                        27 students   1954 B.S. SNU
                                        1957 Ph.D. Indiana
                                        1965-82 Prof. U. Penn

Dwijendra Ray-Chaudhuri ⟶ Jeffrey Kahn ⟶ Jeong Han Kim
1959 Ph.D. UNC   1979 Ph.D. Ohio State   (1962-   )
                                        1985 B.S. Yonsei U
                                        1993 Ph.D. Rutgers
                                        1997 Fulkerson Prize
                                        2006 Prof. Yonsei U

Emil Artin ⟶ Serge Lang ⟶ Minhyong Kim (1963-   )
(1898-1962)   (1927-2005)   1985 B.S. SNU
1921 Ph.D. Leipzig   1951 Ph.D. Princeton   1990 Ph.D. Yale
                                        1995-2005 Arizona
                                        2005-2007 Purdue
                                        2007 Prof. UCL

S.S. Chern ⟶ Alan Weinstein ⟶ Yong-Geun Oh (1961-)
(1911-2004)   1967 Ph.D. Berkeley   1983 B.S. SNU
1936 Ph.D. Hamburg   1988 Ph.D. Berkeley
1983 Wolf Prize   1991-   U. Wisconsin

Robert Gunning ⟶   Yum-Tong Siu ⟶   Jun-Muk Hwang (1963-   )
1955 Ph.D. Princeton   (1943-   )   1986 B.S. SNU
                         1966 Ph.D. Princeton   1993 Ph.D. Harvard
                         1982 Prof. Harvard   1996-99 SNU
                         NAS member   1991-   KIAS
                         Bergman Prize   2006 Korean National Scholar

Louis McAuley ⟶   Ronald Fintushel ⟶   Jongil Park (1963-   )
1954 Ph.D. UNC   (1946-   )   1986 B.S. SNU
                   1975 Ph.D. Binghamton   1996 Ph.D. Michigan State
                   Prof. Michigan State   2004 Prof. SNU

Phillip Griffiths ⟶   Charles H. Clemens ⟶   Yongnam Lee (1964-   )
(1938-   )   1966 Ph.D. Berkeley   1987 B.S. SNU
1962 Ph.D. Princeton   1997 Ph.D. Utah
2008 Wolf Prize   2000-   Prof. Sogang U

Korean School

During Japanese control of Korea from 1910 to 1945, Koreans were prohibited to study mathematics. The only university, Keijo Imperial University (Seoul National University after 1946) founded in 1924, had Faculties of Law, Medicine and Literature, but no mathematics department.

The history of modern mathematics in Korea began in 1946 when the Korean Mathematical and Physical Society was established. The Korean War from 1950 to 1953 completely disrupted real mathematical research because of heavy civilian casualties and destroyment of facilities.

In 1952, the Korean Mathematical Society was separated from the Korean Mathematical and Physical Society. The 41st International Mathematical Olympiad was held in Korea in July 2000.

There are 112 Mathematics departments in Korean universities awarding 2774 bachelors, 170 masters and 78 Ph.Ds. annually. There are also 71 statistics departments awarding 2312 bachelors, 206 masters and 33 Ph.Ds. annually.

Mathematical papers published in international journals by Korean mathematicians increased greatly in the last several years. In 2007, Korea was upgraded to Level IV by the International Mathematical Union.

Current membership in the Korean Mathematical Society reached 2870 including 663 student members.

Rimhak Ree (李林學)

Rimhak Ree was born in 1922 at Hamhung, Korea (now North Korea). He graduated from Keijo Imperial University (now Seoul National University) in 1944 majoring in physics since there was no mathematics department at that time.

He got a job as a supervisor at the aircraft plant in Shenyang in Manchuria which was under Japanese control at that time. After Japan surrendered to the Allies in August 1945, he returned to Seoul. He taught mathematics at Seoul National University from 1946 to 1952.

Ree came across with an article by Max Zorn in a Bulletin of the American Mathematical Society in which he found an unsolved problem. Ree came up with a solution and sent a letter to Max Zorn who submitted it to the American Mathematical

Society. (Ree, Rimhak, On a problem of Max A. Zorn, Bull. AMS 55 (1949) pp. 575–576) This was the first mathematical paper published by a Korean in an international journal.

He was not aware of his paper published until he came to the University of British Columbia in 1953 to study for his doctorate under Stephen Jennings.

Ree received his Ph.D. in 1955 at UBC for a thesis *Witt Algebras* supervised by Jennings.

He remained at the University of British Columbia for the rest of his career. He specialized in finite group theory.

Daniel Gorenstein describes Rees's contribution.[104]

It was Rimhak Ree who first realized the implication of the Lie-theoretic interpretation of Suzuki's family, which he then used as a model for constructing two further families of simple groups related to the exceptional groups $G_2(3^n)$ and $F_4(2^n)$, respectively. Ree's groups constituted the last of the sixteen families of finite simple groups of Lie type.

Centralizers of involutions in Ree's groups of characteristic 3 had an essentially simple structure: $Z_2 \times L_2(q)$, q odd (as well as elementary Abelian Sylow 2-groups of order 8).

In addition, like Suzuki's family, these groups were doubly transitive with a one-point stabilizer containing a regular normal subgroup. It was therefore natural to attempt a Brauer-type characterization of this family in terms of centralizers of involutions of this form. The work was begun by Nathanial Ward, a student of Brauer's and taken up independently by Janko and Thompson.

Together they established the doubly transitive nature of such a group of "Ree type" provided $q \geq 5$.

Ree was elected to the Royal Society of Canada. He wrote a paper with Erdös in December 1958 (The American Mathematical Monthly, Vol. 65, No. 10).

He died on 9 February 2005 in Vancouver, Canada.

Dock Sang Rim (林德相)

Dock Sang Rim was born on 2 October 1929 in Korea. He graduated from Seoul National University in 1954 and received a Ph.D. at Indiana University on 1957 for his thesis *An Axiomatic Approach to Cohomology Theory of Finite Groups* supervised by George Whaples.

Rim was an instructor at Columbia University from 1957 to 1960 and taught at Brandeis University from 1960 to 1965. He became a tenured professor at the University of Pennsylvania in 1965 and served as chairman of the mathematics department from 1975 to 1978.

Stephen Shatz described Rim's work:

He was one of the early experts in the cohomology of groups becoming an expert in homological algebra in general and in its applications to algebraic geometry and number theory, especially in the theory of deformations (initiated in the setting of differential geometry by Kuranishi and carried forward into algebraic geometry by Grothendieck and his Paris School).

Rim died of hepatic cancer on 18 November 1982 in his home in Wynnewood, Pennsylvania, U.S.A.

Figure 278. Rimhak Ree

Figure 279. Jeong Han Kim

Jeong Han Kim (金鼎瀚)

Jeong Han Kim was born on 20 July 1962 in Seoul. He graduated from Yonsei University in 1985 with a B.S. in mathematics. He received a Ph.D. in 1993 from Rutgers University for his thesis *Non-combinatorial approaches to two combinatorial problems* supervised by Jeffrey Kahn.[10]

He worked at the Information Sciences Research Center, AT&T Bell Laboratories from 1993 to 1997. Kim moved to Microsoft Research as a senior researcher in April 1997.

He received the Delbert-Ray Fulkerson Prize from the American Mathematical Society in 1997 for his paper "The Ramsey Number $R(3,t)$ has order of Magnitude $t^2/\log t$" which appeared in Random Structures and Algorithms, Vol. 7, issue 3 in 1995.

Jeong Han Kim's paper solves this Sixty-year old problem by improving the Erdös lower bound to match the upper bound of Ajtai, Komlós, and Szemerédi. The paper is a veritable cornucopia of modern techniques in the probabilistic method; it uses martingales in a sophisticated way to obtain strong larger deviation bounds.[111]

Kim has been Underwood Chair Professor of Mathematics at Yonsei University in Seoul, Korea since 2006. He delivered a lecture "Poisson Cloning Model for Random Graphs" at the International Congress of Mathematicians in 2006 in Madrid, Spain.

Minhyong Kim (金民衡)

Minhyong Kim was born on 4 December 1963 in Seoul, Korea and graduated from Seoul National University in 1985. He received a Ph.D. in 1990 from Yale University for his thesis *Lower Bounds for Lattice Vectors and Arithmetic Intersection Theory* supervised by Serge Lang and Barry Mazur.[10]

Kim taught at the University of Arizona from 1995 to 2005 being promoted to a full professor in 2004. He became a professor of mathematics at Purdue University in 2005. Kim was appointed to the Chair of Pure Mathematics at the University College London in 2007.

His fields of interest are number theory and algebraic geometry. Kim has supervised nine Ph.D. students. He was co-organizer for special session on arithmetical algebraic geometry, AMS-MAA annual meeting in Phoenix, Arzona in January 2004.

Yong-Geun Oh (吳龍根)

Yong-Geun Oh was born on 1 January 1961 in Seoul, Korea and graduated from Seoul National University in 1983. Oh received a Ph.D. in 1988 from the University of California Berkeley for his dissertation *Nonlinear Schrödinger Equations with Potentials: Evolution, Existence, and Stability of Semi-Classical Bound States* under the supervision of Alan Weinstein.[10]

Oh taught two years as Courant Instructor at New York University (1989-1991) and was a member at the Institute for Advanced Study at Princeton (1991-1992). He has been a faculty member at the University of Wisconsin-Madison in 1991 being promoted to full professor in 2001.

He was a section speaker in geometry for the International Congress of Mathematicians in 2006 in Madrid, Spain.

Figure 280.   Jun-Mok Hwang

Jun-Mok Hwang (黃準默)

Jun-Mok Hwang was born on 27 October 1963 in Seoul, Korea. He graduated from Seoul National University in 1986 with a B.S. in physics. Hwang received a Ph.D. from Harvard University in 1993 for his dissertation *Global Nondeformability of the Complex Hyperquadric* supervised by Yum-Tong Siu.[10]

Hwang has been a professor at the Korea Advanced Institute of Science since 1991. He was an assistant professor at Notre Dame University in 1993 and at Seoul National University from 1996 to 1999.

He became a Star Faculty (National Scholar) by the Ministry of Education in Korea in 2006. He was also a section speaker for the International Congress of Mathematicians in 2006 in Madrid, Spain.

Jongil Park (朴鍾逸)

Jongil Park was born on 3 September 1963 in Kangjin, Korea and received a B.S. in mathematics in 1986 and an M.S. in mathematics in 1988 at Seoul National University. He was awarded a Ph.D. in 1996 for his thesis *Seiberg-Witten Invariants of Rational Blow-downs and Geography Problems of Irreducible 4-Manifolds* at Michigan State University supervised by Ronald Fintushel.

Park taught at the University of California, Irvine for one year (1996-1997) and at Konkuk University in Seoul, Korea from 1997 to 2004. He has been at Seoul National University since 2004.

In 2006, Park and Yongnam Lee worked together to construct a simply connected, minimal, complex surfaces of general type with geometric genus $P_g = 0$ and $K^2 = 2$ by modifying Park's symplectic 4-manifold published in 2004. The main techniques used are rational blow-down surgery and Q-Gorenstein smoothing theory, which are totally different from other classical techniques such as a finite group quotient and a double covering, due to Godeaux, Campedelli, Burniat and others. The surface constructed by Park and Lee, so-called Park-Lee surface, opened a new direction in the study of complex surface of general type with $P_g = 0$.[112]

Figure 281. Jongil Park

Figure 282. Yongnam Lee

Yongnam Lee (李龍男)

Yongnam Lee was born on 22 May 1964 in Seoul, Korea and received a B.S. in mathematics in 1987 and an M.S. in mathematics in 1989 at Seoul National University.

He also received a Ph.D. in 1997 at the University of Utah for his thesis *Degeneration of Numerical Godeaux Surfaces* supervised by Charles Herbert Clemens.

He joined the mathematics department of Sogang University, Seoul, Korea in March 2000. Lee is currently the department head.

R. Pardini's comments on Park and Lee's paper "A simply connected surface of general type with $P_g = 0$ and $K^2 = 2$" in the Inventiones Mathematicae, 170, No. 3 (2007) are as follows[116]:

The paper under review contains the construction of a simply connected minimal surface of general type with $P_g = 0$ and $K^2 = 2$, and also of two simply connected minimal surfaces of general type with $P_g = 0$ and $K^2 = 1$. The examples are constructed by a new method, namely as smoothings of singular rational surfaces. The construction of the main example $K^2 = 2$ goes as follows: one takes a certain pencil of cubics in the plane and blows up its base locus, obtaining an elliptic rational surface with a particular configuration of singular fibers. The one blows up further and obtains a rational surface containing 5 disjoints chains of rational curves, which can be blown down to get a singular rational surface X. Every singularity of X is of class T, namely it admits a local Q-Gorenstein smoothing. Using deformation theory, the authors prove that there is indeed a global Q-Gorenstein smoothing of X, namely a one-parameter family of projective surfaces such that: the central fiber is X; the general fibre X_t is smooth and projective; the relative canonical divisor is Q-Cartier. Then it is not difficult to show that the general fibre X_t is a minimal surface of general type with $K^2 = 2$, and $P_g = 0$. Finally, one shows that X_t is simply connected by using standard arguments on Milnor fibers. One should understand that the construction is not only very technical but really subtle and ingenious.

American School

J.J. Sylvester ⟶ George B. Halsted
See British School   (1853-1922)
           1879 Ph.D. Johns Hopkins
           1884-1902 Prof. Texas

H.A. Newton ⟶ Oswald Veblen (1880-1960) ⟶ R.L. Moore (1882-1974)
1850 Ph.D.Yale   1903 Ph.D. Chicago   1905 Ph.D.Chicago
2 students (Yale)   16 students (14 Princeton, 2 Chicago)   1920-69 Texas
              1923-24 AMS President   50 students
E.H.Moore ⟶   1932 Prof. IAS   1937-38 AMS President
(1862-1932)
1885 Ph.D. Yale ⟶ George D. Birkhoff (1884-1944)
31 students   1907 Ph.D. Chicago
(Chicago)   46 students (41 Harvard) ⟶ Alexander Oppenheim
1901-02   1923 Bocher Prize   (1903-1997)
AMS President   1925-26 AMS President   1930 Ph.D. Chicago
                                  1929 Oppenheim Conjecture
                                  See British School
         Leonard E. Dickson ⟶ Abraham A. Albert
         (1874-1954)   (1905-1972)
         1896 Ph.D. Chicago   1928 Ph.D. Chicago
         67 students (Chicago)   29 students (Chicago)
         1928 Frank Cole Prize   1965-66 AMS President
         1917-18 AMS President ⟶ Mina Rees
                                  (1902-1997)
                                  1931 Ph.D. Chicago
                                  1969 CUNY President

         Edward W. Chittenden ⟶ Deane Montgomery
         (1885-1977)   (1909-1992)
         1912 Ph.D. Chicago   1933 Ph.D. Iowa
         1918-54 Iowa   1951-92 IAS
         4 students (Iowa)

Theophil H. Hildebrandt

(1888-1980)

1910 Ph.D. Chicago

11 students (Michigan)

1945-46 AMS President

Mark H. Ingraham ⟶ Georg Whaples ⟶ Dock Sang Rim (1928-1982)

1924 Ph.D. Chicago   1939 Ph.D. Wisconsin   1957 Ph.D. Indiana

14 students   27 students   1965-82 Prof U. Penn

George Birkhoff ⟶ H.C. Marston Morse (1892-1977)

1917 Ph.D. Harvard

5 students

1933 Bocher Prize

1941-42 AMS President

Hassler Whitney ⟶ Herbert E. Robbins

(1907-1989)   (1915-   )

1932 Ph.D. Harvard   1938 Ph.D. Harvard

1982 Wolf Prize   8 students

6 students

David Widder ⟶ Ralph Boas, Jr ⟶ Philip J. Davis (1923-   )

(1898-1990)   (1912-1992)   1950 Ph.D. Harvard

1924 Ph.D. Harvard   1937 Ph.D. Harvard   12 students (Brown)

13 students   1950-80 Northwestern

13 students

Clarence Adams ⟶ Anthony Morse ⟶ Herbert Federer

1922 Ph.D. Harvard   1937 Ph.D. Brown   1944 Ph.D. Berkeley

8 students   15 students   9 students (Brown)

NAS member

Joseph Walsh ⟶ Joseph Leo Doob ⟶ David Blackwell

(1895-1973)   (1910-2004)   (1919-   )

1920 Ph.D. Harvard   1932 Ph.D. Harvard   1941 Ph.D. UIUC

31 students   16 students   65 students (Berkeley)

1949-50 AMS   NAS member   Paul Halmos (1916-2006)

President   1963-64 AMS   1938 Ph.D. UIUC

President   21 students

Robert Carmichael → Harold Mott-Smith → Richard James Duffin ──┐

1911 Ph.D. Princeton   1933 Ph.D. UIUC   1935 Ph.D. UIUC   │

33 students (UIUC)   13 students (Carnegie Mellon) │

                                                                            │
                        ┌───┘
                        └→ Raoul Bott (1923-2005)

                           1949 Ph.D. Carnegie Mellon

                           19 students (Harvard)

                           2000 Wolf Prize

Raoul Bott (1923-2005) → Stephen Smale → Morris Hirsch → William P. Thurston

1949 Ph.D. Carnegie Mellon   (1930-   )   (1933-   )   (1946-   )

19 students   1957 Ph.D. Michigan 1958 Ph.D. Chicago 1972 Ph.D. Berkeley

(17 Harvard)   44 students   23 students   1982 Fields Medal

NAS member   (41 Berkeley)   (Berkeley)   30 students

2000 Wolf Prize   1966 Fields Medal   2003 Cornell Univ

                            NAS member   → Jacob Palis, Jr (1940-   )

                            2007 Wolf Prize   1968 Ph.D. Berkeley

                                                39 students

                                                NAS member

                          → Daniel Quillen (1940-   )

                            1964 Ph.D. Harvard

                            5 students (MIT)

                            1978 Fields Medal

                            NAS member

                          → Robert D. Mac Pherson (1944-   )

                            1970 Ph.D. Harvard

                            24 students

                            NAS member

Georg K. W Hamel   → Richard E. von Mises → Stefan Bergman

(1877-1954)   (1883-1953)   (1895-1977)

1901 Ph.D. Göttingen   1907 Ph.D. Wien   1921 Ph.D. Berlin

1919-1949   19 students   6 students (Stanford)

TH Berlin   See German School

See German School

Samuel Edwards ──────▶ Elliott H. Lieb (1932–   )
(1928–   )   1956 Ph.D. Birmingham
Ph.D. Cambridge   1988 Birkhoff Prize
1966 FRS   9 students (8 Princeton)
1984–95 Cavendish Prof

James Michael ──────▶ Leon M. Simon (1945–   )
1957 Ph.D. Adelaide   1971 Ph.D. Adelaide
                         1994 Bocher Prize
                         7 students

Marshall Stone ──▶ George Mackey ──▶ Andrew Gleason ──
   (1903–1989)   (1916–2006)   (1921–   )
   1926 Ph.D. Harvard 1942 Ph.D. Harvard 1942 BA Yale
   14 students   23 students   16 students (Harvard)
   1943–44 AMS   NAS member   NAS member
G.D. Birkhoff ──▶ President   1981–82 AMS President

Charles. B. Morrey, Jr ▶ Richard Kadison ──▶ James Glimm (1934–)
   1931 Ph.D. Harvard   (1925–   )   1959 Ph.D. Columbia
   15 students (Berkeley) 1950 Ph.D. Chicago 27 students
   1967–68 AMS President 19 students   NAS member
                          NAS member   2007 AMS President

Clarence Moore ──▶ Roger E. Howe (1945–   )
   1960 Ph.D. Harvard 1969 Ph.D. Berkeley
   13 students   15 students (Yale)
   (Berkeley)   NAS member

Richard Palais (1931–   ) ──▶ Karen K. Uhlenbeck (1942–   )
   1956 Ph.D. Harvard   1968 Ph.D. Brandeis
   1960 Prof. Brandeis   17 students
                               NAS member

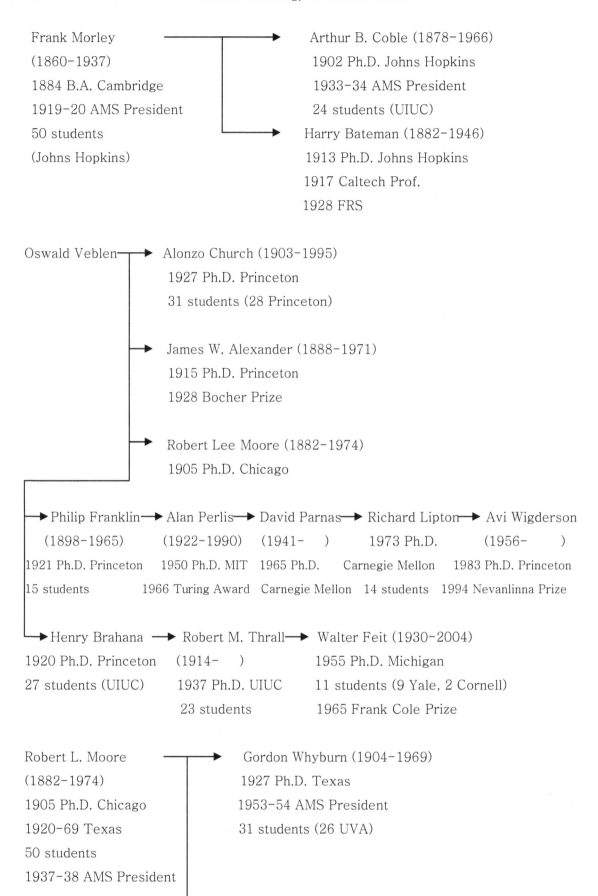

Frank Morley
(1860-1937)
1884 B.A. Cambridge
1919-20 AMS President
50 students
(Johns Hopkins)

Arthur B. Coble (1878-1966)
1902 Ph.D. Johns Hopkins
1933-34 AMS President
24 students (UIUC)

Harry Bateman (1882-1946)
1913 Ph.D. Johns Hopkins
1917 Caltech Prof.
1928 FRS

Oswald Veblen

Alonzo Church (1903-1995)
1927 Ph.D. Princeton
31 students (28 Princeton)

James W. Alexander (1888-1971)
1915 Ph.D. Princeton
1928 Bocher Prize

Robert Lee Moore (1882-1974)
1905 Ph.D. Chicago

Philip Franklin → Alan Perlis → David Parnas → Richard Lipton → Avi Wigderson
(1898-1965)   (1922-1990)   (1941-   )   1973 Ph.D.   (1956-   )
1921 Ph.D. Princeton   1950 Ph.D. MIT   1965 Ph.D.   Carnegie Mellon   1983 Ph.D. Princeton
15 students   1966 Turing Award   Carnegie Mellon   14 students   1994 Nevanlinna Prize

Henry Brahana → Robert M. Thrall → Walter Feit (1930-2004)
1920 Ph.D. Princeton   (1914-   )   1955 Ph.D. Michigan
27 students (UIUC)   1937 Ph.D. UIUC   11 students (9 Yale, 2 Cornell)
                     23 students   1965 Frank Cole Prize

Robert L. Moore
(1882-1974)
1905 Ph.D. Chicago
1920-69 Texas
50 students
1937-38 AMS President

Gordon Whyburn (1904-1969)
1927 Ph.D. Texas
1953-54 AMS President
31 students (26 UVA)

Raymord L. Wilder (1896-1982)

1923 Ph.D. Texas

1926-67 Michigan

1955-56 AMS President

26 students

R.H. Bing (1914-1986)

1945 Ph.D. Texas

1947-73 Wisconsin

1977-78 AMS President

38 students

R.L. Moore ⟶ S. Eldon Dyer (1929-1993) ⟶ Robion C. Kirby

1952 Ph.D. Texas (1938-   )

10 students 1965 Ph.D. Chicago

50 students

John R. Kline ⟶ Arthur N. Milgran

(1891-1955) 1937 Ph.D. U. Penn

1916 Ph.D. U. Penn 3 students

18 students Leo Zippin (1905-1995)

1929 Ph.D. U. Penn

1938-71 CUNY

Alonzo Church ⟶ Alan Turing (1912-1954)

1938 Ph.D. Princeton

See British School

Simon B. Kochen (1934-   )

1958 Ph.D. Princeton

1968 Frank Cole Prize

Stephen C. Kleene (1909-1994)

1934 Ph.D. Princeton

NAS member

Martin Davis (1928-   )

1950 Ph.D. Princeton

1967 Frank Cole Prize

16 students (14 NYU)

John G. Kemeny (1926-1992)

1949 Ph.D. Princeton

1970-81 President Dartmouth

→ Michael Rabin (1931–   ) ⟶ Saharon Shelah (1945–   )

      1956 Ph.D. Princeton   1969 Ph.D. Hebrew U

      14 students   2001 Wolf Prize

→ Hartley Rogers. Jr→ David Park→ Michael Paterson→ Leslie G. Valiant (1949–   )

   (1926–   )   1964 Ph.D. MIT (1942–   )   1974 Ph.D. Warwick

1952 Ph.D. Princeton   1968 Ph.D. Cambridge 1986 Nevanlinna Prize

1964 Prof. MIT   971 Warwick   1991 FRS

                                       2001 FRS   2001– Harvard

Cassius J. Keyser ⟶   Edward Kasner ⟶   Jesse Douglas (1897–1965)

(1862–1947)   (1878–1955)   1920 Ph.D. Columbia

1902 Ph.D. Columbia   1900 Ph.D. Columbia   1930–36 MIT

                         Prof. Columbia   1936 Fields Medal

                                                  1943 Bocher Prize

                       → Eric T. Bell (1883–1960)

                         1912 Ph.D. Columbia

                         1923–53 Caltech

                         1924 Bocher Prize

                         14 students

Josiah Royce   ⟶   Norbert Wiener   ⟶   Norman Levinson

(1855–1916)   (1894–1964)   (1912–1975)

1875 Ph.D. Johns Hopkins   1913 Ph.D. Harvard   1935 Ph.D. MIT

                           18 students   34 students

                                                      1953 Bocher Prize

Philip Hall→ Garrett Birkhoff ⟶ Richard Arens→ Kenneth M. Hoffman (1930–   )

(1908–1982) (1911–1996)   1945 Ph.D. Harvard 1956 Ph.D. UCLA

1951 FRS   1932–33 Cambridge 10 students (UCLA) 11 students(MIT)

1961 Sylvester 1978 Birkhoff Prize

Medal   52 students → George D. Mostow (1923–   )

See British School   (Harvard)   1948 Ph.D. Harvard

                                 1987–88 AMS President

                                 4 students (Yale)

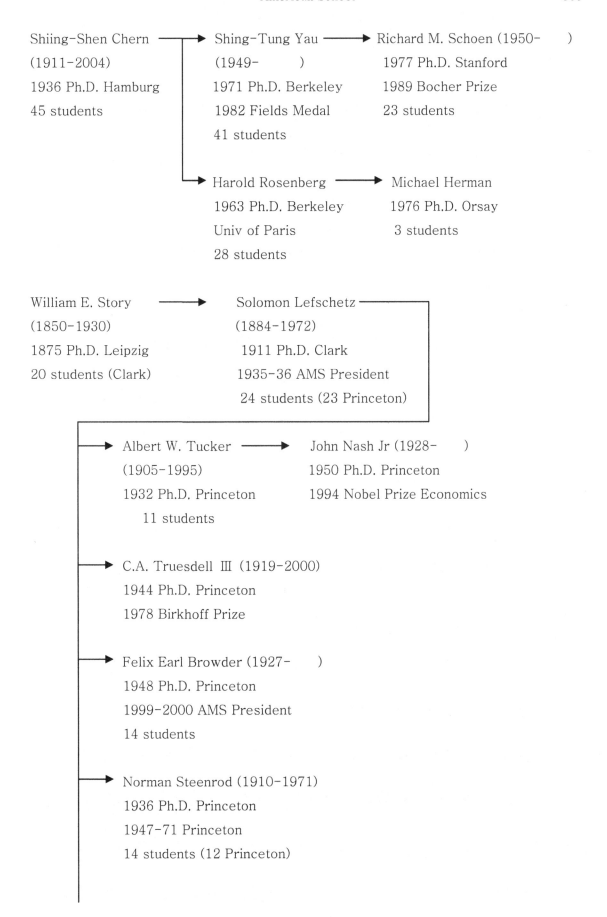

Shiing-Shen Chern ———▶ Shing-Tung Yau ———▶ Richard M. Schoen (1950-   )
(1911-2004)   (1949-   )   1977 Ph.D. Stanford
1936 Ph.D. Hamburg   1971 Ph.D. Berkeley   1989 Bocher Prize
45 students   1982 Fields Medal   23 students
                         41 students

                         Harold Rosenberg ———▶ Michael Herman
                         1963 Ph.D. Berkeley   1976 Ph.D. Orsay
                         Univ of Paris   3 students
                         28 students

William E. Story ———————▶   Solomon Lefschetz
(1850-1930)   (1884-1972)
1875 Ph.D. Leipzig   1911 Ph.D. Clark
20 students (Clark)   1935-36 AMS President
                            24 students (23 Princeton)

        Albert W. Tucker ———————▶ John Nash Jr (1928-   )
        (1905-1995)   1950 Ph.D. Princeton
        1932 Ph.D. Princeton   1994 Nobel Prize Economics
            11 students

        C.A. Truesdell III (1919-2000)
        1944 Ph.D. Princeton
        1978 Birkhoff Prize

        Felix Earl Browder (1927-   )
        1948 Ph.D. Princeton
        1999-2000 AMS President
        14 students

        Norman Steenrod (1910-1971)
        1936 Ph.D. Princeton
        1947-71 Princeton
        14 students (12 Princeton)

Ralph Fox → John W. Milnor → John N. Mather (1942-   )

Ralph Fox
(1913-1973)
1939 Ph.D. Princeton
21 students
(20 Princeton)

John W. Milnor
(1931-   )
1954 Ph.D. Princeton
1962 Fields Medal
1989 Wolf Prize
17 students

John N. Mather (1942-   )
1967 Ph.D. Princeton
2003 Birkhoff Prize
5 students

John R. Stallings, Jr (1935-   )
1959 Ph.D. Princeton
1970 Frank Cole Prize
21 students

Barry Mazur (1937-   )
1959 Ph.D. Princeton
49 students (45 Harvard)
NAS member

John W. Tukey → Frederick Mosteller → Persi W. Diaconis (1945-   )

John W. Tukey
(1915-2000)
1939 Ph.D. Princeton

Frederick Mosteller
(1916-2006)
1946 Ph.D. Princeton
Prof. MIT

Persi W. Diaconis (1945-   )
1974 Ph.D. Harvard
NAS member
37 students

Albert Tucker → David Gale (1921-   )

David Gale (1921-   )
1949 Ph.D. Princeton
14 students
NAS member

M. Ninsky → M. Blum → G. Miller → F. Thomson Leighton → Peter W. Shor

M. Ninsky
(1927-   )
1954 Ph.D.
Princeton

M. Blum
(1938-   )
1964 Ph.D. MIT
23 students
NAS member
1995 Turing Award

G. Miller
1975 Ph.D.
Berkeley

F. Thomson Leighton
1981 Ph.D. MIT
14 students

Peter W. Shor
1985 Ph.D. MIT
1998 Nevanlinna
Prize

Norman Steenrod → George Whitehead, Jr → John C. Moore → William Browder

George Whitehead, Jr
(1918-   )
1941 Ph.D. Chicago
13 students

John C. Moore
(1923-   )
1952 Ph.D. Brown
23 students

William Browder
(1934-   )
1958 Ph.D. Princeton
1989-90 AMS
President
32 students

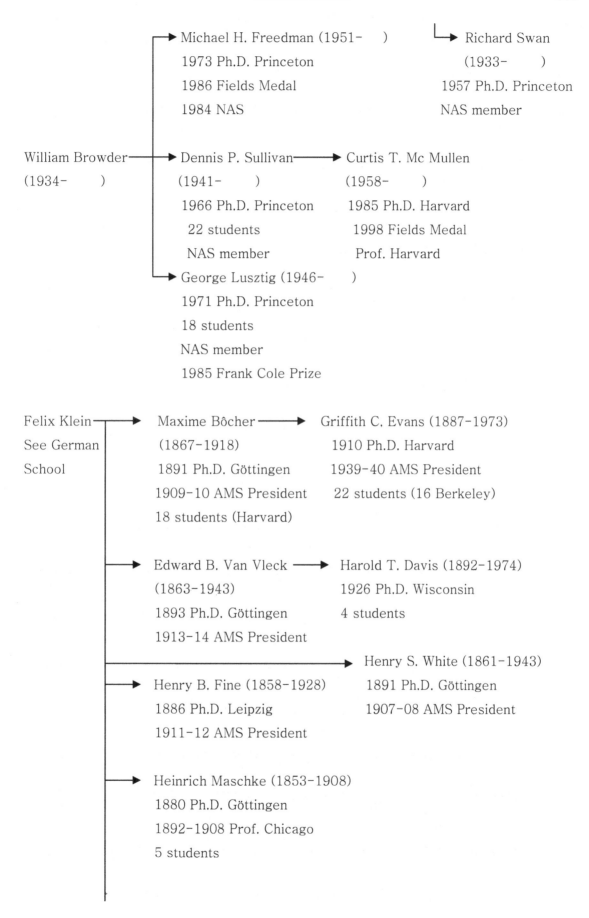

Michael H. Freedman (1951–   )
1973 Ph.D. Princeton
1986 Fields Medal
1984 NAS

Richard Swan
(1933–   )
1957 Ph.D. Princeton
NAS member

William Browder
(1934–   )

Dennis P. Sullivan
(1941–   )
1966 Ph.D. Princeton
22 students
NAS member

Curtis T. Mc Mullen
(1958–   )
1985 Ph.D. Harvard
1998 Fields Medal
Prof. Harvard

George Lusztig (1946–   )
1971 Ph.D. Princeton
18 students
NAS member
1985 Frank Cole Prize

Felix Klein
See German
School

Maxime Bôcher
(1867–1918)
1891 Ph.D. Göttingen
1909–10 AMS President
18 students (Harvard)

Griffith C. Evans (1887–1973)
1910 Ph.D. Harvard
1939–40 AMS President
22 students (16 Berkeley)

Edward B. Van Vleck
(1863–1943)
1893 Ph.D. Göttingen
1913–14 AMS President

Harold T. Davis (1892–1974)
1926 Ph.D. Wisconsin
4 students

Henry S. White (1861–1943)
1891 Ph.D. Göttingen
1907–08 AMS President

Henry B. Fine (1858–1928)
1886 Ph.D. Leipzig
1911–12 AMS President

Heinrich Maschke (1853–1908)
1880 Ph.D. Göttingen
1892–1908 Prof. Chicago
5 students

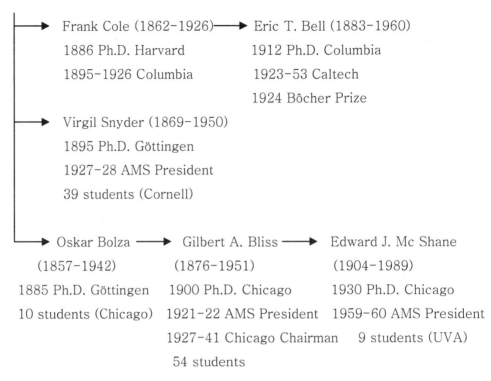

Frank Cole (1862-1926)⟶ Eric T. Bell (1883-1960)
1886 Ph.D. Harvard   1912 Ph.D. Columbia
1895-1926 Columbia   1923-53 Caltech
                             1924 Bôcher Prize

Virgil Snyder (1869-1950)
1895 Ph.D. Göttingen
1927-28 AMS President
39 students (Cornell)

Oskar Bolza ⟶   Gilbert A. Bliss ⟶   Edward J. Mc Shane
(1857-1942)   (1876-1951)   (1904-1989)
1885 Ph.D. Göttingen 1900 Ph.D. Chicago   1930 Ph.D. Chicago
10 students (Chicago) 1921-22 AMS President   1959-60 AMS President
                     1927-41 Chicago Chairman   9 students (UVA)
                     54 students

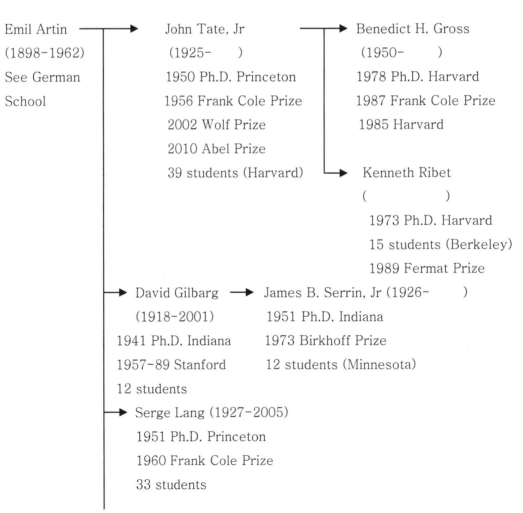

Emil Artin ⟶   John Tate, Jr ⟶   Benedict H. Gross
(1898-1962)   (1925-   )   (1950-   )
See German   1950 Ph.D. Princeton   1978 Ph.D. Harvard
School   1956 Frank Cole Prize   1987 Frank Cole Prize
                     2002 Wolf Prize   1985 Harvard
                     2010 Abel Prize
                     39 students (Harvard)   Kenneth Ribet
                                                (   )
                                                1973 Ph.D. Harvard
                                                15 students (Berkeley)
                                                1989 Fermat Prize

David Gilbarg ⟶   James B. Serrin, Jr (1926-   )
(1918-2001)   1951 Ph.D. Indiana
1941 Ph.D. Indiana   1973 Birkhoff Prize
1957-89 Stanford   12 students (Minnesota)
12 students

Serge Lang (1927-2005)
1951 Ph.D. Princeton
1960 Frank Cole Prize
33 students

↳ Max August Zorn (1906–1993)

1930 Ph.D. Hamburg

1946–1971 Indiana

Hermann Schwarz ⟶ Leon Lichtenstein⟶ Aurel Wintner ⟶ Shlomo Z. Sternberg ⟶

See German School (1878–1933) (1903–1958) ( )

1909 Ph.D. Berlin 1928 Ph.D. Leipzig 1957 Ph.D. Johns Hopkins

7 students 4 students 17 students

(16 Harvard)

↳ Victor Guillemin (1937–)

1962 Ph.D. Harvard

40 students

(39 MIT, 1 Harvard)

Erhard Schmidt ⟶ Salomon Bochner (1899–1982)

See German School 1921 Ph.D. Berlin

36 students (35 Princeton)

⟶ Eberhard Hopf ⟶ Albert C. Schaeffer

(1902–1983) 1936 Ph.D. MIT

1926 Ph.D. Berlin 1948 Bocher Prize

1932–36 MIT

↳ Heinz Hopf ⟶ James Stoker ⟶ Louis Nirenberg (1925–)

(1894–1971) (1905–) 1949 Ph.D. NYU

1925 Ph.D. Berlin 1936 Ph.D. ETH 1959 Bocher Prize

1931 Prof ETH 37 students 44 students

47 students (ETH) (NYU)

See German School

Salomon Bochner ⟶ Gerard Washnitzer ⟶ William E. Fulton (1939–)

1950 Ph.D. Princeton 1966 Ph.D. Princeton

3 students 20 students

⟶ Richard Askey (1933–)

1961 Ph.D. Princeton

13 students (Wisconsin)

Harry (Hillel) Furstenberg (1935-   )

1958 Ph.D. Princeton

2006 Wolf Prize

16 students

Eugenio Calabi (1923-   )

1950 Ph.D. Princeton

5 students

Robert Gunning (1931-) ⟶ Richard Hamilton (1943-)

1955 Ph.D. Princeton   1966 Ph.D. Princeton

1966 Prof Princeton

28 students

Jeff Cheeger (1943-   )

1967 Ph.D. Princeton

13 students

David Eisenbud (1947-   )

1970 Ph.D. Chicago

25 students

2003-04 AMS President

H. Weyl ⟶ Saunders Maclane ⟶ John G. Thompson (1932-   )

(1909-2005)   1959 Ph.D. Chicago

1934 Ph.D. Göttingen   1965 Frank Cole Prize

41 students   1970 Fields Medal

1973-74 AMS   1971-1993 Rouse Ball Prof

President   1992 Wolf Prize

                               2008 Abel Prize

                               22 students

Irving Kaplansky (1917-2006) ⟶ Donald S. Ornstein (1934-   )

1941 Ph.D. Harvard   1957 Ph.D. Chicago

55 students   24 students (22 Stanford)

NAS member   NAS member

1985-86 AMS President   1974 Bocher Prize

Robert M. Solovay   Hyman Bass (1932-   )

1964 Ph.D. Chicago   1959 Ph.D. Chicago

14 students (Berkeley)   25 students (Columbia)

NAS member   NAS member

                               2001-02 AMS President

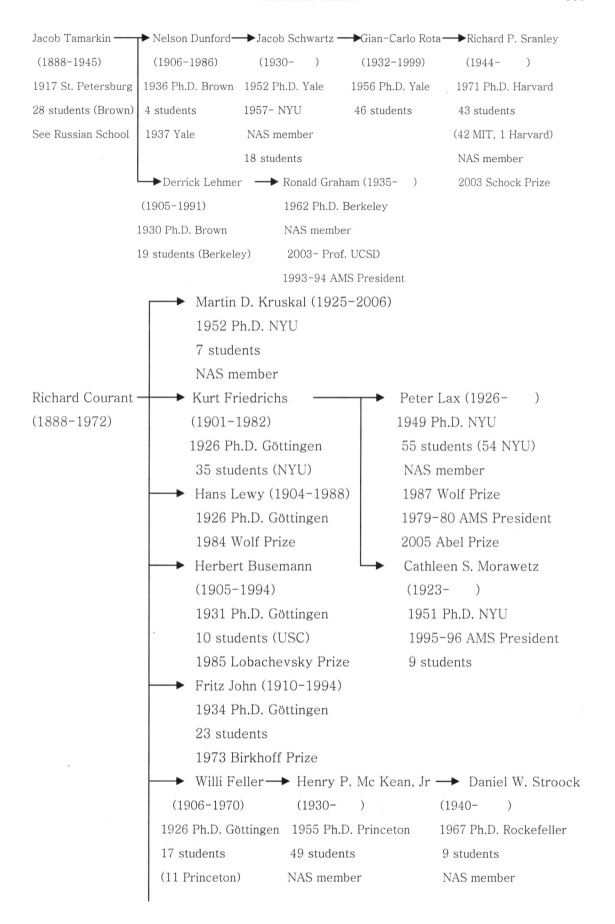

Jacob Tamarkin ——▶ Nelson Dunford ——▶ Jacob Schwartz ——▶ Gian-Carlo Rota ——▶ Richard P. Sranley

(1888–1945)   (1906–1986)   (1930–)   (1932–1999)   (1944–)

1917 St. Petersburg 1936 Ph.D. Brown 1952 Ph.D. Yale 1956 Ph.D. Yale 1971 Ph.D. Harvard

28 students (Brown) 4 students 1957– NYU 46 students 43 students

See Russian School 1937 Yale NAS member   (42 MIT, 1 Harvard)

18 students   NAS member

——▶ Derrick Lehmer ——▶ Ronald Graham (1935–)   2003 Schock Prize

(1905–1991)   1962 Ph.D. Berkeley

1930 Ph.D. Brown NAS member

19 students (Berkeley) 2003– Prof. UCSD

1993–94 AMS President

——▶ Martin D. Kruskal (1925–2006)

1952 Ph.D. NYU

7 students

NAS member

Richard Courant ——▶ Kurt Friedrichs ————————▶ Peter Lax (1926–)

(1888–1972) (1901–1982) 1949 Ph.D. NYU

1926 Ph.D. Göttingen 55 students (54 NYU)

35 students (NYU) NAS member

——▶ Hans Lewy (1904–1988) 1987 Wolf Prize

1926 Ph.D. Göttingen 1979–80 AMS President

1984 Wolf Prize 2005 Abel Prize

——▶ Herbert Busemann ——————————▶ Cathleen S. Morawetz

(1905–1994) (1923–)

1931 Ph.D. Göttingen 1951 Ph.D. NYU

10 students (USC) 1995–96 AMS President

1985 Lobachevsky Prize 9 students

——▶ Fritz John (1910–1994)

1934 Ph.D. Göttingen

23 students

1973 Birkhoff Prize

——▶ Willi Feller ——▶ Henry P. Mc Kean, Jr ——▶ Daniel W. Stroock

(1906–1970) (1930–) (1940–)

1926 Ph.D. Göttingen 1955 Ph.D. Princeton 1967 Ph.D. Rockefeller

17 students 49 students 9 students

(11 Princeton) NAS member NAS member

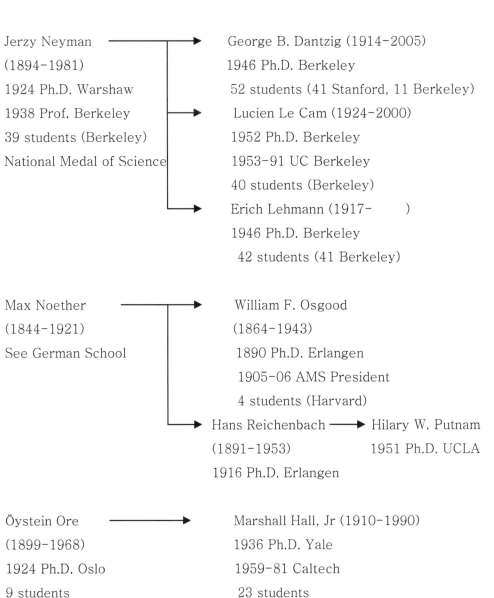

Franz Rellich ➡ Jürgen Moser ➡ Paul H. Rabinowitz
(1906-1955)   (1928-1999)   (1939-   )
1929 Ph.D. Göttingen   1952 Ph.D. Göttingen   1966 Ph.D. NYU
6 students   30 students   11 students (Winconsin)
(Göttingen)   1968 Birkhoff Prize   NAS member
See German School   1994 Wolf Prize   1998 Birkhoff Prize

Joseh Bishop Keller (1923-   )
    1948 Ph.D. NYU
    1996 Wolf Prize
    52 students

Jerzy Neyman ➡ George B. Dantzig (1914-2005)
(1894-1981)   1946 Ph.D. Berkeley
1924 Ph.D. Warshaw   52 students (41 Stanford, 11 Berkeley)
1938 Prof. Berkeley   Lucien Le Cam (1924-2000)
39 students (Berkeley)   1952 Ph.D. Berkeley
National Medal of Science   1953-91 UC Berkeley
                      40 students (Berkeley)
                      Erich Lehmann (1917-   )
                      1946 Ph.D. Berkeley
                       42 students (41 Berkeley)

Max Noether ➡ William F. Osgood
(1844-1921)   (1864-1943)
See German School   1890 Ph.D. Erlangen
                   1905-06 AMS President
                   4 students (Harvard)
Hans Reichenbach ➡ Hilary W. Putnam
(1891-1953)   1951 Ph.D. UCLA
1916 Ph.D. Erlangen

Öystein Ore ➡ Marshall Hall, Jr (1910-1990)
(1899-1968)   1936 Ph.D. Yale
1924 Ph.D. Oslo   1959-81 Caltech
9 students   23 students

Aleksander Rajchman ⟶ Antoni Zygmund

(1890–1940)　　　　　 (1900–1992)

1921 Ph.D. Lwow　　　 1923 Ph.D. Warszawsk

1922–39 Warsaw　　　　 1947–71 Chicago

　　　　　　　　　　　　 40 students

▸Elias Stein (1931–　　)　⟶ ▸Charles Fefferman (1949–　　)

　1955 Ph.D. Chicago　　　　 1969 Ph.D. Princeton

　1999 Wolf Prize　　　　　 1974 Fields Medal

　45 students　　　　　　　 13 students (12 Princeton, 1 Chicago)

　(41 Princeton)

▸ Alberto Calderón　　　　 ▸ Terence Tao (1975–　　)

　(1920–1998)　　　　　　　 1996 Ph.D. Princeton

　1950 Ph.D. Chicago　　　　 2006 Fields Medal

　1979 Birkhoff Prize　　　　 2007 FRS

　1989 Wolf Prize　　　　　 See Australian School

　27 students

▸ Paul Joseph Cohen (1934–2007)

　1958 Ph.D. Chicago

　1964 Bocher Prize

　1966 Fields Medal

　5 students (Stanford)

▸ Guido L. Weiss (1928–　　)

　1956 Ph.D. Chicago

George Chrystal ⟶ Joseph H.M.Wedderburn ⟶ Nathan Jacobson

(1851–1911)　　　　 (1882–1948)　　　　　　 (1910–1999)

1875 B.A. Cambridge　 1903 M.A. Edinburgh　　 1934 Ph.D. Princeton

　　　　　　　　　　　 4 students (Princeton)　　 1971–72 AMS President

　　　　　　　　　　　　　　　　　　　　　　 34 students (Yale)

Oscar Zariski ————————→ David Bryant Mumford

(1899–1986)   (1937–   )

1923 Ph.D. Rome   1961 Ph.D. Harvard

1944 Frank Cole Prize   1974 Fields Medal

1969–70 AMS President   2008 Wolf Prize

1947–69 Harvard   41 students

————————→ Daniel Gorenstein

(1923–1992)

1950 Ph.D. Harvard

NAS member

9 students

————————→ Heisuke Hironaka (1931–)

1960 Ph.D. Harvard

1970 Fields Medal

13 students

————————→ Maxell A. Rosenlicht

(1924–1999)

1950 Ph.D. Harvard

1960 Frank Cole Prize

————————→ Michael Artin (1934–   ) ———→ David Harbater (1952–)

1960 Ph.D. Harvard   1978 Ph.D. MIT

33 students   1995 Frank Cole Prize

NAS member

1991–92 AMS President

————————→ Steven Kleiman ———→ Spencer J. Bloch (1944–   )

(1942–   )   1971 Ph.D. Columbia

1965 Ph.D. Harvard   16 students

28 students   NAS member

Prof. MIT

————————→ Demetrios L. Christodoulou (1951–)

1971 Ph.D. Princeton

1999 Bocher Prize

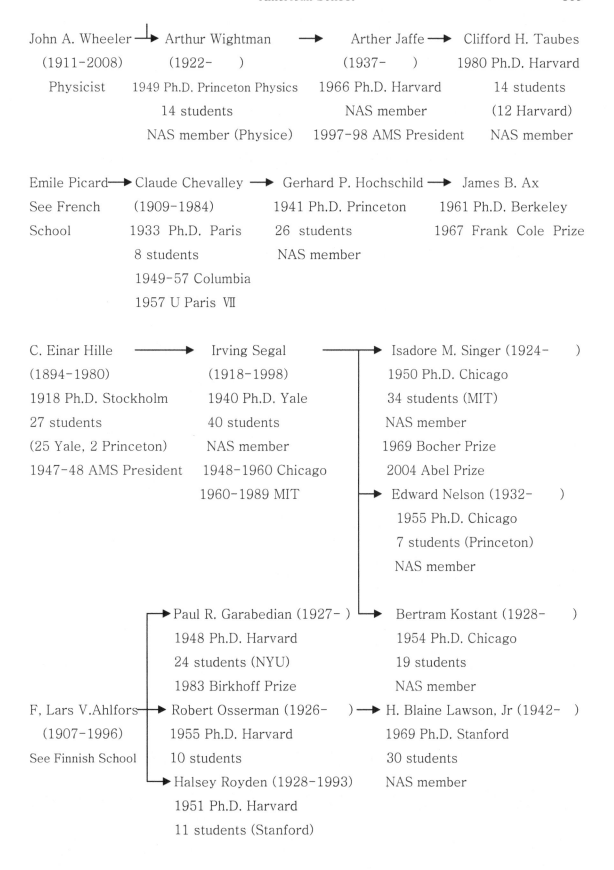

John A. Wheeler ⟶ Arthur Wightman ⟶ Arther Jaffe ⟶ Clifford H. Taubes
(1911-2008)　　　(1922-　　)　　　(1937-　　)　　1980 Ph.D. Harvard
Physicist　　　1949 Ph.D. Princeton Physics　1966 Ph.D. Harvard　　14 students
　　　　　　　　14 students　　　　　　NAS member　　　(12 Harvard)
　　　　　NAS member (Physice)　1997-98 AMS President　NAS member

Emile Picard ⟶ Claude Chevalley ⟶ Gerhard P. Hochschild ⟶ James B. Ax
See French　　(1909-1984)　　1941 Ph.D. Princeton　　1961 Ph.D. Berkeley
School　　　1933 Ph.D. Paris　26 students　　　1967 Frank Cole Prize
　　　　　8 students　　　NAS member
　　　　　1949-57 Columbia
　　　　　1957 U Paris VII

C. Einar Hille ⟶ Irving Segal ⟶ Isadore M. Singer (1924-　)
(1894-1980)　　(1918-1998)　　1950 Ph.D. Chicago
1918 Ph.D. Stockholm　1940 Ph.D. Yale　34 students (MIT)
27 students　　40 students　　NAS member
(25 Yale, 2 Princeton)　NAS member　1969 Bocher Prize
1947-48 AMS President　1948-1960 Chicago　2004 Abel Prize
　　　　　1960-1989 MIT　　Edward Nelson (1932-　)
　　　　　　　　1955 Ph.D. Chicago
　　　　　　　　7 students (Princeton)
　　　　　　　　NAS member

Paul R. Garabedian (1927-)　Bertram Kostant (1928-　)
1948 Ph.D. Harvard　　1954 Ph.D. Chicago
24 students (NYU)　　19 students
1983 Birkhoff Prize　　NAS member

F, Lars V.Ahlfors ⟶ Robert Osserman (1926-　) ⟶ H. Blaine Lawson, Jr (1942-　)
(1907-1996)　　1955 Ph.D. Harvard　　1969 Ph.D. Stanford
See Finnish School　10 students　　30 students
　　　　Halsey Royden (1928-1993)　NAS member
　　　　1951 Ph.D. Harvard
　　　　11 students (Stanford)

Wassilij Höffding ⟶ Donald L. Burkholder (1927-   )

(1914-1991)   1955 Ph.D. UNC

1940 Ph.D. Berlin   19 students (UIUC)

18 students (UNC)   NAS member

? ⟶ Robert Floyd ⟶ Robert E. Tarjan (1948-   )

(1936-2001)   1972 Ph.D. Stanford

1953 B.A. Chicago   1982 Nevanlinna Prize

1969 Prof Stanford   23 students

1978 Turing Award

7 students

Paul Dienes ⟶ Abraham Robinson (1918-1974)

1905 Ph.D. Eötvös   1949 Ph.D. London Birkbeck College

Lorand Univ   1957-62 Hebrew Univ

            1962-67 UCLA

            1967-74 Yale

Stephen Jennings ⟶ Rimhak Ree (1922-2005)

1939 Ph.D. Toronto   1955 Ph.D. UBC

Richard D. Brauer ⟶ Richard H. Bruck ⟶ Michael Aschbacher (1944-   )

(1901-1977)   (1914-1991)   1969 Ph.D. Wisconsin

1926 Ph.D. Berlin   1940 Ph.D. Toronto   5 students

1949 Frank Cole Prize   31 students   NAS member

1957-58 AMS President

Goro Shimura ⟶ Melvin Hochster (1943-   )

(1930-   )   1967 Ph.D. Princeton

1958 Ph.D. Tokyo   31 students

20 students   NAS member

(Princeton)

1977 Frank Cole Prize

C.T. Ionescu Tulcea ⟶ Robert Langlands (1936–) ⟶ James G. Arthur (1944–   )

                                 1960 Ph.D. Yale   1970 Ph.D. Yale

                                 1982 Frank Cole Prize   2005 AMS President

                                 1995 Wolf Prize

                                 NAS member

Enrico Fermi ⟶ Geoffrey Chew ⟶ David Gross ⟶ Edward Witten

(1901–1954)   (1924–   )   (1941–   )   (1951–   )

1922 Ph.D. Pisa   1948 Ph.D. Chicago   1966 Ph.D. Berkeley   1976 Ph.D. Princeton

1938 Nobel Prize   NAS member   1988 Dirac Medal   1985 Dirac Medal

                                 2004 Nobel Prize   1990 Fields Medal

Stanislaw Lesniewski ⟶ Alfred Tarski ⟶ Julia Robinson (1919–1985)

(1886–1939)   (1902–1983)   1948 Ph.D. Berkeley

1912 Ph.D. Lwow   1924 Ph.D. Warszawski   1983 AMS President

                         27 students   NAS member

Donald C. Spencer ⟶ Patrick Gallagher (1935–) ⟶ Dorian M. Goldfeld (1947–   )

(1912–2001)   1959 Ph.D. Princeton   1969 Ph.D. Columbia

1939 Ph.D. Cambridge   22 students (Columbia)   17 students

1948 Bocher Prize   Joseph Kohn (1932–   )   1987 Frank Cole Prize

27 students   1956 Ph.D. Princeton

               16 students

               Phillip A. Griffiths (1938–   )

               1962 Ph.D. Princeton

               1992–2004 Director FAS

               2008 Wolf Prize

               26 students

American School

The first Ph.D. in mathematics in the United States was awarded in 1862 at Yale University. In subsequent years, other university followed suit. Johns Hopkins University in 1878, Harvard University in 1886, University of Chicago in 1896, University of Michigan in 1911, University of Texas in 1916, and M.I.T. in 1925.

By the end of the 19th century, there were 160 Ph.Ds. in mathematics in the United States. Johns Hopkins awarded 30 Ph.Ds, Yale 23, Clark 11, and Harvard 10. The American Mathematical Society was founded in 1888 and the first president was John Howard Van Amringe (1835-1915). As of January 2008 there are 33,000 members.

In 1936, there were 114 holders of European doctorates in mathematics and 1286 Ph.D. degree holders in mathematics awarded in the United States.

From 1862 to 1934, the University of Chicago awarded 237 Ph.Ds, Harvard and Johns Hopkins 103 Ph.D.s respectively, Cornell 89, Yale 79, Illinois 73, Columbia 62, Michigan 55, U Penn 52 and Princeton 48 Ph.Ds.[68]

After Hitler came to power in 1933, many European mathematicians came to the United States. Some of them had difficulties in finding acceptable positions because the US economy continued to feel the lingering effects of the Great Depression. Among the noted immigrants to the United States in those days were[50]

From Germany: Emil Artin, Alfred Brauer, Richard Brauer, Herbert Busemann, Richard Courant, Max Dehn, K.O. Friedrichs, G.P. Hochschild, Hilde Geiringer, Fritz John, Rudolf Carnap, Hans Rademacher, Carl L. Siegel, Richard von Mises, Aurel Wintner, Hermann Weyl, George Lorentz, Hans Lewy, Otto Neugebauer, Emmy Noether, Max Zorn

From Hungary: Paul Erdös, George Pólya, Tibor Radó, Ottó Szász, Gábor Szegö, Theodor von Kármán, John von Neumann, Peter Lax

From Austria: Kurt Gödel, Karl Menger, Abraham Wald, Olga Taussky-Todd

From Czechoslovakia: Charles Loewner

From Yugoslavia: William Feller

From Poland: Nachman Aronszajn, Stefan Bergman, Salomon Bochner, Samuel Eilenberg, Witold Hurewicz, Mark Kac, Jerzy Neyman, Alfred Tarski, Stanislaw Ulam, Antoni Zygmund

From Italy: Oscar Zariski

From Russia: Stefan Warschawski, Alexander Weinstein

From France: Léon Brillouin, Claude Chevalley, Jacques Hadamard, Raphael Salem, André Weil

Eliakim Hastings Moore

Eliakim Moore was born on 26 January 1862 in Marietta, Ohio. He graduated from Yale University in 1883 receiving his B.A. degree. He was awarded a Ph.D. in 1885 at Yale for his thesis *Extentions of certain theorems of Clifford and Cayley in the geometry of n dimensions* supervised by Hubert Anson Newton.

In the late 19th century and early 20th century, many American academics were encouraged to study in Europe after receiving a doctorate in the United States. Moore spent the academic year 1885–86 attending lectures by Kronecker and Weierstrass at the University of Berlin.

Moore took a position as an associate professor of mathematics at Northwestern University.

In 1892 William Rainey Harper, President of the University of Chicago appointed Eliakim Moore as a full professor. Moore brought two German mathematicians Oskar Bolza (1857–1942) and Heinrich Maschke (1853–1908). The University of Chicago provided for the first time the opportunity for American mathematicians to train in a research–intensive environment in the United States.

Among Moore's Ph.D. students at Chicago were Oswald Veblen, George D. Birkhoff, Leonard E. Dickson, Theophil Hildebrandt, Edward Chittenden and Anna Pell Wheeler.[3] Most of them became leaders in American mathematics. R.L. Moore (no relation) was strongly influenced by E.H. Moore although Veblen was his formal adviser at Chicago.

In 1893 Moore married Martha Morris Young, childhood friend. They had two sons, only one of them reached adulthood. In 1896 Moore became head of the mathematics department at Chicago, a post he retained until 1931.

Maschke died in 1908, and his colleague Bolza returned to Germany. For the next decade Chicago maintained an outstanding department but there were signs of in-breeding, and by 1920 the quality began to deteriorate. While Moore's department produced a steady stream of Ph.Ds. they were generally not as impressive as those in the early years.[4]

Eckert Hall to house the mathematics department was constructed in 1930 which was financed by a local businessman. It set a high standard for departmental accommodation.

Moore served as President of the American Mathematical Society from 1901 to 1902 and acted as an editor of the Transactions of the American Mathematical Society from 1899 to 1907. He was elected to the National Academy of Sciences and received many honorary degrees.

Moore died on 30 December 1932 in Chicago, Illinois.

Figure 283. Eliakim H. Moore

(Courtesy of the American Mathematical Society)

Figure 284. Oswald Veblen

(Courtesy of the American Mathematical Society)

Oswald Veblen

Oswald Veblen was born on 24 June 1880 in Decorah, Iowa, USA. His father was a professor of mathematics and physics at the University of Iowa. Veblen received his A.B. in 1898 at the University of Iowa and second B.A. in 1900 at Harvard University.

He was awarded a Ph.D. in 1903 at the University of Chicago for his thesis *A system of axioms for geometry* supervised by E.H. Moore. Veblen became a preceptor at Princeton University in 1905 and was promoted to full professor in 1910.

In 1908 Veblen married Elizabeth Mary Dixon Richardson, a sister of Owen Richardson (1879-1959), Nobel laureate in physics in 1928. Richardson's other sister Charlotte Sarah Richardson married Clinton J. Davisson (1881-1958), Nobel laureate in physics in 1937.

So Veblen became brothers-in-law with Richardson and Davisson, two Nobel laureates. In 1926 Veblen was named Henry B. Fine Professor of Mathematics at Princeton University. Henry B. Fine (1858-1928) played a major role in making Princeton world famous university.

After Henry Fine's tragic death in 1928, his old friends offered funds for the construction and maintenance of a mathematics building in his memory. Veblen took charge of this project. He visited Oxford University in 1928-29 and designed Fine Hall as "Oxford College style". For the interior he worked closely on the furnishings with a high-quality firm of decorators from New York. Faculty members had "studies", not "offices"; some of those were large rooms equipped with fireplaces, carved oak panelling, leather sofas, oriental rugs, and concealed blackboards. Fine Hall included a first-class departmental library, common rooms, and other facilities.[4]

In 1932 Veblen resigned the Henry B. Fine Professorship to become the first professor at the Institute for Advanced Study at Princeton. The original five faculty members in mathematics were Veblen, Einstein, Weyl, von Neumann, and J.W. Alexander. The Institute shared the Fine Hall with the Princeton Mathematics Department until it had its own buildings. Veblen established Princeton as one of the leading centres in the world for topology (analysis situs) research.

After Einstein's theory of general relativity appeared, Veblen turned his attention to differential geometry publishing a few books on that field. He gave a new treatment of spinors, used to represent electron spin.[3] Veblen was President of the American Mathematical Society in 1923-24.

After Hitler came to power in 1933 Veblen helped many European mathematicians to settle in the United States. He was elected to the National Academy of Sciences in 1919 and the *Académie des Sciences* in Paris, the Royal Society of Edinburgh. He was made a Knight of Norway's Royal Order of St. Olaf.

In his last years Veblen became partially blind and he devised a number of aids for blind people.[3]

He died on 10 August 1960 in his summer home in Brooklin, Maine.

Figure 285. George D. Birkhoff

(Courtesy of the American Mathematical Society)

George David Birkhoff

George Birkhoff was born on 21 March 1884 in Overisel, Michigan, USA. He received his A.B. in 1905 and A.M. in 1906 at Harvard University. Birkhoff was awarded a Ph.D.

in 1907 at the University of Chicago for his thesis *Asymptotic properties of certain ordinary differential equations with applications to boundary value and expansion problems* supervised by E.H. Moore.

Birkhoff married Margaret Elizabeth Grafius in 1908 and they had three children. One of their sons Garrett Birkhoff became a famous mathematician.

In 1909 Birkhoff became a preceptor at Princeton University and a full professor in 1911. In 1912 he returned to Harvard as assistant professor, a reflection of the superior academic reputation Harvard enjoyed in those days. He was promoted to full professor in 1919 and named Perkins Professor at Harvard in 1932.

In 1913 Birkhoff proved Poincaré's Last Geometric Theorem, a special case of the 3-body problem. He also wrote extensively on a mathematical theory of aesthetics he had invented.

His main work other than differential equations was on dynamics and ergodic theory. He discovered the "ergodic theorem" in 1931–32, which resolved one of the fundamental problems arising in the theory of gases and statistical mechanics.[3]

In 1923 the American Mathematical Society awarded the first Bôcher Memorial Prize to Birkhoff for his paper, *Dynamical systems with two degree of freedom,* published in 1917. He was President of the American Mathematical Society in 1925–26.

He had the influential position as Dean of the Faculty of Arts and Sciences at Harvard during the period 1935–39.

When many European mathematicians immigrated to the United States after 1933, he found himself in the difficult position of having to balance the claims of young Americans against those of refugees. Birkhoff was compared with Veblen who was more favorable to relocate the European refugees. Consequently Birkhoff was criticized as a great anti-semites. But Birkhoff strongly supported Oscar Zariski's appointment at Harvard in 1947. Zariski was the first Jew to hold a tenure position in the Harvard mathematics department. Birkhoff was elected to the National Academy of Sciences, the *Académie des Sciences* in Paris and many other academies. He championed the cause of Latin American mathematics and persuaded the Guggenheim Foundation to establish a visiting professorship in Latin America. He received thirteen honorary degrees including the University of St. Andrews in 1938.[3]

Towards the end of his life Birkhoff suffered from a weak heart and he died in his sleep at home in Cambridge, Massachusetts on 12 November 1944.[4]

Figure 286. Leonard E. Dickson

(Courtesy of the American Mathematical Society)

Leonard Eugene Dickson

L. E. Dickson was born on 22 January 1874 in Independence, Iowa, USA. He studied at the University of Texas under the influence of George B. Halsted (1853–1922). Dickson received his B.S. in 1893 and M.S. in 1894.

Dickson received the first mathematics doctorate at the University of Chicago in 1896 for his dissertation *The analytic representation of substitutions on a power of a prime number of letters with a discussion of the linear group* supervised by E.H. Moore.

After a short stay at the University of California at Berkeley and the University of Texas, Dickson was appointed assistant professor at the University of Chicago in 1900, promoted to associate professor in 1907 and full professor in 1910. He remained at Chicago for the rest of his career, retiring in 1939.

In 1902 he married Susan McLeod Davis and they had two children.

He wrote his famous book *Linear Groups with an Exposition of the Galois Field Theory* published in 1901 by Teubner Verlag of Leipzig.[3] He also published the three-volume *History of the Theory of Numbers* in 1919–1923 and 16 more books.

Dickson supervised 67 doctoral students at Chicago. This is the second largest number of Ph.D. students in mathematics supervised in the United States. The highest was 72 Ph.Ds. by Wilhelm Magnus. However, Kolmogorov supervised 79 and Hilbert 75.

Dickson was elected to the National Academy of Sciences in 1913 and the *Académie des Sciences* in Paris. He was the first recipient of the Cole Prize by the American Mathematical Society in 1928. He was President of the American Mathematical Society in 1917–18.

Dickson died on 17 January 1954 in Harlington, Texas, USA.

Abraham Adrian Albert

Adrian Albert was born on 9 November 1905 in Chicago, Illinois, USA. Albert received his B.S. in 1926 and M.S. in 1927 at the University of Chicago. He was awarded a Ph.D. in 1928 at Chicago for his thesis *Algebras and their radicals and division algebras* supervised by L.E. Dickson.

In 1927 he married Freda Davis and they had three children. He spent nine months at Princeton University in 1928-29 where Lefschetz suggested to look at open problems in the theory of Riemann matrices. After spending two years at Columbia University he was appointed assistant professor at the University of Chicago, promoted to associate professor in 1937 and full professor in 1941.[3] His main work was on associative algebra, non-associative algebras and Riemann matrices.

He received the Cole Prize in Algebra from the American Mathematical Society in 1939. He also wrote several papers on Jordan Algebras which had been introduced by Pascual Jordan (1902-1980), a German physicist who was one of the founders of quantum mechanics. Albert was elected to the National Academy of Sciences in 1943 and President of the American Mathematical Society in 1965-66.

Albert had intense loyalty to his friends and to his profession. He constantly fought for the improvement of working conditions and student support in his field.

Albert died on 6 June 1972 in Chicago, Illinois, USA.

Figure 287. A. Adrian Albert

(Courtesy of the American Mathematical Society)

Figure 288. Mina S. Rees

Mina Spiegel Rees

Mina Rees was born on 2 August 1902 in Cleveland, Ohio, USA. She graduated Hunter College in 1923 and received her M.A. in 1925 at Columbia University. She was awarded a Ph.D. in 1931 at the University of Chicago for a thesis *Division algebras associated with an equation whose group has four generators* under Dickson's supervision.

From 1932 to 1943 she was a faculty member at Hunter College. During World War II she worked as a technical aid and executive assistant with the Applied Mathematics panel in the Office of Scientific Research and Development.[3] For this work she received the President's Certificate of Merit and King's Medal for Service in the Cause of Freedom from the British Government.[3]

From 1946 to 1953 she worked at the Office of Naval Research, first as Head of the mathematics branch and later as Deputy Science Director. The American Mathematical Society passed a resolution in 1953 recognizing her contribution to post-war development of mathematical research in the United States.

From 1953 to 1961, she returned to Hunter College. In 1961 she moved to the City University of New York where she established graduate research there. She was provost (1968 to 1969) and president (1969 to 1972). The library at the Graduate Center of CUNY is named after her.

She was a member of the National Science Board from 1964 to 1970. She became the first woman president of the American Association for the Advancement of Science in 1971. Rees received the National Academy of Sciences Public Welfare Medal in 1983.[94]

In 1955 Rees married Leopold Brahdy, a physician who died in 1977.[94] She was an accomplished painter.

Rees died on 25 October 1997 in New York City, USA.

Deane Montgomery

Dean Montgomery was born on 2 September 1909 in Weaver, Minnesota, USA. He graduated from Hamline University in Saint Paul, Minnesota in 1929. He was awarded a Master's degree in 1930 and a Ph.D. in 1933 at the University of Iowa for his thesis on point-set topology supervised by Edward W. Chittenden (1885-1977).

In 1933 Montgomery married Katherine (Kay) Fulton and they had two children. From 1935 to 1946, he taught at Smith College and from 1946 to 1948 at Yale University.

In 1948 he moved to the Institute for Advanced Study at Princeton where he stayed until he retired in 1980.[3] His research fields included algebraic and geometric topology. He solved Hilbert's 5th Problem in dimension three in 1948 and he had solved the problem with the assumption of finite dimensionality and this restriction was removed by Yamabe, Montgomery's assistant in 1952.[3] Andrew Gleason is also credited to solve Hilbert's 5th problem.[12]

Montgomery was President of the American Mathematical Society in 1961–62 and he was awarded the Steele Prize by the same society in 1988. He was elected to the National Academy of Sciences in 1955 and he served as President of the International Mathematical Union (1974–1978). He was always seeking out and encouraging young mathematicians.

He died on 15 March 1992 in Chapel Hill, North Carolina near his daughter's home.

Figure 289. Deane Montgomery

(Courtesy of the American Mathematical Society)

Figure 290.   Robert L. Moore

(Courtesy of the American Mathematical Society)

Robert Lee Moore

R.L. Moore was born on 14 November 1882 in Dallas, Texas, USA. Before he entered university, he had learnt university level calculus by independently studying the university textbooks. He entered the University of Texas in 1898 where he took courses by George B. Halsted (1853–1922) and L.E. Dickson (1874–1954). Professor Halsted introduced Moore to the work of Bolyai, Lobachevsky and Hilbert. He graduated from the University of Texas gaining his B.S. and M.A. degrees simultaneously in 1901.[69] Moore, at the age of 18, made his name producing a "delightfully elegant and simple proof" demonstrating redundancy of one of Hilbert's geometry axioms.

E.H. Moore (no relation) heard of this contribution, and decided to award Robert Moore a fellowship to study for his Ph.D. at Chicago.

Meanwhile, Halsted, who had hoped to keep Moore at Texas, nominated him for an instructorship in mathematics in 1902, but another person was appointed. Halsted criticized the university for its decision and, as a result, was dismissed from his professorship in 1902.

After teaching one year at Marshall High School in Texas, Moore arrived in Chicago in September 1903 and received his Ph.D. in 1905 for his thesis *Sets of metrical hypotheses for geometry* supervised by Oswald Veblen.[69]

Moore spent one year at the University of Tennessee, two years (1906-1908) at Princeton University and three years (1908-1911) at Northwestern University before he joined the University of Pennsylvania in 1911 where he spent nine years. Moore started his famous, unique teaching method at the University of Pennsylvania. John Parker states in his book.[69]

1. There would be no textbooks linked to the course and none was to be consulted. Moore quickly developed his personal "radar" for spotting those who had accidentally or otherwise became exposed to the work at hand.

2. Students were asked to prove theorems from given axioms and present their proofs in class without seeking help externally or discussing the problem with each other.

3. The rest of the class would then be encouraged to criticize weaknesses or inaccuracies in the presented proofs.

4. He encouraged a strong spirit of competitiveness among his students and devised various means to promote it.

5. From the outset he placed a good deal of emphasis on logic.

Moore eventually returned to the University of Texas in 1920 as an associate professor at the annual salary of $3,000 and was promoted to full professor three years later. Moore wrote *Foundations of Point Set Topology,* which was published in 1932. The term point-set topology was coined by him.

He was elected to the National Academy of Sciences in 1931 and President of the American Mathematical Society from 1936 to 1938. He taught his last class at the University of Texas in summer of 1969 when he was 86 years old.

He refused to teach black students. There were only one woman, Mary Ellen Rudin, and one Jew E.E. Moise, among his students. He had an ongoing internal feud with his colleague Harry S. Vandiver (1882-1973). Vandiver was a pure mathematician but he had to start a new department of applied mathematics in 1946.[69] Robert Moore supervised 50 doctoral students in Pennsylvania and Texas.[10]

Moore died on 4 October 1974 in Austin, Texas, USA.

Harold Calvin Marston Morse

Marston Morse was born on 24 March 1892 in Waterville, Maine, USA. Marston was his mother's maiden name. He received his B.A. from Colby College in Maine in 1914 and a Master's degree from Harvard University in 1915. He received a Ph.D. from Harvard in 1917 for a thesis *Certain types of geodesic motion of a surface of negative curvature* supervised by G.D. Birkhoff.

During World War I, he fought in the US Army in France and he was awarded the *Croix de Guerre* with Silver Star for his outstanding work in the Ambulance Corps.

Morse taught one year at Harvard (1919-20), five years at Cornell University (1920-25), one year at Brown University (1925-26). He subsequently returned to Harvard, where he stayed until 1935, during which time he was promoted to full professor in 1929.[3]

He married Céleste Phelps in 1922 and they had two children, but ended in divorce in 1930. In August 1932 Céleste Phelps married William F. Osgood (1864-1943), professor of mathematics at Harvard who was 68 years old at the time of marriage. In 1940 Morse married Louise Jeffreys and they had five children.[3]

In 1935 he moved to the Institute for Advanced Study at Princeton as the sixth permanent member in mathematics and remained there until 1962.

In his 1925 paper, *Relations between the critical points of a real function of n independent variables,* he developed Morse theory. In differential topology, the techniques of Morse theory gives a very direct way of analyzing the topology of a manifold by studying differentiable functions on that manifold.

His name can be found in mathematical terms: the Morse lemma, the Morse inequalities, Morse homology and Morse-Bott theory.

Morse received the Bôcher Prize in 1933, the Presidential Certificate of Merit for his war work in 1947, the National Medal for Science for his mathematical contributions and

Legion d'honneur from France. He was President of the American Mathematical Society in 1941-42. He received 20 honorary degrees.

Morse died on 22 June 1977 in Princeton, New Jersey, USA.

Figure 291. H. C. Marston Morse

(Courtesy of the American Mathematical Society)

Figure 292. Joseph L. Walsh

(Courtesy of the American Mathematical Society)

Joseph Leonard Walsh

Joseph Walsh was born on 21 September 1895 in Washington, D.C. He graduated from Harvard University in 1916 and received a Master's degree by the University of Wisconsin in 1917. During World War I, he served in the United States Navy. He was awarded a Ph.D. at Harvard in 1920 for his thesis *On the location of the roots of a Jacobian of two binary forms, and of the derivative of a rational function* supervised by G.D. Birkhoff.

Beginning as an instructor at Harvard in 1921, Walsh became a full professor in 1935. In 1946 he was named Perkins Professor of Mathematics, the position he held for 20 years. He married twice, first to Aline Natalie Burgess in 1931 and second to Elizabeth Cheney Strayhorn in 1946.[3] During World War II, he served as Navy officer reaching the rank of Commander in 1944. He became a Captain in the Naval Reserve in the 1950s. After the war, he returned to Harvard. Upon retiring, he worked at the University of Maryland in College Park until his death.

He had 31 Ph.D. students and published 7 books and 279 papers. The Walsh functions consist of trains of square pulses with the allowed states being -1 and +1 such that transitions may only occur at fixed intervals of a unit time step and the functions satisfy orthogonality relations. Walsh functions are used in electrical engineering.

He was elected to the National Academy of Sciences in 1936 and he served as President of the American Mathematical Society in 1949-50.

He died on 6 December 1973 in College Park, Maryland, USA.

Hassler Whitney

Hassler Whitney was born on 23 March 1907 in New York. His grandfather was William D. Whitney, a famous Sanskrit scholar and his maternal grandfather was Simon Newcomb (1835-1909), the fourth president of the American Mathematical Society. Whitney graduated from Yale University in 1928 and received his doctorate at Harvard in 1932 for a dissertation *The coloring of graphs* under G.D. Birkhoff's supervision.

He married Margaret R Howell in 1930 and they had three children. This marriage ended in divorce and he married Mary Barnett Garfield in 1955 producing two more children. When he was 79 years old, he married to Barbara F Osterman in 1986.

Whitney was appointed instructor at Harvard in 1930 and promoted to full professor in 1946. He moved to the Institute for Advanced Study at Princeton in 1952 where he retired in 1977.

He was elected to the National Academy of Sciences in 1945 and to the *Académie des Sciences* in Paris. He was vice-president of the American Mathematical Society from 1948 to 1950. He was a good mountaineer all his life.

He received the National Medal of Science in 1976 and the Wolf Prize in 1983. His main work was in topology and his name is honored by the term Stiefel-Whitney Characteristic Classes.

Whitney died on 10 May 1989 in Mount Dents Blanches, Switzerland.

Figure 293. Joseph L. Doob

(Courtesy of the American Mathematical Society)

Joseph Leo Doob

Joseph Doob was born on 27 February 1910 in Cincinnati, Ohio, USA. When he was a teenager, he obtained an amateur radio operator licence and built a radio transmitter himself. At Harvard University he intended to major in physics, but changed to mathematics. He received a Bachelor's degree in 1930 and a Ph.D. in 1932 at Harvard University at the age of 22. His dissertation was *Boundary values of analytic functions* under Joseph Walsh's supervision.[3] Ironically, Doob thought that he was ignorant of almost everything in mathematics outside of his thesis because he obtained his Ph.D. in two years.

In June 1931 he married Elsie Haviland Field, a medical student at that time. They had three children. After spending the academic year 1934–35 at Columbia University on a Carnegie fellowship, he was appointed Associate Professor at the University of Illinois in the fall of 1935. At only 25 years of age, he looked like a graduate student, crew cut, shirt sleeves. But Doob was the first well-informed modern mathematician in the mathematics department at Illinois.

During World War II, he worked for the US Navy on mine warfare in Washington, D.C. from 1942 to 1945. He was promoted to full professor at Illinois in 1945. In 1953 he published *Stochastic Processes* and *Classical Potential Theory and its Probabilistic Counterpart* in 1984. Both of them became classic texts.

He was elected to the National Academy of Sciences and the *Académie des Sciences* in Paris. He was president of the American Mathematical Society in 1963–64. He received the National Medal of Science in 1979 and the Steele Prize in 1984.

Even after retiring from the University of Illinois in 1978, his work in mathematics continued. He wrote a well known book on measure theory in 1984 when he was 74 years old.[3] He was commissioner of the Champaign–Urbana Saturday Hike for about 25 years and joined the hikers regulary every Saturday.

Doob had an outgoing, relaxed personality, he seemed to get along with everyone. He was never intellectually dishonest.[70]

Doob died on 7 June 2004 in Clark–Lindsey Village, Urbana, Illinois, USA.

Paul Richard Halmos

Paul Halmos was born on 3 March 1916 in Budapest, Hungary. His mother died when he was six months old and his father, a physician, immigrated to the United States in

1924. Paul came to the United States in 1929 and entered Walter High School in Chicago as 11th grade student skipping four years. He entered the University of Illinois in 1931 at the age of fifteen and graduated in three years in 1934 majoring in philosophy and mathematics.[70]

Halmos became the first Ph.D. student of Joseph Doob and he was awarded his doctorate in 1938 for his thesis *Invariants of certain stochastic transformation: The mathematical theory of gambling systems.*

From 1939 to 1942, he was at the Institute for Advanced Study at Princeton and in his second year he became von Neumann's assistant. In 1942 Halmos published *Finite Dimensional Vector Spaces,* which was based on his lecture "Elementary Theory of Matrices" given at Princeton University.[70] The Institute for Advanced Study at that time shared the Fine Hall with Princeton mathematics department.

From 1943 to 1946, Halmos worked at Syracuse University. In September 1946 he accepted assistant professorship at the University of Chicago for $5,500 annually. Three years later he was promoted to associate professor with $500 raise. In 1956 he was promoted to full professor and his salary in 1960 climbed $14,000.[70]

In 1954 his passport was denied because he refused to cooperate with the American Embassy in Uruguay. They asked Halmos to report on the activities of a student organization, *Federación de Estudiantes Universitarios del Uruguay.*

In the 1950s, the mathematics department at Chicago included luminaries such as Marshall Stone, Adrian Albert, S.S. Chern, Saunders MacLane, Antoni Zygmund and Irving Kaplansky.

Halmos had some difficulties with MacLane who considered Halmos second-rate. Halmos moved to the University of Michigan, where he stayed eight years at an annual salary of $20,000.

In 1968–69 he went to the University of Hawaii as chairman of the mathematics department. Then he moved to Indiana University in 1969 where he stayed until 1985 except two years at the University of Santa Clara in 1975-77 at the annual salary of $41,000.[70]

His main interests were in ergodic theory, operator theory and functional analysis. He wrote many books. Halmos considered himself to belong in the fourth rank among mathematicians. His classifications are[70]:

First Rank: Archimedes, Gauss, Riemann, Poincaré, Hilbert
1.5 Rank: Paul Cohen, Charles Fefferman, Kurt Gödel
Second Rank: Felix Klein, Saunders MacLane

Third Rank: George Mackey, Alfred Tarski, Antoni Zygmund

Fourth Rank: Garrett Birkhoff, Charles Rickart, Jean Dieudonne, Kazimierz Kuratowski,

Paul Halmos

Halmos received the Steele Prize from American Mathematical Society in 1983. He died on 2 October 2006 in Los Gatos, California, USA.

Figure 294. Paul R. Halmos

Figure 295. David H. Blackwell

David Harold Blackwell

David Blackwell was born on 24 April 1919 in Centralia, Illinois, USA. He entered the University of Illinois in 1935 at age 16 and graduated in three years of study. He was awarded a Ph.D. at Illinois in 1941 for a thesis, *Properties of Markov chains,* supervised by Joseph Doobs.[10]

After one year at the Institute for Advanced Study (1941-1942), he worked at the Southern University at Baton Rouge, Louisiana and Clark College in Atlanta, Georgia. He was appointed as an Instructor at Howard University in Washington, D.C. in 1944, and three years later, was promoted to full professor. He moved to the University of California at Berkeley in 1954 strongly supported by Jerzy Neyman. In the ensuing 35 years, until his retirement from Berkeley, in 1989, Blackwell supervised 65 Ph.D. students.[10]

Blackwell published *Theory of Games and Statistical Decisions* in 1954 jointly with Abe Girshick.

One day in 1941 Blackwell explained his Ph.D. thesis to John von Neumann at the Institute for Advanced Study.

Neumann listened for about ten minutes and asked a couple of questions, and then he said-what you have really done is this, and probably this is true, and you could have done it in a somewhat simpler way, and so forth. Blackwell said "He listened to me talk about this rather obscure subject and in ten minutes he knew more than I did."[50]

In 1955 Blackwell was elected President of the Institute of Mathematical Statistics. In 1965 he became a member of the National Academy of Sciences. He received the John von Neumann Theory Prize from the Operations Research Society of America in 1979.

He had found a game theory proof of the Kuratowski Reduction theorem by connecting the areas of game theory and topology.[3]

Blackwell has about 40 acres land in Mendocino County, California where he enjoys outdoor life.[94]

Raoul Bott

Raoul Bott was born on 24 September 1923 in Budapest, Hungary. His mother was Jewish of Hungarian descent. His parents split up and his mother remarried. Bott lived with his mother and step father. When Raoul was eleven years old, his mother died and his step father remarried. Raoul was interested in electricity as a young boy. The step father's family emigrated to Canada and Bott entered McGill University in 1941 studying electrical engineering. He graduated in 1945.

In 1947 he married Phillis, a student of English literature. Bott went to Carnegie Institute of Technology (now Carnegie Mellon University) where he received a Ph.D. for a thesis Electrical Network Theory supervised by Richard Duffin in 1949. Using a theorem by P.I. Richards published in Duke Mathematical Journal No 14 (1947), Bott was able to prove that the Brune function can be synthesized without an ideal transformer.

The positive real function (Brune function) is defined as Re $Z(s) \geq 0$ for Re $s \geq 0$ where $s = \sigma + i\omega$ is a complex variable and $Z(s)$ is real when s is real. s is called the complex frequency in electrical engineering.

P.I. Richards used the transformation

$$R(s) = \frac{kZ(s) - sZ(k)}{kZ(k) - sZ(s)}$$

and showed that if $Z(s)$ is positive real function, $R(s)$ is also positive real for real, positive value of k.

Then $$Z(s) = \frac{1}{\dfrac{1}{Z(k)R(s)} + \dfrac{s}{kZ(k)}} + \frac{1}{\dfrac{k}{Z(k)s} + \dfrac{R(s)}{Z(k)}}$$

which can be synthesized by two network in parallel eliminating the undesirable ideal transformer.[71]

From 1949 to 1951 he worked at the Institute for Advanced Study in Princeton where he learned topology. He was appointed to the University of Michigan in 1951 where his first doctoral student was Stephen Smale, Fields Medalist in 1966. Bott moved to Harvard University in 1959 as a full professor where one of his student was Daniel Quillen, Fields Medalist in 1978.

Bott's name is remembered by the Bott periodicity theorem (1956) and the Morse-Bott functions.

Bott received the National Medal of Science in 1987, the Steele Prize in 1990 and the Wolf Prize in 2000. He was elected to the National Academy of Sciences in 1964 and the *Académie des Sciences* in Paris in 1995.

Bott died on 20 December 2005 in San Diego, California.

Figure 296. Raoul Bott

Figure 297. Stephen Smale

Stephen Smale

Stephen Smale was born on 15 July 1930 in Flint, Michigan, USA. He studied physics at the University of Michigan, but after failing a course in nuclear physics, he switched to mathematics in his senior year receiving his B.S. in 1952 and M.S. in 1953. As a graduate student his poor work habits earned him a C average. T.H. Hildebrandt, the mathematics department chairman, threatened to kick Smale out, so he began to work hard.

His Ph.D. adviser, Raoul Bott, was not a well-known mathematician at that time who joined the Michigan faculty in 1951. Bott provided Smale with an excellent problem: to classify, up to regular homotopy, regular closed curves in an arbitrary manifold. Smale's work on that question resulted in a Ph.D. in 1957 for his thesis *Regular curves in Riemanian manifolds*.

As an Instructor at the University of Chicago in 1958, he proved the counter-intuitively famous theorem that one can turn a 2-sphere in R^3 inside out. In 1904, Poincaré conjectured that any three-dimensional space (closed 3-manifold) in which every loop can be continuously contracted is just a three-dimensional sphere.

Smale proved the higher dimensional Poincaré conjecture in 1961 for n at least 5. Only in 2002 the original Poincaré conjecture was completely solved by the Russian mathematician Grigori Perelman.

Smale spent 1958-60 at the Institute for Advanced Study at Princeton on a National Science Foundation (NSF) Postdoctoral Fellowship. In 1960 he was appointed an associate professor of mathematics at the University of California at Berkeley where he stayed until 1995. From 1961 to 1964 he taught at Columbia University.

Smale became well-known for his political activities. He participated in anti-war demonstrations in California in the 1960s, including co-organizing the 1905 "Vietnam Day" teach-in with activist Jerry Rubin. At one time he was subpoenaed by the House Un-American Activities Committee.

Smale was awarded a Fields Medal at the International Congress of Mathematicians at Moscow in 1966 for his work on the generalized Poincaré conjecture in dimension $n >= 5$. He held a press conference, criticizing the US policy in Vietnam on the steps of Moscow State University where the Congress was held.[89]

In 1995 he moved to the City University of Hong Kong and currently a professor at the Toyota Technological Institute at Chicago where he works with a neuroscientist, Tommy Poggio, on trying to find some kind of mathematical model for human vision in the visual cortex.[72]

In 1996 he received the National Medal of Science from President Clinton. In 1998 he announced a list of 18 problems to be solved in 21st century at the International Mathematical Union in Paris.

He is a member of the National Academy of Sciences.

Smale collected minerals for over 30 years and published a book of his collection. His other hobbies are taking photographs and making an ocean passage to the Marquessas islands in South Pacific in his 43-foot ketch with a two-mathematician crew.[89]

In 2007, Smale received the Wolf Prize in mathematics to became one of the twelve Fields Medalists to win both prizes.

Figure 298. Daniel G. Quillen

Figure 299. William P. Thurston

Daniel Grey Quillen

Daniel Quillen was born on 27 June 1940 in Orange, New Jersey, USA. He received his bachelor's degree at Harvard University in 1961 and a Ph.D. in 1964 for his thesis *Formal properties of over-determined systems of linear partial differential equations* under Raoul Bott's supervision.

Quillen was appointed to the faculty of Massachusetts Institute of Technology, but spent many years undertaking research at other institutions including work at Oxford University in England. He was strongly influenced by Alexandre Grothendieck and Michael Atiyah.

Quillen received a Fields Medal at the International Congress of Mathematicians in Helsinki in 1978 as the principal architect of the higher algebraic K-theory, a new tool that successfully used geometric and topological methods and ideas to solve major problems in algebra.[3]

He is a member of the National Academy of Sciences. He and his wife Jean, a violinist, have five children.

William Paul Thurston

William Thurston was born on 30 October 1946 in Washington, D.C, USA. He graduated from New College of California in San Francisco in 1967. He received his doctorate in 1972 at the University of California at Berkeley for a thesis, *Foliations of 3-Manifolds which are Circle Bundles,* supervised by Morris Hirsch and Stephen Smale. During his graduate studies at Berkeley, he was a conscientious objector.[89]

After working the academic year 1972-73 at the Institute for Advanced Study at Princeton, he was appointed an assistant professor at MIT in 1973. One year later he was appointed professor of mathematics at Princeton University. Since 2003 he is a professor at Cornell University.

Thurston received a Fields Medal at the International Congress of Mathematicians in Warsaw in 1983 for the depth and originality of his contributions to mathematics. His ideas revolutionized the study of topology in 2 and 3 dimensions, and introduced useful connections among analysis, topology and geometry.[3]

He was awarded the Oswald Veblen Geometric Prize of the American Mathematical Society in 1976, the Alan T. Waterman Award in 1979. He won the first AMS Book Prize for his *Three Dimensional Geometry and Topology.* Thurston is a member of the National Academy of Sciences.

Stefan Bergman

Stefan Bergman was born into a Jewish family on 5 May 1895 in Czestochowa, Poland. He received his Ph.D. in 1921 at the University of Berlin for his thesis *Über die Entwicklung der harmonischen Functionen der Ebene und des Raumes nach Orthogonal Funktionen* supervised by von Mises.[10]

He was dismissed from his post at the University of Berlin by the Nazis in 1933 and stayed in Russia until 1937. In 1939 he went to the United States with von Mises as his sponsor. He spent most of his remaining career at Stanford University.

Bergman is best known for his kernel function, which he invented in 1922 in Berlin, and the Bergman projection, as well as his research in several complex variables.

When his wife died, the terms of her will established the Bergman Prize, which was administered by the American Mathematical Society. David Catlin received the first Bergman Prize in 1989.

Bergman died on 6 June 1977 in Palo Alto, California, USA.

Figure 300.   Marshall H. Stone

(Courtesy of the American Mathematical Society)

Marshall Harvey Stone

Marshall Stone was born on 8 April 1903 in New York City. His father Harlan Fiske Stone (1872-1946) was a famous lawyer serving the dean of Columbia Law School, Attorney General in Coolidge Administration (1923-1925), and Associate Justice (1925-1941), and Chief Justice of the US Supreme Court (1941-1946).

Marshall Stone entered Harvard University in 1919 intending to study at Harvard law school. He completed his undergraduate work in three years, during which time his interest moved to mathematics. Stone remained at Harvard, where he earned a Ph.D. in 1926 for his thesis *Ordinary linear homogenous differential equations of order n and the Related expansion problems* under G.D. Birkhoff's supervision.[3]

In 1927 he married Emmy Portman. They had three children but the marriage ended in divorce in 1962. He remarried Ravijojla Kostic.

Stone was at Harvard from 1927 to 1946 except two years at Yale in 1931-33. He published his 662 page book *Linear Transformations in Hilbert Space and Their*

Applications to Analysis in 1932. In the same year Banach published *Théorie des Opérations linéaires.*

He was appointed head of the mathematics department at the University of Chicago in 1946. Stone greatly improved its reputation by inviting André Weil, Saunders MacLane, Antoni Zygmund and S.S. Chern to join the department. Stone's chairmanship period of 1946-52, became affectionately known as "Stone age". He left Chicago in 1968 and moved to the University of Massachusetts where he stayed until 1980.

He was elected to the National Academy of Sciences in 1938 and he served as president of the American Mathematical Society in 1943-44. Stone was president of the International Mathematical Union in 1952-54.

Stone also proved results on spectral theory, arising from group theoretical methods in quantum mechanics conjectured by Hermann Weyl. Stone is best known for Stone-von Neumann uniqueness theorem, Stone-Cech compactification theory, Stone-Weierstrass theorem.

He enjoyed cooking and travel. He died on 9 January 1989 in Madras, India while travelling there.

Figure 301. Andrew M. Gleason

(Courtesy of the American Mathematical Society)

Figure 302. George W. Mackey

George Whitelaw Mackey

George Mackey was born on 1 February 1916 in St. Louis, Missouri, USA. He graduated from the Rice Institute (now Rice University) in Houston, Texas majoring in physics in 1938. He received a Ph.D. at Harvard University in 1942 for his thesis *The subspaces of the conjugate of an abstract linear space* supervised by Marshall Stone.[3]

He was appointed as an Instructor at Harvard in 1943 and promoted to full professor of mathematics at Harvard, where he retired in 1985.

Mackey wrote many books, *The Mathematical Foundations of Quantum Mechanics* in 1963, *Lectures on the Theory of Functions of a Complex Variable* in 1967, *The Theory of Unitary Group Representations* in 1976, *Unitary Group Representations in Physics, Probability, and Number Theory* in 1978 and *The Scope and History of Commutative and Noncommutative Harmonic Analysis* in 1992.

Clifford Taubes, chairman of Harvard University's mathematics department, said Mackey was a very honest person intellectually.[3] He became a member of the National Academy of Sciences in 1962. He was vice president of the American Mathematical Society in 1964-65 and received the Society's Steele Prize in 1975.

Mackey married Alice Willard in 1960 and they had one daughter.

He died on 15 March 2006 in Belmont, Massachusetts, USA.

Andrew Mattei Gleason

Andrew Gleason was born on 4 November 1921 in Fresno, California, USA. He graduated from Yale University in 1942. He received an honorary MA degree from Yale University in 1953.[89] He worked as a code breaker for the United States Navy during World War II. In 1946 Harvard University appointed him as a Junior Fellow, which is a very prestigious academic position similar to the research fellow of Colleges at Cambridge University. Gleason did not need to get a doctorate. Several Junior fellows won Nobel Prizes.

During the Korean War (1950-1953) Gleason went back to Navy service. After the war, he returned to Harvard where he was steadily promoted, becoming a full professor in 1957.

He married Jean Berko, a psychologist, in 1959 and they had three daughters.

In 1969 he was appointed the Hollis Professor of Mathematics and Natural Philosophy at Harvard, where he retired in 1992.

Gleason's paper in 1952 *Groups Without Small Subgroups,* gave a complete solution to Hilbert's fifth problem which States "A connected locally compact group G is a projective limit of a sequence of Lie groups; and, if G has no small subgroups, then it is a Lie group".[3] Gleason, Montgomery, Zippin and Yamabe are credited as solvers to the fifth problem.[12]

In 1966 Gleason published *Fundamentals of Abstract Analysis,* which was highly praised by Jean Dieudonné. Gleason served as president of the American Mathematical Society in 1981–82. He is member of the National Academy of Sciences.

Karen Keskulla Uhlenbeck

Karen Uhlenbeck was born on 24 August 1942 in Cleveland, Ohio, USA. She graduated from the University of Michigan in 1964. Karen married the son of George Uhlenbeck (1900–1988), a discoverer of the election spin.

She received a Ph.D. at Brandeis University in 1968 for her thesis *Calculus of Variations and Global Analysis* supervised by Richard Palais.

After temporary positions at Massachusetts Institute of Technology and the University of California, Berkeley, she worked for twelve years at the University of Illinois (1971–1983) and five years at the University of Chicago (1983–1988).

Since 1988 she has been at the University of Texas at Austin where she holds the Sid W Richardson Foundation Regents Chair in Mathematics.

In 1990 she was a Plenary Speaker at the International Congress of Mathematicians in Kyoto. She was the second woman (after Emmy Noether in 1932) to give a Plenary Lecture at the Congress.

Uhlenbeck is a leading expert on partial differential equations and her work provided analytic tools to use instantons as an effective geometric tool.[3]

Figure 303.　Karen K. Uhlenbeck　　　　**Figure 304.　Frank Morley**

(Courtesy of the American Mathematical Society)

Frank Morley

Frank Morley was born on 9 September 1860 in Woodbridge, Suffolk, England. He graduated from King's College, Cambridge with a B.A. in 1884. He taught at a secondary school for three year before he emigrated to the United States in 1887. He worked at Haverford College in Pennsylvania from 1887 to 1900.

In 1889 he married Lilian Janet Bird, a musician and poet. They had three sons. The youngest son, Frank Vigor Morley (1899-1985), was a mathematician who collaborated with his father over twenty years[3]

Morley was appointed Professor of Mathematics at Johns Hopkins University in 1900 where he supervised 49 Ph.D. students.[10]

Morley was the editor of the American Journal of Mathematics for 30 years and he served as president of the American Mathematical Society in 1919-20.

He died on 17 October 1937 in Baltimore, Maryland, USA.

Harry Bateman

Harry Bateman was born on 29 May 1882 in Manchester, England. He was Senior Wrangler in the Mathematical Tripos Examinations of 1903 when he was awarded his B.A. at Trinity College, Cambridge. He won a Smith's prize in 1905 for an essay on differential equations and became a Fellow of Trinity College.

In 1906 he was appointed a lecturer at Liverpool University and a Reader in mathematical physics at the University of Manchester in 1907. In the same year, Ernest Rutherford was appointed to Chair of Physics at Manchester. In 1910 Bateman emigrated to the United States and spent the years 1912-1917 at Johns Hopkins University as a research fellow.

He received a Ph.D. at John Hopkins in 1913 for his thesis *The quartic curve and its inscribed configurations* under Frank Morley's supervision. At the time Bateman already had over 60 publications.[3]

Bateman was appointed in 1917 as a professor in the Division of Mathematics, Theoretical Physics and Aeronautics at Throop College of Technology (now California Institute of Technology).

Bateman was one of the first to apply Laplace transforms to integral equations in 1906. In 1910 he solved systems of differential equations discovered by Rutherford which describe radio-active decay.

He wrote many books and around 200 papers over the course of his 40 year career. Bateman was elected to the Royal Society of London in 1928 and to the National Academy of Sciences in 1930. In 1912 he married Ethel Horner Dodd. They had one son and one adopted daughter.

Bateman was a top chess player. Bateman died of a coronary thrombosis on 21 January 1946 on a train near Milford, Utah on his way to New York to receive an award from the Institute of Aeronautical Science.[3] After his death Arthur Erdélyi and the staff of the Bateman Manuscript Project published *Higher Transcendental Functions* in three volumes and *Tables of Integral Transforms* in two volumes.

Figure 305.　Harry Bateman

Figure 306.　James W. Alexander

James Waddell Alexander

James Alexander was born on 19 September 1888 in Sea Bright, New Jersey, USA. He received a B.S. degree in 1910 and an M.S. degree in 1911 at Princeton University, where he was a student of Veblen.

His mother had the same family name Alexander and her father John Waddell Alexander was the President of the Equitable Life Insurance Company. James Alexander's father John White Alexander was a famous artist.

James studied at the Universities of Paris and Bologna in 1912 before he submitted his Ph.D. dissertation *Functions which map the interior of the unit circle upon simple regions* at Princeton University in 1915 supervised by T.H. Gronwall, Swedish-American mathematician.[11]

In 1917 he married Natalia Levitzkaya, a Russian woman he met in Paris. They had two children. He was a lieutenant of the United States Army at the Aberdeen Proving Ground during World War I.

Alexander became assistant professor at Princeton in 1920, associate professor in 1926 and full professor in 1928. In 1924 he persuaded Veblen to bring Solomon Lefschetz to Princeton from Kansas. With Alexander and Lefschetz, Princeton became one of the best research centers on topology attracting many of the leading researchers from Europe: Aleksandrov, Cech, Hopf and Hurewicz.[11]

In 1933 Alexander became one of the original five permanent members of mathematics at the Institute for Advanced Study in Princeton where he retired in 1951. He never drew a salary from the Institute because he was a millionaire through inherited wealth.

During World War II, he worked at the US Army Air Force Office of Scientific Research and Development. He was known as a socialist and the atmosphere of the McCarthy era pushed him into greater seclusion.

The last time Alexander was seen in public was July 1954 when he signed a letter supporting J. Robert Oppenheimer, who had lost his security clearance.[3]

He was a noted mountaineer. The Alexander Chimney in the Rocky Mountain National Park is named after him. He liked to climb the University buildings and always left his office window on the top floor of Fine Hall at Princeton University open so that he could enter by climbing the building.

After his wife Natalia died in 1967 his health slowly declined. Alexander died of pneumonia in the Princeton Hospital on 23 September 1971.

His name can be found in many mathematical terms: Alexander horned sphere, Alexander polynomial, Alexander Cochain, Alexander-Spanier cohomology, Alexander duality and Alexander trick.

Alonzo Church

Alonzo Church was born on 14 June 1903 in Washington, D.C., USA. He graduated from Princeton University in 1924 and he was awarded a Ph.D. in 1927 for a thesis *Alternatives to Zermelo's assumption* supervised by Veblen.

Church spent two years as a National Research Fellow, one year at Harvard and a year at Göttingen and Amsterdam. He became Assistant Professor of Mathematics at Princeton in 1929, Associate Professor in 1939 and Full Professor in 1947.

In 1961 he became professor of mathematics and philosophy at Princeton where he retired in 1967. Then he was appointed as Kent Professor of Philosophy and professor of mathematics at the University of California at Los Angles (UCLA) in 1967 where he retired in 1990 at the age of 87.[3]

His main field of interest was in mathematical logic. He published the classic book *Introduction to Mathematical Logic* in 1956, which was a revised and enlarged edition of his 1944 book.

He was elected to the National Academy of Sciences in 1978 and received three honorary degrees.

Alan Turing, Simon B. Kochen and Stephen Kleene were among his 31 doctoral students.

Church died on 11 August 1995 in Hudson, Ohio.

Figure 307. Alonzo Church

Figure 308. R.H. Bing

(Courtesy of the American Mathematical Society)

R.H. Bing

R.H. Bing was born on 20 October 1914 in Oakwood, Texas, USA. His father, Rupert Henry Bing, was Superintendent of the Oakwood School District in Texas and his mother was a primary school teacher. Somehow his name was "R. H", which is not an abbreviation.

His mother greatly influenced his early education. After his father died in 1919, the family had to live very frugally. Bing went to the Southwest Texas State Teacher's College to become a teacher. He graduated in 1935 completing his B.A. in two and half years. Bing taught four and a half year at three different schools.[3]

He married Mary Blanche Hobbs in 1938 and they had one son and three daughters. He was awarded a Ph.D. at the University of Texas in 1945 for a thesis *Concerning simple plane webs* under R.L. Moore's supervision.[10] Bing did not like Moore's close supervision because Bing felt Moore wanted his own way rather than allowing Bing to write his thesis as he chose.

He was a faculty member at the University of Wisconsin from 1947 to 1973, during which time he was appointed Rudolph E Langer Professor in 1968. From 1973 to 1985 he taught at the University of Texas.

He published *The Geometric Topology of 3-Manifolds* in 1983. Bing was elected to the National Academy of Sciences in 1965. He was president of the American Mathematical Society in 1977-78.[3] He had unfailing sense of humor and was an exemplary man.

Bing died of cancer on 28 April 1986 in Austin, Texas, USA.

Gordon Thomas Whyburn

Gordon Whyburn was born on 7 January 1904 in Lewisville, Texas, USA. He studied at the University of Texas majoring in chemistry received his B.A. in 1925 and M.A. in 1926. In response to encouragement from R.L. Moore, Whyburn switched to mathematics and was awarded a Ph.D. in mathematics in 1927 for his thesis *Concerning continua in the plane* supervised by R.L. Moore. Gordon's brother William Marvin Whyburn (1901-1972) also received a Ph.D. in mathematics at Texas under Hyman Joseph Ettlinger's supervision.

Gordon married Lucille Smith also from Lewisville in 1925. He spent the academic year 1929-30 in Europe, working with Hans Hahn in Vienna and visited Kuratowski and Sierpinski in Warsaw.

After three years at Johns Hopkins University as Associate Professor, he was appointed Professor and Chairman of the Department of Mathematics at the University of Virginia in 1934 where he stayed until his death. He supervised 31 doctoral students (26 at Virginia).[10] Whyburn was president of the American Mathematical Society in 1953-54. He published his famous text *Analytic Topology* in 1942 and *Topological Analysis* in 1958.

He suffered a heart attack in 1966 and died on 8 September 1969 in Charlottesville, Virginia, USA.

Raymond Louis Wilder

Raymond Wilder was born on 3 November 1896 in Palmer, Massachusetts, USA. The family loved music and Raymond learned to play the piano and the cornet. During his undergraduate years at Brown University he served in the US Navy as an ensign during World War I. He received his B.S. at Brown in 1920 and a master's degree in actuarial mathematics in 1921.[3]

He married Una Maude Greene in 1921, and they had three daughters and one son. He went to University of Texas in 1921 to become an instructor. R.L. Moore declined to give Wilder permission to take his topology course because of Wilder's background in actuarial mathematics.

However, after Wilder solved one of hardest problems Moore posed to the class, Wilder became Moore's research student. Wilder received his Ph.D. in 1923 with his dissertation *Concerning continuous curves*. Wilder was Moore's first Texas doctorate.[3]

After two years (1924–1926) at Ohio State University, he joined the faculty at the University of Michigan where he retired in 1967. Then he moved to the University of California at Santa Barbara where he remained for the rest of his life.

He published *Topology of Manifolds* in 1949 and *Introduction to the Foundations of Mathematics* in 1952. He was also interested in cultural anthropology. He was president of the American Mathematical Society in 1955–56. Wilder was elected to the National Academy of Sciences in 1963.

He died on 7 July 1982 in Santa Barbara, California. His wife survived him for nineteen years, dying at the age of 100.[3]

Figure 309. Gordon T. Whyburn

(Courtesy of the American Mathematical Society)

Figure 310. Raymond L. Wilder

(Courtesy of the American Mathematical Society)

Leo Zippin

Leo Zippin was born in 1905. He received a Ph.D. in 1929 at the University of Pennsylvania for a thesis *A study of continuous curves and their relation to the Janiszewki-Mullikin theorem* under John R. Kline's supervision.[10]

Zippin joined in the faculty of Queens College, the City University of New York in 1938 where he retired in 1971. He helped solve the fifth problem of Hilbert.[11] He died on 11 May 1995 in Manhattan, New York.

Robion Cromwell Kirby

Robion Kirby was born in 1938 and received his Ph.D. in 1965 at the University of Chicago for a thesis entitled *Smoothing locally flat imbeddings* supervised by S. Eldon Dyer.[10]

Kirby received the Oswald Veblen Prize in Geometry by the American Mathematical Society in 1971. He is currently professor of mathematics at the University of California, Berkeley where specializes in low-dimension topology. He has supervised 50 doctoral students at Berkeley and UCLA. He was elected to the National Academy of Sciences in 2001.

Martin Davis

Martin Davis was born in 1928 in New York City. He was awarded a Ph.D. in 1950 at Princeton University for his thesis *On the theory of recursive unsolvability* under Alonzo Church's supervision.

In 1967 he received the Frank Cole Prize from the American Mathematical Society. He has supervised 16 Ph.D. students mostly at New York University. He is co-inventor of the Davis-Putnam and the DPLL algorithms. He is also known for his model of Post-Turing machines.[73]

Stephen Cole Kleene

Stephen Kleene was born on 5 January 1909 in Hartford, Connecticut, USA. After graduating from Amherst College, he received a Ph.D. at Princeton University in 1934 for a thesis, *A theory of positive integers in formal logic,* under Church's supervision.

He joined the faculty at the University of Wisconsin at Madison in 1935 and was promoted to full professor in 1948. He retired after 44 years of service to Wisconsin. He was elected to the National Academy of Sciences and awarded the Steele Prize by the American Mathematical Society in 1983.

His research focused on the theory of algorithms and recursive functions as well as mathematical intuitionism. Kleene's best known books are *Introduction to Mathematics* (1952) and *Mathematical Logic* (1967).

He died on 25 January 1994 in Madison, Wisconsin, USA.

Figure 311.   Stephen C. Kleene

Saharon Shelah

Saharon Shelah was born on 3 July 1945 in Jerusalem. He received a Ph.D. in 1969 at the Hebrew University under Michael O. Rabin's supervision. Shelah is currently a professor at the Hebrew University and also at Rutgers University in New Jersey, USA.

As of 2006, he had published over 900 mathematical papers together with over 200 co-authors.[3] He was awarded the Bolyai Prize by the Hungarian Academy of Sciences in 2000 for his work, *Cardinal Arithmetic,* published by Oxford University Press in 1994, in which he presented his pcf theory.

He also received the Wolf Prize in mathematics in 2001. His main field of interest is in mathematical logic, in particular model theory and set theory.[74]

Leslie Gabriel Valiant

Leslie Valiant was born on 28 March 1949. He graduated from King's College, Cambridge and Imperial College in London. He received a Ph.D. in 1974 at the University of Warwick for a thesis, *Decision procedures for families of deterministic pushdown automata,* under Michael Paterson's supervision.[10]

He is currently the T. Jefferson Coolidge Professor of Computer Science and Applied Mathematics at Harvard University. He is renowned for his work in theoretical computer science and he also works in computational neuroscience focusing on understanding memory and learning.

Valiant was elected a Fellow of the Royal Society of London in 1991 and a member of the National Academy of Sciences.

He received the Nevanlinna Prize in 1986.

Edward Kasner

Edward Kasner was born in 1878. He graduated from the City College of New York in 1897 and received a Ph.D. at Columbia University in 1900 for a thesis entitled *The invariant theory of the inversion group* under Cassius J. Keyser's supervision. Kasner was the first Jew appointed to a faculty position in the sciences at Columbia University.

In 1938 Kasner asked his nine-year-old nephew Milton Sirotta to name a very large number, one followed by a hundred zeros(10^{100}). Milton answered "googol."

The Internet search engine "Google" originated from a misspelling of "googol." The Googolplex is the number represented by a 1 followed by a googol of zeros.[75]

$$googolplex = 10^{10^{100}}$$

One of Kasner's students at Columbia, Jesse Douglas won a Fields Medal in 1936.

Jesse Douglas

Jesse Douglas was born on 3 July 1897 in New York City. He graduated from the City College of New York with honors in mathematics in 1916 and received a Ph.D. at Columbia University in 1920 for a thesis *On certain two-point properties of general families of curves* under Edward Kasner's supervision.

From 1926 to 1930 he visited Princeton, Harvard, Chicago, Paris and Göttingen on a National Research Fellowship. During this period, he solved the Plateau Problem which had been posed by Lagrange in 1760. Douglas presented details of his work in *Solution of the Problem of Plateau* in the Transactions of the American Mathematical Society in 1931. He was awarded the first Fields Medal at the International Congress of Mathematicians in Oslo in 1936. Douglas also received the Bôcher Prize by the American Mathematical Society in 1943.

Douglas married Jessie Nayler in 1940 and they had one son. From 1930 to 1937, he was a faculty member at Massachusetts Institute of Technology and from 1942 to 1954, he taught at Brooklyn College and Columbia University.

In 1955 he was appointed professor at the City College of New York where he stayed the rest of his life. In the same year his wife died.[3]

He died on 7 October 1965 in New York City.

Figure 312.   Jesse Douglas

Figure 313.   Eric T. Bell

Eric Temple Bell

Eric Temple Bell was born on 7 February 1883 in Peterhead, near Aberdeen, Scotland. In 1884 his family moved to San Jose, California. After his father's death, Eric returned to Britain in 1896 with his mother and older brother. From 1898 Bell attended Bedford Modern School, near Cambridge. In 1902 Bell alone returned to the United States.

He graduated Stanford University with honors in mathematics in 1907 completing undergraduate work in two years. He received a Master's degree at the University of Washington in Seattle in 1908. He was awarded a Ph.D. in 1912 at Columbia University for a thesis *The cyclotomic quinary quintic* nominally supervised by Cassius J. Keyser. It took only one year for him to get a Ph.D.

He taught at the University of Washington from 1912 to 1926 then at the California Institute of Technology from 1926 to 1959.

In 1914 he married Jessie Lillian Smith Brown and they had one son, who became a doctor.

Bell was awarded the Bôcher Prize in 1924 by the American Mathematical Society for his memoir *Arithmetical Paraphrases*. He was elected to the National Academy of Sciences in 1927.

His books *Algebraic Arithmetic* (1927) and *The Development of Mathematics* (1940) are well known. He also wrote *Men of Mathematics* (1937) and *Mathematics, Queen and Servant of Science* (1951).

Bell wrote sixteen science fiction novels under the name John Taine and several volumes of poetry. A well written biography about Bell, *"The Search for E.T. Bell: Also known as John Taine"* by Constance Reid was published in 1996.

Bell died on 21 December 1960 in Watsonville, California.

Figure 314. Garrett Birkhoff

Garrett Birkhoff

Garrett Birkhoff, a son of G.D. Birkhoff, was born on 19 January 1911 in Princeton, New Jersey, USA. He studied at Harvard University attending lectures by Oliver Kellogg, E.C. Kemble, Marston Morse, Joseph Walsh and Hassler Whitney. After graduating from Harvard in 1932 he studied with Philip Hall at the University of Cambridge publishing a joint paper on group theory in 1936.

Birkhoff became a Junior fellow at Harvard University in 1933. Junior fellows receive the salary of assistant professors but with no teaching obligations to allow them to engage in research for three years. Most of them already have Ph.Ds, but Birkhoff was appointed without a doctorate.

In 1936 he was appointed as an instructor at Harvard. He wrote his famous book *Survey of Modern Algebra* jointly with Saunders Mac Lane in 1941.[3]

During World War II Birkhoff worked at the Ballistic Research Laboratory at the Aberdeen Proving Ground and also for the United States Navy. His war works resulted in two books.

He was appointed George Putnam Professor of Pure and Applied Mathematics at Harvard in 1969 where he retired in 1981. He never earned a Ph.D. When someone asked why he did not have a Ph.D., his reply was "Who would examine me?"[50]

He was elected to the National Academy of Sciences and he received the Birkhoff Prize in honor of his father in 1978. He had supervised 50 doctoral students and published 229 papers.[3]

He died on 22 November 1996 in Water Mill, New York, USA.

Figure 315. Norbert Wiener

Figure 316. Norman Levinson

Norbert Wiener

Norbert Wiener was born on 26 November 1894 in Columbia, Missouri where his father Leo Wiener was professor of modern language at the University of Missouri. Leo

Wiener later became a professor of Slavic language at Harvard University. Norbert was mostly educated at home and he graduated in 1906 from Ayer High School at the age of eleven. He was awarded a B.A. degree in 1909 at Tufts College (now Tufts University), graduating in three years. Then he studied mathematics and philosophy at Cornell University and Harvard University where he received his Ph.D. in 1913 for a thesis, *A comparison between the treatment of the algebra of relatives by Schroeder and that by Whitehead and Russell,* at the age of eighteen.[10] Wiener's book *Ex-Prodigy* (1953), published by MIT Press elaborates his life when he was young.

From 1913 to 1915, he studied in Europe with Russell, Hardy, Hilbert and Edmund Landau. During World War I, Wiener also worked at the Aberdeen Proving Ground on ballistics. After the war ended in 1918, he was appointed as an instructor at Massachusetts Institute of Technology where he spent the rest of his life. He was promoted to full professor in 1932. Wiener collaborated with Max Born on quantum mechanics during Born's visit to MIT from November 1925 to January 1926.

In 1926 Wiener married Margaret Engemann, a former student of his father. They had two daughters. Wiener exemplified the stereotypical absent-minded professor, although his behavior approached problematic proportions. For example, shortly after the Wiener family moved to a new home in Cambridge, Massachusetts, Wiener unwittingly returned to his former house and asked a little girl if she knew where the Wieners had moved. 'Yes daddy' the girl said, 'I am your daughter and mom sent me to take you to our new home.'[4]

In another instance, Wiener drove Professor Dirk Struik and two graduate students from Cambridge, Massachusetts to Yale University to attend a mathematics meeting. When they arrived New Haven, it was raining. Wiener took the student's hat and placed it on his head thinking it was his hat. Struik and the two students returned to Cambridge by a different means because of Wiener's erratic driving.

After the conference, Wiener returned to Cambridge by bus having forgotten that he had driven down to New Haven. When he went to his garage the next morning to drive to work, he found it was not there and reported it as stolen to police.[95]

Wiener received the Bôcher Prize in 1933 for his memoir *Tauberian theorems*. He published *The Fourier Integral, and Certain of Its Applications* in 1933.

During World War II he worked on gunfire control. Sometimes he has been described as "the father of automation."

After the war, he published *Cybernetics: Control and Communication in the Animal and the Machine (1948)* and *Extrapolation, Interpolation, and Smoothing of Stationary Time*

Series (1949) which was originally classified. The term 'Cyber Space' came from 'Cybernetics'.

Wiener felt uneasy about the quality of his mathematical output and pressed his colleagues to confirm that his productivity was not declining. He would ask "tell me, am I slipping?" or "Am I really a good mathematician?"

He would normally fall asleep in many seminars, but after the talk is over, he would wake up asking very relevant questions. Wiener received the National Medal of Science in January 1964 from President Lyndon Johnson.

His name is honored by the Wiener filter, Wiener process, Paley-Wiener theorem, Wiener-Khinchin theorem, Wiener-Hopf theorem, Abstract Wiener Space and Wiener's tauberian theorem.

He died of a heart attack on a visit to Stockholm on 18 March 1964 at the age of sixty-nine.

Norman Levinson

Norman Levinson was born on 11 August 1912 in Lynn, Massachusetts, USA. His father was a very poor Russian Jewish immigrant working at a shoe factory.

Levinson entered Massachusetts Institute of Technology in 1929 and received his S.B. and S.M. degree in Electrical Engineering in 1934. By this time, he had taken twenty graduate courses in mathematics at MIT. After Levinson showed some aptitude for mathematics, Wiener gave him the Paley-Wiener manuscript to read. Levinson found an error and proved a lemma to fix it. Later Wiener paid several visits to Levinson's family to introduce himself and to assure them of Levinson's future in mathematics.[35]

Norbert Wiener arranged for Levinson to receive an MIT Redfield Proctor Traveling Fellowship so that he could study at the University of Cambridge. During Levinson's time there, he considered Littlewood to be a better mathematician than Hardy.

When he returned to MIT in 1935 he was awarded a Sc.D. degree, which is essentially the same as a Ph.D., for his thesis *Non-vanishing of a function.*[3] Then he spent one and a half years at the Institute for Advanced Study in Princeton working with von Neumann.

Wiener recommended Levinson to be hired as an Instructor at MIT, but Vannevar Bush, MIT provost, turned him down mainly due to anti-Semitism. G.H. Hardy, on a visit to MIT, went with Wiener to Bush's office. Hardy is reported to have said: "Tell me, Mr. Bush, do you think you are running an engineering school or a theological seminar? Is this the Massachusetts Institute of Theology? If it isn't why not hire Levinson?"[3]

Levinson was appointed as an Instructor at MIT in February 1937 where he was promoted to full professor in 1949. He married Zipporah (Fagi) Wallman on 11 February 1938 and they had two daughters. He received the Bôcher Prize in 1954 for his contributions to non-linear differential equations.

Levinson published *Theory of Ordinary Differential Equations* jointly with Earl Coddington in 1955, which became a classic. He received Chauvenet Prize of the Mathematical Association of America in 1971 for his paper on number theory.

Levinson joined the American Communist Party believing that it was against anti-Semitism and discrimination against blacks. But he left the Party after he came to understand the direction of Stalin's Communism in the Soviet Union. The House Un-American Activities Committee called for Levinson to name other members of the American Communist Party, but he declined.

He died of a brain tumor on 10 October 1975 in Boston, Massachusetts, USA.

Figure 317. Shiing-shen Chern

Figure 318. Shing-Tung Yau

Shiing-shen Chern (陳省身)

Shiing-shen Chern was born on 26 October 1911 in Jiaxing, Zhejiang Province, China. He graduated from Nankai University in Tientsin and undertook graduate study at Tsing Hua University in Beijing, where his teacher was Dr. Dan Sun, a former student of E.P. Lane at the University of Chicago. Chern received a fellowship from the Boxer

Indemnity Fund in 1934 and went to the University of Hamburg. Chern was awarded a doctorate at Hamburg in 1936 under Blaschke's supervision.[94] Then he studied under Elie Cartan in Paris.

After returning from Paris he became professor of mathematics at Tsing Hua University in 1937. Because Japan invaded China in 1937, the universities in Beijing moved to rural southwestern China where the Southwest Associated University was formed. Chern taught there from 1938 to 1943. C.N. Yang and T.D. Lee were students there at that time.

In 1948 Chern went to the Institute for Advanced Study at Princeton at the invitation of Weyl and Veblen.

Marshall Stone then invited Chern to join the University of Chicago in 1949 where he stayed until 1960. Chern joined Stone's other recruits, Weil, MacLane and Zygmund in building the mathematics department at Chicago.

In 1960 he moved to the University of California, Berkeley where he supervised 31 Ph.D. students in addition to 10 doctoral students at Chicago. He was the first director of the Mathematical Sciences Research Institute at Berkeley from 1981 to 1984. He permanently moved back to Nankai University in 1999.

He was elected to the National Academy of Sciences in 1961 and Royal Society of London in 1985. He received the National Medal of Science in 1975 and the Wolf Prize in 1984. Chern also received the AMS Steele Prize in 1983 and the Lobachevsky Prize in 2002. He gave an intrinsic proof of the n-dimensional Gauss-Bonnet theorem. His name is honored in the Chern classes, Chern-Weil homomorphism, Chern-Simons invariants, and Chern-Moser invariants.

He married Shih-Ning Cheng in 1940 and they had one son and one daughter. His wife died in 2000 in Tientsin (Tianjin). Chern died on 3 December 2004 in Tianjin, China.

Shing-Tung Yau (丘成桐)

Shing-Tung Yau was born in Shantou, Guangdong Province, China on 4 April 1949. His name in pinyin is Qiu Chengtong. He graduated from the Chinese University of Hong Kong in 1969 and received a Ph.D. at the University of California, Berkeley in 1971 for a thesis *On the fundamental group of compact manifolds of non-positive curvature* supervised by Chern.[10]

After spending one year at the Institute for Advanced Study at Princeton, he was appointed assistant professor at the State University of New York at Stony Brook in

1972. In 1974 he moved to Stanford University where he proved the Calabi conjecture in 1976. Yau returned to IAS Princeton in 1979. From 1984 to 1987 he was at the University of California at San Diego after which he moved to Harvard University.

Yau was awarded a Fields Medal in 1983 at the International Congress of Mathematicians in Warsaw for his contributions to partial differential equations, to the Calabi conjecture in algebraic geometry, to the positive mass conjecture of general relativity theory, and to the real and complex Monge-Ampère equations. He received the Oswald Veblen Prize in Geometry in 1981 and the National Medal of Sciences in 1997. He was elected to the National Academy of Sciences in 1993.

Yau, jointly with Karen Uhlenbeck, solved higher dimensional versions of the Hitchin-Kobayashi conjecture in 1986.

His book with Richard Schoen, *Lectures on Differential Geometry* (1994) is a popular textbook. Recently, he was involved in the controversy related to Perelman's proof of the Poincaré conjecture.

Yau's name can be found in mathematical terms in Tian-Yau Conflict, Calabi-Yau manifold, positive energy theorem, and Schoen-Yau conjecture.

Figure 319. William E. Story

Figure 320. Solomon Lefschetz

(Courtesy of the American Mathematical Society)

Richard Melvin Schoen

Richard Schoen was born on 23 October 1950. He received a Ph.D. at Stanford University in 1977 for a thesis *Existence and regularity theorems for some geometric variational problems* supervised by Shing-tung Yau.

In 1979, together with Yau, Schoen proved the fundamental positive energy theorem in general relativity. This work has applications to the formation of black holes.

Schoen received the Bôcher Prize in 1989. He is currently Robert M. Bass Professor of Humanities and Sciences at Stanford University.

William Edward Story

William Story was born on 29 April 1850 in Boston, Massachusetts. He graduated with honors from Harvard University in 1871. Story studied in Berlin and Leipzig where he received a doctorate in 1875 for a thesis, *On the algebraic relations existing between the polars of the binary quintic,* under Carl Neumann's supervision.

When J.J. Sylvester sought to strengthen the research department of mathematics at Johns Hopkins University in 1876, he appointed Story as a faculty member. Story married Mary Deborah Harrison in 1878 and they had one child.

Story moved to Clark University in Worcester, Massachusetts in 1887. Clark University was established as a graduate institution attracting many scientists such as Albert A. Michelson, Arthur G. Webster, Oskar Bolza, Henry Seely White. When the University of Chicago was founded in 1890, President William Harper recruited fifteen professors from Clark University, including Michelson and Bolza. The research support at Clark declined sharply, and Arthur G. Webster, professor of physics at Clark and third president of the American Physical Society, committed suicide in 1923.

From 1892 to 1919 Clark University awarded 26 Ph.Ds. in mathematics, sixteen of them supervised by William Story.[3] His most famous student was Solomon Lefschetz.

Story was elected to the National Academy of Sciences in 1908. He died on 10 April 1930 in Worcester, Massachusetts, USA.

Solomon Lefschetz

Solomon Lefschetz was born into a Russian Jewish family on 3 September 1884 in Moscow, Russia. The family moved to Paris where French became Lefschetz's first language. He studied at the *Ecole Centrales des Arts et Manufacures* where he took courses from Emile Picard and Paul Appell, graduating in 1905. He emigrated to the United States in November 1905 and became an engineer at Westinghouse Electric Company in Pittsburgh from 1907. He was able to speak six languages.

He lost both hands and forearms in a laboratory accident at Westinghouse and could not continue working as an engineer. He had two prostheses in place of his hands, but they did not move or function in any way. Over each he wore a black leather glove. Someone had to push a piece of chalk into Lefschetz's hand each morning and remove it at the end of the day. He could not operate a doorknob, so his office door was equipped with a lever.[35] But he could write reasonably well. When he had to take a subway train, he would ask strangers to take a token out of his pocket and put it in the turnstile.

He decided to become a mathematician and went to Clark University in 1910 where he received a Ph.D. in 1911 for a thesis *On the existence of loci with given singularities* supervised by William Story. He married Alice Hays, his fellow student at Clark in 1913.

He taught at the University of Nebraska in Lincoln (1911–1913) and the University of Kansas in Lawrence (1913–1924). The majority of Lefschetz's contribution to algebraic geometry occurred at Kansas.

James W. Alexander persuaded Veblen to invite Lefschetz to Princeton in 1924 as a visiting professor for a year. Lefschetz was offered a permanent post at Princeton as associate professor in 1925 which he gladly accepted.

Lefschetz was the editor of the *Annals of Mathematics* from 1928 to 1958 bringing it up to the most revered mathematical journal in the world. He served as president of the American Mathematical Society from 1935 to 1936.

The word 'topology' comes from the title of a monograph written by Lefschetz in 1930. Before that it was called 'analysis situ'.

He received the Bôcher Prize in 1924. Lefschetz became Henry Fine Research Professor in 1933 when Veblen moved to the Institute for Advanced Study. He was elected to the National Academy of Sciences and the Royal Society of London in 1961. In 1965 he received the National Medal of Science awarded by President Johnson. He received the Order of the Aztec Eagle in recognition of his achievements in establishing a lively Mexican research school in mathematics in Mexico.

Lefschetz ran the mathematics department in an autocratic fashion. If it voted with him, fine; if not, he did what he wanted anyway. Many members of the department did not even bother to attend faculty meeting.[76] Lefschetz retired from Princeton in 1953 and Albert Tucker succeeded him as Chairman.

After retiring from Princeton in 1953, he established the Research Institute for Advanced Studies (RIAS) in Baltimore funded by industry. RIAS gained an international reputation for its work on mathematical theory of control and stability. In 1964 the major portion of RIAS was transferred to Brown University to become the Lefschetz Center for Dynamical Systems.[4]

When Zariski asked Lefschetz "How do you draw the line between algebra and topology?" Lefschetz answered, "Well, if it's just turning the crank, it's algebra, but if it's got an idea in it, it's topology!"[94]

Lefschetz died on 5 October 1972 in Princeton, New Jersey.

Figure 321.   Princeton University Department of Mathematics 1951

(Courtesy of Princeton University)

Figure 322.   Norman E. Steenrod

Norman Earl Steenrod

Norman Steenrod was born on 22 April 1910 in Dayton, Ohio.

He finished elementary and secondary school in nine years. Before entering university he worked as a tool designer for two years. Using the skills he had learnt at this time he could work at the Chevrolet plant as a die designer.

At the University of Michigan at Ann Arbor, he studied physics, philosophy and economics, but he also took a topology course given by Raymond Wilder.[3]

After graduating from the University of Michigan in 1932, he wrote his first paper on topology, which led to offers of fellowships from three universities including Harvard, Princeton and Duke University.

At Princeton, Steenrod worked for his doctorate supervised by Lefschetz. He received a Ph.D. in 1936 for a thesis titled *Universal homology groups*.

Steenrod married Cardyn Witter in 1938 and they had two children.

Steenrod worked three years at the University of Chicago from 1939 until 1942 and five years at the University of Michigan from 1942 until 1947 before he moved to Princeton in 1947 where he remained for the rest of his life.

He published *The Topology of Fibre Bundles* in 1951, which was highly praised by mathematicians and *Foundations of Algebraic Topology* in 1952 jointly with Samuel Eilenberg.

He also wrote *Cohomology Operations* in 1962 from a lecture note. He was elected to the National Academy of Sciences. He is best known for the Steenrod algebra, Eilenberg–Steenrod axioms and his research on fibre bundles.

Steenrod died on 14 October 1971 in Princeton, New Jersey.

Figure 323.   Albert W. Tucker

Figure 324.   John F. Nash

Albert William Tucker

Albert Tucker was born on 28 November 1905 in Ontario, Canada.

After graduating from the University of Toronto in 1928, he went to Princeton University where he received a Ph.D. for a thesis titled *An abstract approach to manifolds* under Lefschetz's supervision in 1932.

He was appointed to the faculty at Princeton where he stayed until 1970 serving as chairman of mathematics department for about 20 years.[77]

He served as president of the Mathematical Association of America in 1961 and 1962. He is best known for the Karush–Kuhn–Tucker conditions in non-linear programming, Tucker tableaux and Prisoner's Dilemma.

John Forbes Nash, Nobel laureate in economics 1994 was one of his students.

Tucker died on 25 January 1995 in Hightstown, New Jersey, USA.

John Forbes Nash

John Nash was born on 13 June 1928 in Bluefield, West Virginia. He entered the Carnegie Institute of Technology (now Carnegie-Mellon University) in 1945 where John Synge recognized Nash's mathematical talents. Nash received a B.A. and an M.A. in mathematics in 1948.

While studying for his doctorate at Princeton in 1949 he wrote a paper that, 45 years later, was to win a Nobel Prize for economics.[3]

He received a Ph.D. at Princeton with a thesis titled *Non-cooperative games* in 1950 supervised by Albert Tucker.

In 1951, he was appointed as an instructor at Massachusetts Institute of Technology where he was promoted to assistant professor in 1953 and associate professor in 1957.

His famous implicit function theorem was published in the paper *The Imbedding Problem for Riemannian Manifolds* in 1956.[3]

He met Eleanor Stier and they had a son, John David Stier, born on 19 June 1953. In 1954, Nash lost his job at the RAND Corporation after being arrested for being homosexual. He was dismissed from RAND.

In February 1957, Nash married Alicia Larde, one of his students at MIT. Around that time, Nash began to exhibit erratic and unusual behavior. Norbert Wiener was one of the first to recognize that Nash was suffering from a mental disorder. Alicia divorced Nash in 1962.

Nash spent several years intermittently at the Institute for Advanced Study at Princeton. In the 1990s, Nash made a recovery from the schizophrenia he had suffered from the 1950s.

He was awarded the 1994 Nobel Prize in Economic Science for his work on game theory jointly with Harsanyi and Selten. He also received the AMS Steel Prize and Von Neumann Theory Prize by the Operations Research Association of America in 1999. Sylvia Nasar wrote *A Beautiful Mind*, a story of John Nash which was produced into a successful movie.

Felix Earl Browder

Felix Browder was born in Moscow in the Soviet Union on 31 July 1927. His brother William Browder is also a mathematician. Felix Browder graduated from Massachusetts Institute of Technology with a S.B. degree in 1946 and he received a Ph.D. at Princeton University in 1948. His dissertation was *The topological fixed point theory and its applications in functional analysis* supervised by Lefschetz.[10]

Browder taught at MIT from 1948 until 1951, Boston University from 1951 to 1953 and Brandies in 1955 and 1956. In 1956 he was appointed a full professor at Yale. In 1963 he moved to the University of Chicago where he stayed 23 years, ten years as department chairman. Since 1986, he has been at Rutgers University.

He was awarded the National Medal of Science in 1999 by President Clinton for his creation of nonlinear functional analysis and its applications to partial differential equations.

Browder is a member of the National Academy of Sciences. He also served as president of the American Mathematical Society in 1999–2000.

He married Eva Tislowitz on 5 October 1949 and they have two sons. His own library at home has over 30,000 volumes.

Figure 325.   Felix E. Browder

(Courtesy of the American Mathematical Society)

Clifford Ambrose Truesdell III

Clifford Truesdell was born on 18 February 1919 in Los Angeles, California.

Before going to university he spent two years at Oxford and traveling elsewhere in Europe improving his knowledge of Latin and Greek. He became proficient in German, French and Italian.

He graduated from the California Institute of Technology gaining a B.S. degree in physics and mathematics in 1941 and an M.S. in mathematics in 1942. He was profoundly influenced by his teacher Harry Bateman at Caltech. Truesdell received a Ph.D. at Princeton University in 1944 for his thesis *The membrane theory of shells of revolution* supervised by Lefschetz.

From 1944 to 1951, he worked at the MIT Radiation Laboratory, the US Naval Ordnance Laboratory in White Oak, Maryland, and the US Naval Research Laboratory in Washington, D.C. He was professor of mathematics at Indiana University from 1950 until 1961 and professor of rational mechanics at the Johns Hopkins University from 1961 to 1989.

He received the Euler Medal of the USSR Academy of Sciences in 1958 and 1983. Truesdell also received the Birkhoff Prize in Applied Mathematics. He founded two great journals: *Archive for Rational Mechanics and Analysis* in 1957 and *Archive for History of Exact Sciences* in 1960.

Truesdell died on 14 January 2000 in Baltimore, Maryland.[78]

Figure 326. Ralph H. Fox

Figure 327. John W. Milnor

Ralph Hartzler Fox

Ralph Fox was born on 24 March 1913 in Morrisville, Pennsylvania.

He studied at Swarthmore College for two years while learning piano at the Leefson Conservatory of Music in Philadelphia. He received a master's degree from Johns Hopkins University and a Ph.D. at Princeton University for his thesis entitled *On the*

Lyusternick-Schnirelmann category supervised by Lefschetz in 1939. In later years, he disclaimed all knowledge of Lyusternik-Schnirelmann.

He is best known for Fox *n*-coloring of knots and the phrases *slice know, ribbon knot* and *Seifert circle*, which all appeared for the first time under his name. He also popularized the playing of the Asian game of Go at both Princeton University and the Institute for Advanced Study.[79]

He supervised 21 doctoral candidates mostly at Princeton, including John Milnor, John Stalling, Jr. and Barry Mazur.

Fox died on 23 December 1973 in Philadelphia, Pennsylvania.

Figure 328.   Barry C. Mazur

John Willard Milnor

John Milnor was born on 20 February 1931 in Orange, New Jersey.

He received his A.B. in 1951 and a Ph.D. in 1954 at Princeton University for his thesis titled *Isotopy of links* supervised by Ralph Fox.

Milnor was a faculty member at Princeton from 1953 to 1988 during which time he was appointed to the Henry Putnam Chair in 1962. Since 1988 he has been at the State University of New York at Stony Brook.

He was awarded a Fields Medal at the 1962 International Congress of Mathematicians for his proof that a 7-dimensional sphere can have several differential structures. This work opened up the new field of differential topology.[3]

Milnor received the National Medal of Science in 1967 and became a member of the National Academy of Sciences.

He received the AMS Steele Prize in 1982 and the Wolf Prize in 1989. Milnor is one of only twelve Fields Medalists who have also received the Wolf Prize.

Milnor is married to Margaret Dusa Waddington McDuff, Distinguished Professor of Mathematics at Stony Brook and they have one son.

Barry Charles Mazur

Barry Mazur was born on 19 December 1937 in New York City. He graduated from the Bronx High School of Science. He attended Massachusetts Institute of Technology, but could not graduate because he had not fulfilled the requirement of the Reserve Officer Training Corps (ROTC) Program.[3]

Princeton University accepted him as a graduate student and he received a Ph.D. in 1959 for his thesis *On imbedding of spheres* supervised by Ralph Fox and R.H. Bing.[10]

Mazur was a Junior Fellow at Harvard University from 1959 until 1962. Junior Fellows receive salaries commensurate with assistant professors, but do not have any teaching obligations. Consequently, it is considered a very prestigious academic position.

In 1962, he was appointed assistant professor at Harvard, associate professor in 1965, full professor in 1969 and William Petscheck Professor of Mathematics in 1982.

He was named Gerhard Gade University Professor at Harvard in 1998, the highest professorial chair at Harvard.

Mazur's paper *Modular Curves and the Einstein Ideal* in 1978 laid the foundation for many of the most important results in algebraic geometry including the proof of the Main Conjecture of Iwasawa theory, the proof of Taniyanca-Shimura conjecture and Wile's proof of Fermat's Last Theorem.[3]

Mazur received the Veblen Prize in 1966 for his work on the generalized Schoenflies theorem, the Cole Prize for number theory in 1982, the Steel Prize in 2000 from the American Mathematical Society. He also received the Chauvenet Prize in 1994 from the Mathematical Association of America. He was elected to the National Academy of Sciences in 1982.

His name can be found in mathematical terms such as the Mazur manifold, the Mazur swindle and the Mazur torsion theorem.

In 1960 he married Grace Dane and they have one child. She later became a famous novelist.

William Browder

William Browder, Felix Earl Browder's younger brother, was born on 6 January 1934.

He graduated from the Massachusetts Institute of Technology in 1954 and received a Ph.D. at Princeton University in 1958 for a dissertation entitled *Homology of loop spaces* supervised by John C. Moore.

Since 1964, he has been professor of mathematics at Princeton University, recognized as a leading topologist of his generation.

He is one of the pioneers, with Sergei Novikov, Dennis Sullivan and Terry Wall, of the surgery theory method for classifying high-dimensional manifolds.[80]

He served as president of the American Mathematical Society in 1989-90, ten years before his brother.

Figure 329. William Browder

(Courtesy of the American Mathematical Society)

Michael Hartley Freedman

Michael H. Freedman was born on 21 April 1951 in Los Angeles, California. He received his doctorate at Princeton University in 1973 with his dissertation entitled *Codimension-two surgery* supervised by William Browder.

After working three years at the University of California at Berkeley and the Institute for Advanced Study at Princeton, he became an assistant professor at the University of California at San Diego in 1976. In 1985 he was appointed to the Charles Lee Powell Chair of Mathematics at the same institution. Since 1997, he has been a senior research scientist at the Microsoft Corporation in addition to his professorship. He was elected to the National Academy of Sciences in 1984.

Freedman was awarded a Fields Medal in 1986 for his work on proving the Poincaré hypothesis for 4-dimensional topological manifolds.

In 1987 President Ronald Reagan presented him the National Medal of Science at the White House.

He married Leslie Blair Howland in September 1983.

Dennis Parnell Sullivan

Dennis Sullivan was born in 1941 in Port Huron, Michigan.

He received a Ph.D. at Princeton University in 1966 for a thesis entitled *Triangulating homotopy equivalences* supervised by Willian Browder.

Sullivan was a permanent member of the *Institute des Hautes Études Scientifiques* in France for about 20 years.

Currently he holds the Albert Einstein Chair at the City University of New York.

He founded rational homotopy theory along with Daniel Quillen. Sullivan is best known for the Sullivan conjecture and the Parry-Sullivan invariants.

He received the AMS Oswald Veblan Prize in Geometry (1971), the Prix Élie Cartan of the Académie des Sciences (1981), the King Faisal International Prize for Science (1994), the National Medal of Science (2004) and the AMS Steele Prize (2006).[81]

Curtis Tracy McMullen

Curtis McMullen was born on 21 May 1958. He graduated from Williams College in 1980, completed the Certificate of Advanced Study in Mathematics at Emmanuel College, Cambridge in 1981. He received a Ph.D. at Harvard University in 1985 for a thesis entitled *Families of rational maps and iterative root-finding algorithms* under Dennis Sullivan's supervision.[10]

He was on the faculty at Princeton University from 1987 until 1990 and at the University of California, Berkeley from 1990 to 1997, before joining Harvard University in 1997.[82]

McMullen won the Fields Medal in 1998 for his work in complex dynamics and hyperbolic geometry.

He has written many books: Logic Minimization Algorithms for VLSI Synthesis (1984), Complex Dyamics and Renormalisation (1994), Hyperbolic Manifolds, Discrete Groups and Ergodic Theory (1996) and Theichmuller Theory (1993) among others.

George Lusztig

George (Gheorghe) Lusztig was born in Timisoara, Romania in 1946. He did his undergraduate work at the University of Bucharest before he left for the United States.

He received a Ph.D. at Princeton University in 1971 for a thesis entitled *Novikov's higher signature and families of elliptic operators* supervised by William Browder.[10]

Lusztig worked for seven years from 1971 to 1978 at the University of Warwick, England where he was promoted to professor of mathematics in 1974. He supervised five doctoral students at Warwick.

In 1978, he moved to the Massachusetts Institute of Technology where currently he is Norbert Wiener Professor.

He is best known for the Deligne-Lusztig variety and the Kazhdan-Lusztig polynomials.

Lusztig was elected to the National Academy of Sciences in 1992. He received the AMS Cole Prize in Algebra in 1985 and the AMS Steele Prize in 2008.[83]

Figure 330.   Maxime Bôcher

(Courtesy of the American Mathematical Society)

Maxime Bôcher

Maxime Bôcher was born on 28 August 1867 in Boston, Massachusetts. His father was a professor of modern languages at MIT and later professor of French at Harvard.

Bôcher received a bachelor's degree at Harvard University in 1888 and a doctorate at the University of Göttingen in 1891. His dissertation was entitled *Über die Reihenentwicklungen der Potential Theorie* (Development of the Potential Function into Series) supervised by Felix Klein.[3]

Bôcher married Marie Niemann in July 1891 in Göttingen and they had three children.

He was appointed as an instructor in 1892 at Harvard and promoted to a full professor in 1904. He published *Introduction to Higher Algebra* in 1906, which was translated into German and Russian. His book, *An Introduction to the Study of Integral Equations* (1909), was reprinted in 1971.

He was elected to the National Academy of Sciences in 1909 and he served as president of the American Mathematical Society in 1909–1910. He founded and was the first Editor-in-Chief of the *Transactions of the American Mathematical Society* for five years.[3]

Bôcher died on 12 September 1918 in Cambridge, Massachusetts.

The American Mathematical Society established the Bôcher Memorial Prize with an original endowment of $1,450.00. The first recipient was George D. Birkhoff in 1923. It is now awarded every three years.

Figure 331.   Griffith Conrad Evans

(Courtesy of the American Mathematical Society)

Griffith Conrad Evans

Griffith Evans was born on 11 May 1887 in Boston, Massachusetts, where his father was a mathematics teacher.

Evans graduated from Harvard University in 1907 and also received his Ph.D. there in 1910 for a dissertation titled *Volterra's integral equation of the second kind with discontinuous kernel* supervised by Bôcher.

With a Sheldon Travelling Fellowship, he studied with Volterra in Rome most of 1910 to 1912 except one summer with Max Planck in Berlin.[3]

Evans was at the Rice Institute (now Rice University) in Houston, Texas from 1912 to 1934. During World War I, he was a captain in the Air Branch of the US Army Signal Corps (1918-1919).

In the summer of 1934, he became Chairman of the mathematics department, University of California, Berkeley from where he retired in 1955. University management gave him the explicit mandate to revitalize the mathematics department at Berkeley. He appointed fifteen faculty members from 1934 to 1949, including Jerzy Neyman, Hans Lewy and Alfred Tarski.

Evans undertook work related to the war effort at the Aberdeen Proving Ground and the Applied Mathematics Panel of the National Defense Research Council (NDRC) during World War II. For his war work Evans was awarded the Distinguished Assistance Award of the War Department and the Presidential Certificate of Merit.

He was elected to the National Academy of Sciences in 1933 and President of the American Mathematical Society in 1939-40.

The new mathematics building at Berkeley was named Evans Hall in 1971.

He died on 8 December 1973 in Berkeley, California.

Edward Burr Van Vleck

Edward Van Vleck was born on 7 June 1863 in Middletown Connecticut. His father was a vice-president of the American Mathematical Society in 1904 and a professor at Wesleyan University in Middletown, Conneticut.

Edward Van Vleck graduated from Wesleyan University in 1884 and studied at Johns Hopkins University as a graduate student.

He received a doctorate in 1893 at the University of Göttingen for a thesis *Zur Kettenbruchentwicklung Laméscher und ähnlicher Integrale* under Felix Klein's supervision.

After returning to the United States, he married Hester Laurence Raymond in 1893.

Their son, John Hasbrouck Van Vleck was born on 13 March 1899. John Van Vleck received the Nobel Prize for Physics in 1977 jointly with P.W. Anderson and Neville Mott for their fundamental theoretical investigations of the electronic structure of magnetic and disordered systems.

Edward Van Vleck taught at Wesleyan University from 1895 to 1906 and the University of Wisconsin (1893-1895 and 1906-1929). He served as president of the American Mathematical Society in 1913-14 during which time he ensured the Society remained a single national society instead of breaking into smaller local societies.[3]

Edward Van Vleck was a respected gentleman and his son had the same character. Edward Van Vleck's hobby was collecting Japanese art prints.

He died on 2 June 1943 in Madison, Wisconsin. Twenty years after his death, a building at the University of Wisconsin was named for him.

Oskar Bolza

Oskar Bolza was born on 12 May 1857 in Bergzabern, Rhenish Palatinate (now Germany). He studied physics at the University of Berlin then changed to mathematics. He passed the *Lehramtsprüfungen* (secondary school teaching certificate examination) to teach mathematics at secondary schools in 1882.

Bolza received his doctorate in 1886 from the University of Göttingen for his thesis *Über die Reduction hyperelliptischen Integrale Erster Ordnung und Erster Gattung auf Elliptische, Insbesondere über die Reduction Durch Eine Transformation Vierten Grades* under Felix Klein's supervision.[3]

Bolza emigrated to the United States in 1888. After a short-term appointment at Johns Hopkins University, Bolza became a professor at the newly founded Clark University in Worcester, Massachusetts. Unfortunately, problems that mounted at the new university culminated in a vote of no confidence for President G. Stanley Hall. William Rainey Harper, President of the new University of Chicago recruited fifteen professors from Clark, including Bolza who moved to the University of Chicago in 1892.

Bolza persuaded E.H. Moore, the head of mathematics department at Chicago to invite Heinrich Maschke. The three of them were influential in building a strong research school in mathematics. Between 1892 and 1934, the University of Chicago awarded 237 Ph.D.s in mathematics, highest in the United States.[84] From 1892 to 1910, Bolza supervised nine Ph.D.s at Chicago.

His text, *Lectures on the Calculus of Variations,* published in 1904, became a classic in its field.

Bolza served as vice-president of the American Mathematical society in 1904. Unfortunately Maschke died in 1908 and Bolza went back to Germany in 1910. He became an honorary professor at the University of Freiburg, where he retired in 1933.[3]

Bolza died on 5 July 1942 in Freiburg im Breisgau, Germany.

Figure 332. Edward B. Van Vleck

(Courtesy of the American Mathematical Society)

Figure 333. Oskar Bolza

(Courtesy of the American Mathematical Society)

Heinrich Maschke

Heinrich Maschke was born on 24 October 1853 in Breslau, Germany (now Wroclaw, Poland). He passed the *Lehramtsprüfungen* (secondary school teaching certificate examination) in 1878 and received a Ph.D in 1880 at the University of Göttingen supervised by Felix Klein. Mashke became a close friend of Oskar Bolza.

Maschke taught at a secondary school many years before he joined the University of Chicago where Bolza was a professor. Maschke supervised five doctoral students from 1892 to 1907.[10]

He was a teacher of great ability and many students attended his lectures. He served as vice-president of the American Mathematical Society in 1907.

Unfortunately he died on 1 March 1908 in Chicago, Illinois following an emergency surgery.[3]

Henry Burchard Fine

Henry Fine was born on 14 September 1858 in Chambersburg, Pennsylvania, USA. He graduated from the College of New Jersey (changed to Princeton University in 1896) in 1880 where George B. Halstead inspired Fine to study mathematics.

Figure 334.   Heinrich Maschke

Figure 335.   Henry B. Fine

(Courtesy of the American Mathematical Society)

He went to the University of Leipzig in 1884 attending lectures by Felix Klein and Carl Neumann among others. Fine received a Ph.D. at Leipzig for his dissertation, *On the singularities of curves of double curvature,* in May 1885 supervised by Klein.[3]

When Fine returned to the United States, he was appointed assistant professor at Princeton.

He married Philena Fobes in September 1888 and they had three children.

He wrote several elementary texts and papers on differential equations.

Woodrow Wilson, president of Princeton University, appointed Fine as Dean of the Faculty in 1903, which is the second most influential position at Princeton. In 1909, he became Dean of the Departments of Science. During his time as an administrator, Fine lifted the academic standards at Princeton.

He served as president of the American Mathematical Society in 1911-12. He played the flute in the College orchestra and was a good amateur musician.

His life ended tragically. His wife died in April 1928. Not long thereafter, on 21 December 1928, an automobile struck Fine as he biked around Princeton. He died the next morning without having recovered consciousness.[3]

The Princeton mathematics building is named after him.

Figure 336. George B. Halsted

George Bruce Halsted

George Halsted was born on 23 November 1853 in Newark, New Jersey. He graduated from the College of New Jersey (later Princeton University) in 1875 and received his doctorate at Johns Hopkins University in 1879 for his thesis, *Basis for a dual logic,* under J.J. Sylvester's supervision. Halsted also studied with Carl W. Borchart (1817–1880) at the University of Berlin. Borchart was a member of the Berlin Academy, which entitled him to give lectures at the University of Berlin.

From 1879 to 1884, Halsted was at Princeton, first as a tutor and later an instructor. While there, Halsted encouraged Henry Fine to study mathematics.

Halsted was appointed as a professor in the Department of Pure and Applied Mathematics at the University of Texas in 1884, which had only begun operating the year before.

L.E. Dickson and R.L. Moore were Halsted's students at Texas. Halsted's main interests were the foundations of geometry and he introduced non-euclidean geometry into the United States. Halsted had proposed R.L. Moore for an instructorship in mathematics in 1901 when Moore received his B.S. and M.A. degrees simultaneously at the age of 18.

But someone else was appointed. Halsted criticized the University administration and the board of regents. Halsted was dismissed from his professorship in 1902.

After leaving Texas, he worked at St. John's College in Annapolis, Maryland, Kenyon College in Gambier, Ohio and Colorado College of Education (now University of Northern Colorado in Greenley). Due to his outspoken character, he was not popular.

He spent most of his last years of his life working as an electrician in the electrical supply store he founded in Greenley, Colorado.

He also criticized the careless way that mathematics was described in many textbooks.[3]

Halsted died on 16 March 1922 in New York City.

Frank Nelson Cole

Frank Cole was born on 20 September 1861 in Ashland, Massachusetts.

He graduated from Harvard University in 1882 and he spent the years 1883–85 at the University of Leipzig studying under Felix Klein. When Cole returned to the United States, he presented his thesis, *A contribution to the theory of the general equation of the sixth degree,* to Harvard University in 1886. The topic had been suggested by Klein who was effectively Cole's Ph.D. supervisor.

In July 1888, Cole married Martha Marie Streiff, whom Cole met in Göttingen. They had one daughter and three sons.

From 1888 to 1895 he worked at the University of Michigan.

In 1895, he was appointed professor of mathematics at Columbia University where he stayed until his death. Eric T. Bell attended Cole's Theory of Invariants course and he expressed indebtedness to Professor Cole in his 1921 paper, but Bell didn't clearly indicate the name of his doctoral supervisor.[85] Cassius J. Keyser was Bell's second supervisor.

Cole's most important work was at the American Mathematical Society (AMS). He was Secretary of AMS from 1896 until 1920 and Editor-in-Chief of the Bulletin of AMS from 1897 to 1926. He was vice president of the Society in 1921.

His fellow members of the AMS collected money in appreciation of his twenty-five years of service to the Society. Instead of accepting the money, he let the Society set up the Frank Nelson Cole Prizes in algebra and number theory. The Cole Prizes are now highly prestigious awards.

Cole translated Eugen Netto's (1846-1919) book into English and it was published in 1892 as *The Theory of Substitutions and Its Applications to Algebra.*

Cole died on 26 May 1926 in New York City.

Figure 337.   Virgil Snyder   Figure 338.   Henry S. White

(Courtesy of the American Mathematical Society)

Virgil Snyder

Virgil Snyder was born on 9 November 1869 in Dixon, Iowa. He studied at Iowa State College (now Iowa State University) and Cornell University (1890–1892). He received a Ph.D. in 1894 at the University of Göttingen for a thesis titled *Über die linearen Komplexe der Lie'schen Kugelgeometrie* under Felix Klein's supervision.

He spent his entire professional career at Cornell University, first as an instructor and then as a full professor in 1910. He retired in 1938.

He supervised 33 Ph.D. students at Cornell.[10]

Snyder was an editor of the Bulletin of the American Mathematical Society (AMS) from 1904 to 1920 and president of AMS in 1927–28.

His hobby was mountain climbing and long walks.[3]

Snyder died on 4 January 1950 in Ithaca, New York.

Henry Seely White

Henry White was born on 20 May 1861 in Cazenovia, New York. He studied at the Wesleyan University, graduating in 1882.

John Monroe Van Vleck, father of Edward Van Vleck, encouraged White to study mathematics at the graduate level. He received a Ph.D. at Göttingen for his dissertation, *Abelsche Integrale auf Singularitätenfreien, Einfach überdeckten. Vollständigen Schnittkurven eines Beliebig Ausgedehnten Raumes* in 1891 under Klein's supervision.[3]

White returned to the United States in March 1890 before receiving his doctorate. He was appointed assistant in mathematics at Clark University in Worcester, Massachusetts in the fall of 1890.[86]

From 1889 to 1892, Clark University was the preeminent school of mathematics in North America. The faculty included William Story, Oskar Bolza, Henry Taber, Joseph de Perott (1854-1924) and Henry White. Contrary to the original promises, Clark University was impoverished. From 1892 until 1900, Jonas Gilman Clark, the founder of Clark University gave no more money to the University. During this period, there was only $32,000 per year to support the university. The combined salary for five faculty members in mathematics was $7,200.[86]

Bolza left Clark for the University of Chicago and White left for Northwestern University in 1892. White was promoted to full professor in 1894.

In 1905, White moved to Vassar College in Poughkeepsie, New York where he was elected to the National Academy of Sciences in 1915 and president of the American Mathematical Society in 1907-08.

In 1890, he married Mary Willard Gleason, a composer which was fitting because he loved music and they had three children. White died on 20 May 1943 in Poughkeepsie, New York.

Gilbert Ames Bliss

Gilbert Bliss was born on 9 May 1876 in Chicago, Illinois. He graduated from the University of Chicago in 1897 and received his Ph.D. at Chicago in 1900 for a thesis, *The geodesic lines in the anchor rind*, supervised by Bolza.[3] Bliss spend the academic year 1902-03 at the University of Göttingen.

Before he came back to the University of Chicago in 1908 succeeding Maschke, he taught at the University of Missouri and Princeton University. Bliss was chairman of the mathematical department at Chicago from 1927 to 1941. Unfortunately, Chicago was no longer the preeminent research center of mathematics during this period. After Marshall Stone came to Chicago in 1946, the reputation of Chicago improved greatly.

Bliss worked in the Range Firing Section of the Aberdeen Proving Ground in Maryland during World War I.

He was elected to the National Academy of Sciences in 1916 and president of the American Mathematical Society from 1921 to 1922.[3]

Bliss published a major book entitled *Lectures on the Calculus of Variation* in 1946.

He died on 8 May 1951 in Harvey, Illinois.

Figure 339.   Gilbert A. Bliss

(Courtesy of the American Mathematical Society)

Edward James McShane

Edward McShane was born on 10 May 1904 in New Orleans, Louisiana. He graduated from Tulane University in 1925 receiving two bachelor's degrees in mathematics and engineering. He was awarded a Ph.D. at the University of Chicago in 1930 for his dissertation, *Semi-continuity in the calculus of variations and absolute minima for isoperimetric problems,* supervised by Gilbert Bliss.[10]

He married Virginia Haun in September 1931 and they had two daughters and one son. McShane and his wife spent the academic year 1932-33 at Göttingen, where he observed the Nazis rise to power.

In 1935, he was appointed as a professor at the University of Virginia in Charlottesville where he stayed until his retirement in 1974.

During World War II he also worked at the Aberdean Proving Ground in 1942-45.

He wrote three books on integration and published *Stochastic Calculus and Stochastic Models* in 1975.

He was president of the Mathematical Association of America (MAA) in 1953-54 and received the MAA Chauvenet Prize in 1953. He also served as president of the American Mathematical Society in 1959-60.

McShane died on 1 June 1989 in Charlottesville, Virginia.

Figure 340.   Edward J. McShane

(Courtesy of the American Mathematical Society)

John Torrence Tate, Jr.

John Tate, Jr. was born on 13 March 1925 in Minneapolis, Minnesota, as a son of John Torrence Tate (1889-1950), a famous physicist who received his Ph.D. at the University of Berlin under James Franck in 1914.

John Tate, Jr. received a Ph.D. at Princeton University in 1950 for a thesis titled *Fourier analysis in number fields and Hecke's zeta functions* supervised by Emil Artin.

Tate was a professor at Harvard University for 36 years from 1954 until 1990 and is now at the University of Texas at Austin. He had supervised 39 Ph.D. students at Harvard. Tate received the Frank Cole Prize in 1956 and the Wolf Prize in 2002.

Benedict Gross, one of his former students, also received the Frank Cole Prize in 1987. Tate is best known for the Sato-Tate conjecture, Sato-Tate measure, Tate module and Néron-Tate height.

David Gilbarg

David Gilbarg was born in 1918 in Brooklyn, New York. He graduated from City College of New York in 1937. He completed his Ph.D. in 1941 at Indiana University for a dissertation, *On the structure of the group of p-adic 1-units,* supervised by Emil Artin.

During World War II, he worked at the National Bureau of Standards and at the Naval Ordnance Laboratory. His war work motivated a shift in his interests from algebra to the fields of fluid dynamics and nonlinear partial differential equations.

After working at Indiana University, he was appointed professor of mathematics at Stanford University in 1957 when the faculty included George Pólya, Gabor Szegö, Charles Loewner, Stefan Bergman and Max Schiffer. Gilbarg was Executive Head of the mathematics department during the period 1959–70 and he retired in 1989.

During Gilbarg's tenure as Executive Head, he brought aboard Paul Cohen, Lars Hörmander, Kunichiro Kodaira, Donald Ornstein, Donald Spencer, Ralph Phillips and Hans Samuelson, which ensured Stanford's place as a leading center of research in Geometry, Topology, Partial Differential Equations and Classical analysis.

Gilbarg wrote *Elliptic Partial Differential Equations of Second Order* in 1977 jointly with Neil S. Trudinger, his former student.

Gilbarg died of natural causes on 20 April 2001 in his residence in Palo Alto, California.

James B. Serrin, Jr

James Serrin, Jr. was born in 1926. He received his Ph.D. at Indiana University in 1951 for a thesis titled *The existence and uniqueness of flows solving four free boundary problems* under Gilbarg's supervision.[10]

Serrin married Barbara West in 1952 and they have three daughters.

His entire career from 1954 to 1995 was spent at the University of Minnesota except many visiting appointments elsewhere. Serrin became Regent's professor of Mathematics at Minnesota in 1969.

His areas of interest are differential equations and foundations of thermodynamics and countinuum mechanics. He received the Birkhoff Prize in Applied Mathematics from the American Mathematical Society in 1973 and he was elected to the National Academy of Sciences in 1980.

Salomon Bochner

Salomon Bochner was born into an orthodox Jewish family on 20 August 1899 in Podgorze, near Krakow, Austo-Hungarian Empire (now Poland). After graduating a gymnasium in Berlin, he studied at the University of Berlin where he completed his Ph.D. in 1921 for his thesis, *Über Orthogonale Systeme Analytischer Funktionen,* under Schmidt's supervision.[10]

In 1924, he worked with Harald Bohr on almost periodic functions in Copenhagen and he travelled to England to work with Hardy and Littlewood.

He received habilitation in 1927 at the University of Munich and lectured there from 1930 to 1933. He published a book in 1932 entitled *Vorlesungen über Fouriersche Inegrale,* which became a classic.[3]

After Hitler came to power in January 1933, Bochner accepted a post as an associate at Princeton University and was promoted to assistant professor in 1934. He married Naomi Weinberg in November 1938 and they had one daughter. He had become a US citizen in the same year. Bochner was appointed as Henry Burchard Fine Professor in 1959 and retired in 1968. He became Edgar Odell Lovett Professor of Mathematics at Rice University in 1969 and held this position until 1982.[3]

Bochner wrote *Several Complex Variables* in 1948 jointly with W.T. Martin and *Harmonic Analysis and the Theory of Probability* in 1955.

He was elected to the National Academy of Sciences in 1950 and vice-president of the American Mathematical Society in 1957-58. He received the Steele Prize in 1979. He supervised 36 Ph.D. students.

Bochner died on 2 May 1982 in Houston, Texas.

Figure 341.   Salomon Bochner

Figure 342.   Eberhard F. F. Hopf

Eberhard Friedrich Ferdinand Hopf

Eberhard Hopf was born on 4 April 1902 in Salzburg, Austria. He completed his Ph.D. at the University of Berlin under Schmidt's supervision in 1926. His dissertation was entitled *Über die zusammenhänge zwischen gewissen höeren*

Differezenquotieten reeller Funktionen einer reelen Variablen und deren Differenzierbarkeitseigenschaften.[10]

Observers often confuse Eberhard Hopf with Heinz Hopf (1894–1971) because the latter also received his Ph.D. under Schmidt in the same year.

Eberhard Hopf received his habilitation in Mathematical Astronomy in Berlin in 1929 and held the post of privatdozent until 1937.

Hopf spent the years 1931 to 1936 at the Massachusetts Institute of Technology when the Wiener–Hopf equation was derived.[87]

$$\int_{-\infty}^{\infty} g(u) R_{s+n}(u-\tau) d\,u = R_{s+n,s}(a+\tau) \qquad \tau \geq 0$$

where $g(u)$ = optimum weighting function

R_{s+n} = autocorrelation function $s(t)+n(t)$

$R_{s+n,s}$ = crosscorrelation between $s(t)+n(t)$ and $s(t)$

$s(t)$ is the signal variable and $n(t)$ is the noise

From 1937 to 1944, he held a chair (Dritter Lehrstuhl) at the University of Leipzig. Then he moved to the University of Munich, where he stayed until 1948. Hopf was never forgiven by many people for his moving to Germany in 1936 during the Nazis regime.

He came back to the United States in 1948 and became a US citizen in 1949.

He was appointed a professor at Indiana University in 1949 where he retired in 1972.

Hopf published *Mathematical Problems of Radiative Equilibrium* in 1934 and *Ergodentheorie* in 1937. He received the AMS Steele Prize in 1981.

Hopf died on 24 July 1983.

Figure 343.   Louis Nirenberg

Louis Nirenberg

Louis Nirenberg was born on 28 February 1925 in Hamilton, Canada. He received his B.S. at McGill University and his Ph.D. at New York University's Courant Institute of Mathematical Sciences in 1949. Years later, he served as director of the Courant Institute from 1970-72.

Nirenberg is a member of the National Academy of Sciences, *Académie des Sciences de Paris*, and Honorable Member of European Academy of Sciences.

In 1959, he received the AMS Bôcher Prize for outstanding contributions to mathematical analysis. In 1982, he was the first recipient of the Crafoord Prize from the Royal Swedish Academy of Sciences. Nirenberg received the National Medal of Science in 1995 for his fundamental contributions to mathematical analysis. He developed intricate interactions between mathematical analysis, differential geometry, complex analysis and applied them to the theory of fluid flow and other physical phenomena.[88]

Nirenberg has supervised 44 doctoral dissertations mostly at New York University.[10]

Hillel (Harry) Furstenberg

Hillel Furstenberg was born in 1935 in Berlin, Germany and soon emigrated to the United States. He received his B.A. and M.Sc. at Yeshiva University in 1955. Furstenberg received a Ph.D. at Princeton University in 1958 for his thesis *Prediction theory* supervised by Bochner.

After several years at the University of Minnesota, he joined the mathematics faculty at the Hebrew University of Jerusalem in 1965.

He is member of the Israel Academy of Sciences and Humanities as well as the US National Academy of Sciences.

Furstenberg's main interest is application of probability theory and ergodic theory to number theory and Lie groups. He received the Wolf Prize in 2006.

Saunders MacLane

Saunders MacLane was born on 4 August 1909 in Norwich, Connecticut. His father was a congregational minister.

He graduated from Yale University in 1930 and he continued his study at the University of Chicago where he took courses from Dickson and Bliss. E.H. Moore persuaded him to study for a doctorate at Göttingen.

He went to Göttingen with a fellowship from the Institute of International Education in 1931. The full professors there were Herman Weyl, Edmund Landau, Richard Courant and Gustav Herglotz. The associate professors (extraordinary professor) were Emmy Noether, Otto Neugebauer and Paul Bernays. Hans Lewy and Franz Rellich were Privatdozents.[89]

MacLane received a Ph.D. in July 1933 for his thesis titled *Abgekürzte Beweise im Logikkalkul* (Abbreviated Proofs in the Logical Calculus). His original advisor was Paul Bernays, but Bernays was dismissed due to the Nazis racist law. As a result, MacLane was examined by Hermann Weyl and Gustav Herglotz. MacLane married Dorothy Jones whom he met in Chicago in July 1933 in the Rathaus in Göttingen. She typed all his books and most of his papers.

After he returned to the United States, he worked at Yale University (1933–1934), Harvard University (1934–1936), Cornell University (1936–1937), then again at Harvard. He wrote his famous text, *A survey of modern algebra,* jointly with Garrett Birkhoff in 1941.

In 1947, he was appointed professor of mathematics at Chicago. At that time, Chicago had Marshall Stone (Chairman), Abraham Albert, Irving Kaplansky, Otto Schilling and Andre Weil, S.S. Chern and Antoni Zygmund.

Figure 344. Saunders MacLane

(Courtesy of the American Mathematical Society)

Figure 345. John G. Thompson

MacLane was elected to the National Academy of Sciences. He was president of the Mathematical Association of America (MAA) in 1951-52 and president of the American Mathematical Society (AMS) in 1973-74.

He received the MAA Chauvenet Prize in 1941 and the AMS Steele Prize in 1986.

MacLane died on 14 April 2005 in San Francisco, California.

John Griggs Thompson

John Thompson was born on 13 October 1932 in Ottawa, Kansas. He graduated from Yale University in 1955 and received a Ph.D. at the University of Chicago in 1959 for his thesis titled *A proof that a finite group with a fixed-point-free automorphism of prime order is nilpotent* supervised by MacLane. His thesis solved one of the conjectures of Georg Frobenius (1849-1917) that had remained unsolved for about 60 years.[3]

From 1962 to 1968, he was a professor at the University of Chicago. He went to the University of Cambridge to become a visiting professor and a fellow of Churchill College in 1968. Then he was appointed Rouse Ball Professor of Pure Mathematics at Cambridge in 1971, the post he held until 1993.[42]

Since 1993, he has been a Graduate Research Professor at the University of Florida.

Thompson was elected to the National Academy of Sciences in 1971 and a Fellow of the Royal Society of London in 1979.

He married Diane Oenning in 1960 and they have one son and one daughter. He received the AMS Cole Prize in 1965 and a Fields Medal 1970 for proving jointly with Walter Feit that all non-cyclic finite simple groups have even order. The extension of this work by Thompson determined the minimal simple finite groups, in other words, the simple finite groups whose proper subgroup are solvable.

Walter Feit also received the AMS Cole Prize in 1965. In 1985, Thompson received the Sylvester Medal from the Royal Society. He received the Wolf Prize and the Poincare Medal in 1992. Thompson is one of the twelve Fields Medalists who have also received the Wolf Prize. In 2000, he received the National Medal of Science from President William Clinton. Thompson also received the Abel Prize in 2008.

Irving Kaplansky

Irving Kaplansky was born on 22 March 1917 in Toronto, Canada shortly after his parents emigrated from Poland. His mother opened a bakery in Toronto and everybody in the family worked there. When he was young he took piano lessons for eleven years. His parents wanted him to be a concert pianist. He became a mathematician and "the perfect accompanist".

He graduated from the University of Toronto receiving a B.A. in 1938 and a master's degree in 1940. At Toronto he took courses from Richard Brauer (1901-1977) and greatly influenced by him. After winning a William Lowell Putnam Fellowship, he studied for a doctorate at Harvard University under MacLane. He received his Ph.D. in 1941 for a thesis entitled *Maximal fields with valuations*.[3] Kaplansky was MacLane's first Ph.D. student.

From 1941 to 1944, he was a Benjamin Peirce Instructor of Mathematics at Harvard University.

In 1945, he moved to the University of Chicago where he stayed until 1984. He served as chairman of the mathematics department from 1962 to 1967. Saunders MacLane, Kaplansky's former teacher and colleague at Chicago, advised him "Always behave as though you'll have to explain your actions to a Senate Investigating Committee the next day."[89] Kaplansky was appointed George Herbert Mead Distinguished Service Professor in 1969, the post he held until his retirement in 1984.

Then he moved to California to become the director of the Mathematical Sciences Research Institute (MSRI) in Berkeley in 1984.

Kaplansky was elected to the National Academy of Sciences and served as president of the American Mathematical Society in 1985-86. In 1989, he received the AMS Steele Prize in recognition of cumulative influence extending over a career, including the education of 55 doctoral students. He wrote many books including *Infinite Abelian Groups* in 1954.

Kaplansky died on 25 June 2006.

Gian-Carlo Rota

Gian-Carlo Rota was born on 27 April 1932 in Vigevano, Italy.

Since his father was a prominent anti-fascist, his family had to leave Italy for Ecuador in 1945. He came to the United States in 1950. He graduated from Princeton University in 1953 and completed his Ph.D. at Yale University in 1956 as the first graduate student of Jacob T. Schwartz. His thesis was titled *Extension theory of differential operators*.[3]

He married Teresa Rondón in 1956.

From 1957 to 1959, he was a Benjamin Peirce Instructor of Mathematics at Harvard University. Then he joined the faculty at the Massachusetts Institute of Technology where he remained for the rest of his career, except two years at Rockefeller University in 1965-67.

In 1972, he was appointed professor of applied mathematics and philosophy at MIT. He served as a consultant to the Los Alamos Scientific Laboratory from 1966 to 1971.[3] He had supervised 46 Ph.D. dissertations.[10]

He was elected to the National Academy of Sciences in 1982 and received the AMS Steele Prize in 1988 for his paper *On the foundations of combinatorial theory*.

He served as vice-president of the American Mathematical Society in 1995-97.

Rota died in his sleep and was found on 18 April 1999 in Cambridge, Massachusetts.

Figure 346. Irving Kaplansky

(Courtesy of the American Mathematical Society)

Figure 347. Gian-Carlo Rota

David Eisenbud

David Eisenbud was born on 8 April 1947. He received a Ph.D. at the University of Chicago in 1970 for a thesis entitled *Torsion module over Dedekind prime rings* supervised by Saunders MacLane.

He taught at Brandeis University in Waltham, Massachusetts from 1970 to 1997. Then, he moved to the University of California Berkeley in 1997 as a professor and director of the Mathematical Sciences Research Institute (MSRI). He went back to full-time teaching and research at Berkeley in 2007. From 2003 to 2005 he was president of the American Mathematical Society.

His name can be found in the mathematical terms Eisenbud-Goto conjecture and Buchsbaum-Eisenbud criterion.

Figure 348. David Eisenbud

(Courtesy of the American Mathematical Society)

Donald Samuel Ornstein

Donald Ornstein was born in 1934. He completed his Ph.D. at the University of Chicago in 1957 for a thesis titled *Dual vector spaces* under Kaplansky's supervision

He won the 1974 Bôcher Prize for his work on the isomorphism of Bernoulli shifts. He was elected to the National Academy of Sciences in 1981.

During his professorships at Stanford University and the University of Chicago, he supervised 24 Ph.D. dissertations.

Hyman Bass

Hyman Bass was born in 1932. He received a Ph.D. at the University of Chicago in 1959 for a thesis titled *Global dimensions of rings* supervised by Irving Kaplansky.

From 1959 to 1998, he was a professor of mathematics at Columbia University where he supervised 25 Ph.D. students.

His research interests include algebraic K-theory, commutative algebra and algebraic geometry, algebraic groups and geometric methods in group theory.

Since 1999, he has been the Roger Lyndon Collegiate Professor of Mathematics and Mathematics Education at the University of Michigan.

He is a member of the National Academy of Sciences and served as president of the American Mathematical Society in 2001-02.

Figure 349. Hyman Bass

(Courtesy of the American Mathematical Society)

Figure 350. Kurt O. Friedrichs

Kurt Otto Friedrichs

Kurt Friedrichs was born on 28 September 1901 in Kiel, Germany.

He studied at the University of Düsseldorf, University of Greifswald, University of Freiburg and University of Graz.

Then, he went to Göttingen in 1922 where he was impressed by Carl Ludwig Siegel and Emil Artin, and became good friends with Hans Lewy, who was three years younger than Friedrichs.

Friedrichs received a doctorate at Göttingen in 1925 for his dissertation, *Die Randwert-und Eigenwertprobleme aus der Theorie der Elastischen Platten,* supervised by Courant. Friedrichs became Courant's assistant.

He received his habilitation at Göttingen in 1929 and became a professor at the *Techniche Hochschule in Braunschweg* in 1931.[1]

Friedrichs met Nellie Bruell, a young Jewish assistant at the Techniche Hochschule Braunschweig in February 1933. The Nazis law forbade marriage between Aryans and non-Aryans, so Friedrichs decied to emigrate to the United States.

Friedrichs arrived in New York on 4 March 1937 and married Nellie Bruell. They had five children.

Courant and Friedrichs wrote the classic text *Supersonic Flow and Shock Waves* in 1948. Friedrichs became interested in quantum theory fields in the 1950s.

He was elected to the National Academy of Sciences (NAS) in 1959. He received the NAS Award in applied mathematics in 1972 and the National Medal of Science in 1976.

He supervised 35 Ph.D. students at New York University. Friedrich died on 31 December 1982 in New Rochelle, New York.

Hans Lewy

Hans Lewy was born on 20 October 1904 in Breslau, Germany (now Wroclaw, Poland). He received a Ph.D. at the University of Göttingen in 1926 for his thesis titled *Über einen Ansatz zur numerischen Lösung von Randwertproblem* supervised by Courant.[10] He became a privatdozent in 1927 and held the post until 1933.[1]

In the academic year 1929-30, he was at the University of Rome and in 1930-31 at the University of Paris on a Rockefeller Foundation Fellowship.

Lewy was impressed by Vito Volterra, Tullio Levi-Civita, Jacques Hadamard and Henri Lebesgue during this period.[89]

Lewy went to the United States in 1933 and was appointed to Brown University at the annual salary of $3,000.[89] Then, in 1935, he went to the University of California at Berkeley as a lecturer. One year before Lewy joined the mathematics department in 1935, Griffith Evans was selected as the department head by physicists and chemists because senior professor of mathematics were reluctant to invite a new department head. Griffith Evans established the "publish or perish" idea at Berkeley and upgraded the department greatly.

During World War II, Lewy worked at the Aberdeen Proving Ground. He was promoted to full professor at Berkeley in 1945.

Lewy married Helen Crosby in June 1947 and they had one son. Lewy played the violin very well. At the age of fifteen he was a soloist with the Bautzen Symphony Orchester. Later he played in the Göttingen Orchestra too.[89]

In 1950, he refused to sign a loyalty oath imposed on the faculty by the University of California Board of Regents.

Lewy and other professors were dismissed, but a court order reinstated them.

He died on 23 August 1988 in Berkeley, California.

Figure 351. Hans Lewy

Figure 352. Martin D. Kruskal

Martin David Kruskal

Martin Kruskal was born on 28 September 1925 in New York City. He graduated from the University of Chicago in 1945 and received a Ph.D. at New York University for his thesis, *The bridge theorem for minimal surfaces,* in 1952 supervised by Courant.[10]

In 1951, he joined in Princeton's Project Matterhorn which is now called the Princeton Plasma Physics Laboratory (PPPL). Kruskal published several papers related to plasma physics and nonlinear plasma oscillations. He also published *Maximal extention of Schwarzschild's metric* in 1960 in which he introduced Kruskal coordinate. That year, he was appointed as a lecturer in astronomy at Princeton.

With George Szekeres, he introduced the Kruskal–Szekeres coordinates for the Schwarzchild geometry, the spherically symmetric solution to the Einstein field equation.

In 1961, he was promoted to professor of astronomy and continued his work in plasma physics. Kruskal was appointed as director of the applied mathematics at Princeton, the position he held for twenty years.

He then became a professor of mathematics in 1979 and left Princeton in 1989. That year, he was appointed to the newly created David Hilbert Chair of Mathematics at Rutgers University.

He married Laura in 1950 and they had three children. Kruskal is most famous for his role in starting the "Soliton solution", considered one of the most important mathematical advances of the last half of the 20th Century. He and Norman Zabusky formed nonlinear waves that behave in many ways like linear waves, which they termed solitons.[91]

Kruskal was elected to the National Academy of Sciences in 1980 and a Fellow of the Royal Society of London in 1997. He won the National Academy of Sciences Award in Applied Mathematics and Numerical Analysis in 1989 and the National Medal of Science in 1993 by President Clinton. Kruskal also received the AMS Steele Prize in 2006.

He died on 26 December 2006 in Princeton, New Jersey.

His brother William Kruskal (born 1928) discovered the multidimensional scaling and Kruskal's algorithm.

Kruskal's son Clyde is also a mathematician.

Fritz John

Fritz John was born on 14 June 1910 in Berlin, Germany. He studied from 1929 to 1933 at the University of Göttingen where he was influenced by Herglotz, Courant and Lewy.

He was awarded a Ph.D. at Göttingen in 1934 for a thesis titled *Bestimmung einer Funktion aus iherm intergralen über gewisse Mannigfatigkeiten* under Courant's supervision.

Because of the Nazi's racist laws, he left Germany for England. He spent one year at St. John's College, Cambridge as a research scholar where he published papers on the Radon transform.[3]

He was appointed as assistant professor at the University of Kentucky at Lexington in 1935 where he stayed until 1946.

He became a naturalized US citizen in 1941.

During World War II, he worked at the Ballistic Research Laboratory at the Aberdeen Proving Ground in Maryland from 1943 until 1945.

He was appointed as an associate professor at New York University in 1946 and was promoted to a full professor in 1951.

In 1978, he took the Courant Chair at the Courant Institute of Mathematical Sciences at New York University where he retired in 1981.[3]

His later research interests were in convex geometry, ill-posed problems, the numerical treatment of partial differential equations, quasi-isometry and blow-up in nonlinear wave propagation.

He received the Birkhoff Prize in applied mathematics in 1973 and the AMS Steele Prize in 1984.

John died on 10 February 1994 in New York City.

William (Willy) Feller

William Feller (originally Vilibald Feller) was born on 7 July 1906 in Zagreb, Croatia. His mother was a Catholic, but his father was born to a Polish Jewish family. Feller graduated from the University of Zagreb in 1925 and received his Ph.D. at Göttingen in 1926 for a thesis entitled *Über algebraisch rektifizierbare transzendente Kurven* under Courant's supervision.[10] He was only twenty years old when he earned his doctorate. He spent two more years at Göttingen before he was appointed a privatdozent at the University of Kiel in 1929.

After Hilter came to power in 1933, Feller left Germany[1] for Copenhagen where he remained until 1934. Then, he moved to the University of Stockholm working with Harald Cramér and Marcel Riesz.

In 1938, Feller married Clara Mary Nielsen, his student at Kiel. They had no children.

The following year, Feller and his wife emigrated to the United States and he became an associate professor of mathematics at Brown University in Providence, Rhode Island. He was naturalized as a US citizen in 1944. From 1945 to 1950 he taught at Cornell University.

In 1950, Feller was appointed Eugene Higgins Professor of Mathematics at Princeton University. At that time, Princeton had Lefschetz, Church, Tucker, Artin. Because of the august faculty, Feller considered himself to be one of the lowest ranking members of the department.[3]

Feller made contributions to the mathematical theory of Brownian motion and diffusion processes. His most important work was *Introduction to Probability Theory and its Application* in two volumes published in 1950 and 1961.

Many mathematical terms are named after him; Feller process, Feller explosion test, Feller-Brown movement, Feller property and Feller-Lindberg theorem.

He was elected to the National Academy of Sciences and received the 1969 National Medal of Science, but he died shortly before the presentation was to be made. His wife received the medal on his behalf.

He died on 14 January 1970 in New York City.

Figure 353. Fritz John

Figure 354. Willy Feller

Peter David Lax

Peter Lax was born into a Hungarian Jewish family in Budapest on 1 May 1926. Lax learned mathematics from Albert Korodi, his mother's younger brother who was an engineer by profession. He was also tutored by a mathematician Rózsa Péter (1905–1977).[89] His father was a doctor and the American consul in Budapest was one of his patients. The Lax family left Budapest on 15 November 1941. By the time they arrived in New York, Hungary and the United States had declared war.

Von Neumann had been notified by Dénes König that a really outstanding young mathematician was coming from Hungary. Dénes König also wrote to Gabor Szegö and Otto Szász on Lax's behalf.

Peter Lax's mother and Szegö's wife were first cousins. Lax graduated Stuyvesant High School in New York in 1943 and entered New York University where he took graduate courses in mathematics while he was an undergraduate student.

He was drafted into the United States Army in 1944 and worked at Los Alamos Laboratory participating in the Manhattan Project. He graduated from New York University in 1947.

In 1948, Lax married Anneli Cahn, whom he met in an NYU graduate course in complex analysis. Anneli also received her Ph.D. at NYU in 1955 supervised by Courant. She was the long-time technical editor of the Mathematical Associate of America's *New Mathematical Library*.

Lax received his Ph.D. in 1949 from New York University for a thesis titled *Nonlinear system of hyperbolic partial differential equation in two independent variables* under Friedrichs' supervision.

Lax was appointed as an assistant professor at New York University in 1951 and was promoted to full professor in 1958.

His first paper was *On a conjecture of Erdös*.

He was director of the Courant Institute of Mathematical Sciences from 1972 to 1980 and served as president of the American Mathematical Society in 1979-80.

He wrote *Linear Algebra* in 1997, *Functional Analysis* in 2002, *Hyperbolic Partial Differential Equations* in 2006 and many other books.

Lax was elected to the National Academy of Sciences, the *Académie des Sciences* in Paris and the Russian Academy of Sciences.

He was awarded the National Medal of Science in 1986, the Wolf Prize in 1987, the AMS Steele Prize in 1992 and the Abel Prize in 2005.

Figure 355. Peter D. Lax

(Courtesy of the American Mathematical Society)

Figure 356. Cathleen S. Morawetz

(Courtesy of the American Mathematical Society)

Cathleen Synge Morawetz

Cathleen Morawetz was born on 5 May 1923 in Toronto, Canada as a daughter of John Lighton Synge (1897-1995), a mathematician and Eleanor Mabel Allen Synge. They are both Irish Nationalists rebelling against British rule.[89]

In 1945, Morawetz graduated from the University of Toronto, where she took her father's course.

She met Herbert Morawetz, who had come to Canada in 1939 as a refugee from Czechoslovakia. Herbert's father was the president of the jute cartel and owned a big factory in Czechoslovakia. Herbert graduated the University of Toronto with a master's degree in chemical engineering. Cathleen married him on 28 October 1945 while she was a graduate student at Massachusetts Institute of Technology. They have three daughters and one son.

She received a master's degree at MIT in 1946. She had a hard time getting a job because she was a woman and Herbert had a similar problem because he was Jewish.

Kathleen was hired by Courant at New York University to solder connections on a machine to solve linear equations. Later she edited Courant and Friedrich's book *Supersonic Flow and Shocks waves.* She received a Ph.D. at New York University in 1951 for a thesis titled *Contracting spherical shocks treated by a perturbation method* supervised by Friedrichs.[10]

She became Director of the Courant Institute in 1984 and served as president of the American Mathematical Society in 1995-96. In 1998, she was awarded the National Medal of Science. She was the first woman member of the Applied Mathematics Section of the National Academy of Sciences. Morawetz also received the AMS Steele Prize in 2004 and the Birkhoff Prize in 2006.

One day she was called to jury's duty. When she answered her profession as Distinguished Professor at New York University, she was not selected for the trial.

Next time, she replied that her profession was a teacher, and she was selected for the jury.

Jürgen Kurt Moser

Jürgen Moser was born on 4 July 1928 in Königsberg, Germany (now Kaliningrad, Russia). In early 1945, the Soviet Army began a siege of Königsberg which lasted for two months during which time most of Moser's class was killed. Moser's elder brother was killed in the fighting. Königsberg became a part of the Soviet Union in 1945 and was renamed Kaliningrad in 1946. Königsberg produced so many mathematicians and physicists such as Hilbert, Minkowski, Bessel, Jacobi, Franz Neumann, Carl Neumann, Gustav Kirchhoff, Otto Hesse, Rudolf Clebsch and F.J. Richelot.

Moser escaped to Göttingen in 1947 after risking his life to cross the border. He was awarded a Ph.D. at Göttingen in 1952 for his thesis *Störungstheorie des Kontinuierlichen Spektrums für gewöhliche Differentialgleichungen Zweiter Ordnung* supervised by Franz Rellich.

Moser spent the academic year 1953–54 at the Courant Institute, New York University working with Peter Lax. Then he became an assistant to Carl Siegel at Göttingen in 1954–55.

He emigrated to the United States in 1955 to marry Gertrude Courant on 10 September 1955 in New York.

After working two years at New York University from 1955 until 1957, he was appointed as an associate professor at the Massachusetts Institute of Technology in 1957. He went back to NYU where he served as director of the Courant Institute from 1967 to 1970.

Figure 357. Jürgen Moser

In 1980, he became a professor at *Eidgenössische Technische Hochschule in Zurich* where he remained until his retirement in 1995.

He introduced the "Moser twist stability theorem" in his 1962 paper *On invariant curves of area-preserving mapping of an annulus*. When combined with the work of Kolmogorov and Arnold, it became the KAM Theory, which provided a new approach to stability problems in celestial mechanics.[3]

He received the Birkhoff Prize in 1968 and he was elected to the National Academy of Sciences in 1973. He served as president of the International Mathematical Union from 1983 to 1986. Moser received the L. E. J. Brouwer Medal from the Dutch Scientific Society and the Georg Cantor Medal from the German Mathematical Society in 1992. He also received the Wolf Prize in 1994.

Moser died of prostate cancer on 17 December 1999 in Zurich, Switzerland.[3]

Jerzy Neyman

Jerzy Neyman was born on 16 April 1894 in Bendery, Russia. The family considered itself to be Polish, but Poland did not exist as a nation when Neyman was born. His name in Russian was Yuri Czeslawovich Neyman.[3]

He graduated from Kharkov University in 1917 with a diplom degree in mathematics. He received a Candidate of Science degree in 1920 and became a lecturer at Kharkov. He married Olga Solodovnikova in May 1920. In 1921, he went to Poland and received a doctorate in 1924 at the University of Warsaw for a thesis on application of probability to agricultural experimentation. His thesis was examined by Sierpinski and Mazurkiewicz.[3]

From there, Neyman went to the University College, London on a Rockefeller Fellowship in 1925. He became friendly with Egon Pearson there and they later published many papers together.

In the academic year 1926–27, Neyman was in Paris attending lectures by Borel, Lebesgue and Hadamard. Neyman considered Lebesgue as his Doktorvater.[93] Neyman returned to Paris in 1927 and obtained his habilitation in June 1927 and began lecturing at the University of Warsaw as a docent. He went to London in 1934 to teach at the University College and remained there until 1938 at the annual salary of 500 pounds.

In August 1938, Neyman arrived Berkeley to accept a lectureship ($4,500 per year) at the University of California, Berkeley. At that time Griffith Evans was trying to upgrade the Berkeley mathematics department. Neyman tried hard to build a world-leading school of mathematical statistics at Berkeley. He successfuly created a separate statistics department at Berkeley on 1 July 1955. Neyman supervised 39 doctoral candidates.

George B. Dantzig, Lucien Le Cam and Erich Lehmann were some of Neyman's students.

David Blackwell was recruited from Howard University to Berkeley by Neyman.

Neyman was elected a Fellow of the Royal Society of London in 1979. He received the Guy Medal of the Royal Statistical Society in 1966 and the National Medal of Science from President Lyndon Johnson in 1969.

Nyman helped Antoni Zygmund escape from Poland to the United States in 1940. Neyman could speak Polish, Russian, Ukrainian, French, German, Latin and English.

Neyman died on 5 august 1981 in Oakland, California.

Figure 358. Jerzy Neyman

Figure 359. George Dantzig

George Dantzig

George Dantzig was born on 8 November 1914 in Portland, Oregon.

His father, Tobias Dantzig was born in Russia and studied mathematics in Paris under Poincaré. His father gave him thousands of geometry problems which helped to develop his analytic power. His mother also studied mathematics at Sorbonne. They emigrated to the United States where Tobias Dantzig received a Ph.D. at the University of Indiana in 1917 and became a faculty member at the University of Maryland.

Tobias Dantzig published his famous book *Number: the Language of Science* in the late 1920s. His mother became a linguist at the Library of Congress in Washington D.C.

George graduated from the University of Maryland in 1936 and married Anne Shmuner in the summer of 1936. He moved to the University of Michigan to study under G.Y. Rainich and took courses from R.L. Wilder, T.H. Hildebrandt and H.C. Carver. Dantizig received his master's degree there in 1937.

From 1937 to 1939 he worked at the US Bureau of Labor Statistics as a Junior statistician, reading statistics paper by Jerzy Neyman. Dantzig then went to Berkeley to study for a doctorate under Neyman.[89]

I arrived late one day to one of Neyman's classes during my first year in Berkeley. On the blackboard there were two problems which I assumed had been assigned for homework. I copied them down. A few days later I apologized to Neyman for taking so long to do the homework – the problems seemed to be a little harder to do than usual. I asked him if he still wanted the work. He told me to throw it on his desk.

The problems on the blackboard which I had solved thinking they were homework were in fact two famous unsolved problems in statistics.

A year later when I began to worry about a thesis topic, Neyman just shrugged and told me to wrap two problems in a binder and he would accept them as my thesis.[89]

Because of the World War II, Dantzig's graduate study was interrupted. He worked for the US Army Air Force as a civilian from 1941 to 1946. At that time Air Force belonged to the US Army. In 1944, Dantzig was awarded the War Department Exceptional Civilian Service Medal.

In 1946, Dantzig received a Ph.D. at Berkeley for his thesis titled *I Complete form Neyman-Pearson fundamental lemma, II on the Non-existence of tests of student's hypothesis having power functions independent of Sigma*, which were the two problems in Neyman's class.

A minister in Indiana told the story of Dantzig's two homework problems in his sermon as an example of positive thinking. If Dantzig had known that the problems were not homework but were in fact two famous unsolved problems in statistics he probably would have discouraged and would never have solved them.[89]

He was offered an academic position at the University of California Berkeley at only $1400 per year, so he turned it down. Instead, he became a mathematical adviser to the US Air Force Comptroller. His colleagues challenged him to see what can be done to mechanize the planning process; that is, to find a more rapid way to complete a time-staged deployment, training and logistical supply program.

In response Dantzig invented the simplex method of optimization in 1947. Wassily Leontief first formulated the Inter-industry Model of the American Economy around 1932 which was a steady-state model. Dantzig's model was a highly dynamic model or a time-staged dynamic linear program with a staircase matrix structure.

Leontief was awarded the Nobel Prize in Economic Science in 1973. The term "linear programming" was proposed by Tjalling Koopmans. Dantzig published his famous book *Linear Programming and Extensions* in 1963.

Leonid V. Kantarovich and Tjalling Koopmans received the Nobel Prize in Economic Sciences in 1975 for their contributions to the theory of optimum allocation of resources.

Koopmans regretted that Dantzig was unable to share the honors.

After working at the RAND Corporation for eight years, Dantzig was appointed as a professor at Berkeley in 1960.

In 1966, he moved to Stanford as professor of operations research and computer science where he remained for the rest of his career.[3]

He was a member of the National Academy of Engineering and the National Academy of Science. He received the National Medal of Science in 1976 from President Ford, and the NAS Award in applied mathematics and numerical Analysis in 1977. Dantzig also received the Von Neumann Theory Prize in Operations Research in 1975.

He died on 13 May 2005 in Palo Alto, California.

Figure 360. Lucien Le Cam

Oystein Ore

Oystein Ore was born in 9 October 1899 in Kristiania (now Oslo), Norway. He graduated from the University of Kristiania for a dissertation, *Zur Theorie der algebraischen Körper,* in 1924.

He was appointed as an assistant professor of mathematics at Yale University in 1927 and was promoted to full professor two years later. He married Gudrum Lundevall in August 1930 in Larvik, Norway. They had two children.

In 1931, Ore was named Sterling Professor at Yale, a position he held until his retirement in 1968.[3]

During World War II, he played a major role in helping Norway for which he was decorated with the Knight Order of St. Olaf in 1947.

He wrote ten books including *Number Theory and Its History* (1948), *Theory of Graphs* (1962), *Graphs and Their Uses* (1963), *The Four-Color Problem* (1967) and *Invitation to Number Theory* (1969).

Marshall Hall, Jr. was one of his outstanding students.

He had a deep interest in painting and sculpture as well as on ancient maps.

Ore died in 13 August 1968 in Oslo, Norway where he was scheduled to lecture at a mathematical meeting.

Lucien Le Cam

Lucian Le Cam was born on 18 November 1924 in Croze, Creuse, France. In May 1944, he joined an underground group to fight against the Germans. In December 1944, he took the entrance examination for *Ecole Normale Superieure* but flunked the oral part of the exam.

He decided to go to the University of Paris. In Le Cam's words.[89]

You went to the window and said, "I want to register at the Université." "Do you have your high school diploma?" "Yes, but I don't have a copy of it."

"Well, that's all right. We can check. Which kind of courses are you going to take? Humanities or sciences?" "Well, I will take this and that course." So they wrote that down and said, "Can you pay the fee?" "Well, I don't know. How much is it?" "Oh, five hundred francs." That was about one dollar at that time.

Le Cam received his degree, Licence es Sciences (equivalent to bachelor's degree), in 1945. Then he worked for the next five years at Electricité de France, where the three main subjects were.[89]

(1) Where and how to construct new hydroelectric plants

(2) How to operate dams for the best results

(3) How to evaluate the probabilities of drastic droughts or destructive floods

Jerzy Neyman invited Le Cam to Berkeley in 1950 as an instructor. Le Cam received a Ph.D. at Berkeley in 1952 for his thesis titled *On some asymptotic properties of maximum likelihood estimates and related Baye's estimates* supervised by Neyman.[10]

Le Cam married Louise Romig in April 1952 and they had two sons and one daughter, Linda.

Le Cam was appointed assistant professor in 1953 at Berkeley and his first Ph.D. student, Julius Blum received his degree in the same year. Over the next 38 years until his retirement in 1991, Le Cam supervised 40 Ph.D. students.

He was a principal architect of the modern asymptotic theory of statistics. He wrote many books including *Locally Asymptotically Norman (LAN) Families of Distributions* (1960), *Asymptotic Methods in Statistical Decision Theory* (1986) and *Asymptotics in Statistics–Some Basic Concepts* (1990) with his former student Grace Lo Yang.

In 1972, a malignant tumor developed on the left leg of Le Cam's daughter Linda whose leg had to be removed at the hip. Dr. Vera Byers, immunologist, helped to cure Linda and LeCam provided the statistical assistance. Linda was cured and later got married. [89]

Le Cam died on 25 April 2000.

Antoni Szczepan Zygmund

Antoni Zygmund was born on 26 December 1900 in Warsaw, Russia (now Poland). At the age of 18, he entered the University of Warsaw (Uniwersytet Warzawski), which was founded in 1816. He received a Ph.D. in 1923 for a dissertation on the Riemannian theory of trigonometric series under Aleksander Rajchman (1890–1940) and Stefan Mazurkiewicz (1888–1945). Rajchman died in Dachau concentration camp in 1940.

From 1922 to 1929, Zygmund taught at the Polytechnic School of Warsaw (Politechnika Warszawski, Technical University of Warsaw). In 1926, he received habilitation and began teaching at the University of Warsaw.

Zygmund married Irena Parnowska in February 1925. She later became a mathematics teacher.

Zygmund spent the academic year 1929–30 in England as a Rockefeller Fellow at Oxford and Cambridge where he wrote ten papers.

From 1930 to 1939, he taught at the University of Stefan Batory in Vilnius. The City of Vilnus became a part of Lithuania in 1939 and Zygmund had to leave because Lithuania became a part of Soviet Union.

In 1940, he emigrated to the United States thanks to the efforts of J.D. Tamarkin, Norbert Wiener and Jerzy Neyman. Zygmund stayed at Mount Holyoke College for five years. At the invitation of Marshall Stone in 1947, Zygmund went to the University of Chicago, where he remained until he retired in 1971.

Stone brought many distinguished mathematicians to Chicago including S. MacLane, S.S. Chern and A.Weil. Adrian Albert, E.P. Lane and L.M. Graves were already there.

The junior faculty members were I. Kaplansky, P. Halmos and I.E. Segal.

Zygmund published *Trigonometric Series* in 1935, the second edition in 1959 and the third edition in 2002. He also wrote *Analytic Functions* in 1938 with his friend Stanislaw Saks.

Zygmund supervised 40 Ph.D. students (3 in Vilnius, 1 in Pennsylvania and 36 in Chicago). Elias Stein, Paul Cohen, Alberto Calderón and Guido Weiss are some of his former students. Reference 96 lists his Ph.D. students as well as students of Zygmund's students.

He was elected to the Polish Academy of Sciences in 1959 and to the National Academy of Sciences in 1961. He received the AMS Steele Prize in 1979 and the National Medal of Science in 1986 from President Reagan for his creation and leadership of the strongest school of analytical research in the contemporary mathematical world.[3] R.R. Coifman and R.S. Strichartz stated.[96]

He was gentle, generous and friendly. One day he passed a lounge in the University of Chicago that was filled with a loud crowd watching TV, he asked one of his students what was going on. He explained they were watching the World Series and how it worked. Zygmund thought about it for a few minutes and commented, "I think it should be called the World Sequence."

Zygmund died on 30 May 1992 in Chicago, Illinois.

Figure 361. Antoni S. Zygmund

Figure 362. Alberto P. Calderón

Alberto P. Calderón

Alberto Calderón was born on 14 September 1920 in Mendoza, Argentina. He graduated from the University of Buenos Aires in 1947 with a degree in civil engineering.

He became an assistant to a professor of Electric Circuit Theory in the School of Engineering at his alma mater.[3]

Antoni Zygmund met Calderón when he visited Buenos Aires in 1948. Seeing Calderon's potential, Zygmund brought him to the University of Chicago where Calderón received a Ph.D. in 1950 supervised by Zygmund. His Ph.D. dissertation was in three parts; I. On the Ergodic Theorem, II. On the Behavior of Harmonic Functions at the Boundary, III. On the Theorem of Marcinkiewicz and Zygmund.[10]

Josef Marcinkiewicz (1910-1940) was a Ph.D. student of Zygmund at the University of Vilnius (Wilno).

After working at Ohio State University (1950-1953), the Institute for Advanced Study at Princeton (1954-1955), and the Massachusetts Institute of Technology (1955-1959), Calderon was appointed professor of mathematics at the University of Chicago.

He remained at Chicago until he retired in 1985 except three years at MIT from 1972 until 1975. Upon returning to Chicago, he was appointed university professor of mathematics.

With his mentor Zygmund, Calderon formulated a theory, now known as the Calderon-Zygmund theory, of what are called singular integrals. Singular integrals are mathematical objects that look infinite, but when interpreted properly are finite and well-behaved. Calderon showed how the singular integrals could be used to solve equations in geometry and analyze functions of complex variables. He also showed how singular integrals could be used to study partial differential equations. He was one of the central links between two major areas of mathematical analysis, namely Fourier analysis and partial differential equations.[97]

Together with Zygmund, Calderon founded the "Chicago School of Analysis", the most influential school in analysis in the twentieth century.

Calderon married Mabel in 1950 and they had two children. Mabel died in 1985 and Calderón married Alexandra Bellow, professor of mathematics at Northwestern University, in 1989.

Calderon was a member of the National Academy of Sciences (1968), the French Academy of Sciences (1984), and many other academies.

Calderon received the AMS Bôcher Prize in 1979, the AMS Steele Prize in 1989, the National Medal of Science in 1991, and Argentina's Consagracion Nacional Prize in 1989, as well as the Wolf Prize in 1989.

He had supervised 27 Ph.D. students (16 Chicago, 5 MIT, 5 Buenos Aires).[10]

Calderón died on 16 April 1998 at Northwestern Memorial Hospital in Chicago, Illinois after a brief illness.

Figure 363. Paul Joseph Cohen

Paul Joseph Cohen

Paul Cohen was born on 2 April 1934 in Long Branch, New Jersey, USA. His father was Abraham Cohen and his mother was Minnie Kaplan and they came over to the United States in their teens from Poland, which was a part of Russia at that time. Paul was the youngest among four children. He graduated from Stuyvesant High School, one of the two famous math-science schools, at the age of sixteen.

He read Birkhoff and MacLane's Algebra and Titchmarsh's book on the theory of functions in high school.

After studying three years at Brooklyn College without graduating, he went to the University of Chicago for graduate study. He received his M.S. degree in 1954 and a Ph.D. in 1958 for a thesis titled *Topics in the theory of uniqueness of trigonometric series* under the supervision of Zygmund.[3]

Zygmund made everybody feel extremely comfortable. He and his students talked to each other a great deal and had very good relationships.

In 1957, before Cohen was awarded his doctorate, he taught at the University of Rochester for a year. Then, he spent one year at the Massachusetts Institute of Technology and two years at the Institute for Advanced Study at Princeton. In 1961, he moved to Stanford University where he was promoted to full professor in 1964.

In April 1963, Cohen solved Hilbert's first problem which states:

Cantor's problem of the cardinality of the continuum.

Is the continuum the cardinality next to the denumerable set – and can the continuum be considered as well ordered?[6]

Cohen hand-carried a draft proof to Gödel in Princeton. In a letter dated 20 June 1963, Gödel approved Cohen's proof of Hilbert's first problem. It is remarkable because Cohen was an analyst to solve this difficult problem in logic.

He was awarded a Fields Medal at the International Congress of Mathematicians in Moscow in 1966 for proving the independence in set theory of the axiom of choice and of the generalized continuum hypothesis.

In 1964, he received the AMS Bôcher Prize and the National Medal of Science in 1967. He was a member of the National Academy of Sciences.

He met Christina Karls during a cruise from Stockholm to Leningrad in the summer of 1962. They married on 10 October 1963 and had three sons.

Cohen spoke English, Swedish, French, Spanish, German and Yiddish.

In 1972, he became the first holder of the Marjorie Mhoon Fair Professorship in Quantitative Science at Stanford University.

He died in 23 March 2007 at Stanford Hospital of a rare lung disease.

Elias Menachem Stein

Elias Stein was born on 13 January 1931 in Belgium. His family fled to the United States to escape Nazism. He graduated from Stuyvesant High School in New York in 1949 one year before Paul Cohen.

Stein received a Ph.D. at the University of Chicago in 1955 for a thesis entitled *Linear operators on Lp spaces* supervised by Zygmund. He taught for the next three years at the Massachusetts Institute of Technology and the following five years at the University of Chicago before being appointed a full professor at Princeton University.

His main field of research has been harmonic analysis. His name can be found in mathematical terms such as Stein interpolation, Stein maximal principle, Stein complementary series representations, Nikishin-Pisier-Stein factorization in operator theory, the Tomas-Stein restriction theorem in Fourier analysis, the Kunze-Stein phenomenon, the Coltar-Stein lemma and the Fefferman-Stein theory of the Hardy space H^1.

He has supervised 45 Ph.D. students, including two Fields medalists, Charles Fefferman and Terence Tao.

Stein received the Schock Prize from the Swedish Academy of Sciences in 1993, the Wolf Prize in 1999 and the National Medal of Science in 2002. He was elected to the National Academy of Sciences in 1974. Stein has an astonishing ability to find connections between several branches of mathematics.

Figure 364. Elias M. Stein   Figure 365. Charles L. Fefferman

Charles Louis Fefferman

Charles Fefferman was born on 18 April 1949 in Washington, D.C.

He graduated from the University of Maryland in 1966, at the age of 17 with two bachelor's degrees in mathematics and physics.

Fefferman received his Ph.D. at Princeton University in 1969 for a thesis entitled *Inequalities for strongly regular convolution operators* supervised by Elias Stein. Fefferman became a faculty member at the University of Chicago in 1970 and one year later, he was named to full professor there, earning him the distinction of becoming the youngest full professor in the United States at the age of 22.[3]

Felix Klein became a full professor at the age of 23 and Augustus De Morgan at the age of 22 at the University College, London.

In 1973, Fefferman returned to Princeton, where in 1984, he was appointed Herbert Jones Professor.

He received a Fields Medal at the International Congress of Mathematicians at Helsinki in 1978 for several innovations that revised the study of multidimensional complex analysis by finding correct generalization of classical (low-dimensional) results.

In 1992, he also received the Bergman Prize, named in honor of Stefan Bergman (1895-1977).

Figure 366.   Nathan Jacobson   **Figure 367.   Derrick H. Lehmer**

(Courtesy of the American Mathematical Society)

Nathan Jacobson

Nathan Jacobson was born into a Jewish family on 8 September 1910 in Warsaw, Russian Empire (now Poland). His family emigrated to the United States during World War I. He graduated from the University of Alabama in 1930 and received his Ph.D. at Princeton University in 1934 for a thesis entitled *Non-commutative polynomials and cyclic algebras* supervised by Joseph Wedderburn.[3]

He spent the 1935-36 academic year at Bryn Mawr College to fill the post of Emmy Noether, who died from a post-surgical infection on 14 April 1935. Then he went to the University of Chicago for the following year.

During World War II he taught at the US Navy School for pilots. He married Florence (Florie) Dorfman, a Ph.D. student under Adrian Albert, in August 1942.

Jacobson taught at Johns Hopkins University from 1943 to 1947 after which he moved to Yale University, where he remained until he retired in 1981.

He wrote sixteen algebra books. He was elected to the National Academy of Sciences and he served as president of the American Mathematical Society in 1971-72. Jacobson received the AMS Steele Prize in 1998.

From 1972 to 1972, Jacobson and Pontryagin were both vice-presidents of the International Mathematical Union.[3] Pontryagin attacked Jacobson being a mediocre scientist and an aggressive Zionist.

He died on 5 December 1999 in Hamden, Connecticut, USA.

Derrick Henry Lehmer

Derrick Lehmer was born on 23 February 1905 in Berkeley, California where his father Derrick Norman Lehmer ("DNL"), was a professor of mathematics at the University of California Berkeley. Derrick (Dick) graduated from U.C. Berkeley in 1927 with a B.A. in physics.

He was then awarded a Ph.D. at Brown University in 1930 for a thesis entitled *An extended theory of Lucas functions* supervised by Tamarkin.

Unfortunately, it was a difficult time to obtain a permanent academic position due to the Great Depression. After spending a few years as a post doctoral fellow at the California Institute of Technology, Stanford University and the Institute for Advanced Study at Princeton, he was appointed to the faculty at Lehigh University in Bethlehem, Pennsylvania.

In 1940, he moved to the University of California, Berkeley.

During World War II, the Lehmers worked at Aberdeen Proving Ground where Dick helped set up and operate the Electronic Numerical Integrator and Calculator (ENIAC) to compute trajectories for ballistics problems. The Lehmers were able to use the computer to solve certain number theory problems.[3]

Because Dick Lehmer refused to sign the loyalty oath in early 1950s, he lost his position at Berkeley but he was fortunate to take up the post of director of numerical analysis at the National Bureau of Standards. Ultimately, Lehmer was reinstated at Berkeley by a court order.

Figure 368.   Ronald L. Graham

(Courtesy of the American Mathematical Society)

Lehmer was a pioneer in the application of digital computers to the solution of problems in number theory. He was the first mathematician to solve the Riemann Hypothesis by using a computer to check of the roots of the Riemann equation lie on the critical line.[3]

He died on 22 May 1991 in Berkeley, California.

Ronald Lewis Graham

Ronald Graham was born on 31 October 1935 in Taft, California. At the age of fifteen, he entered the University of Chicago winning a Ford Foundation scholarship. He received a B.S. in physics at the University of Alaska while serving in the US Air Force.[94]

He then received a Ph.D. at the University of California, Berkeley in 1962 for a thesis *On finite sums of rational numbers* supervised by Lehmer. Graham is now Irwin and Joan Jacobs Professor at the Department of Computer Science and Engineering of the University of California, San Diego as well as Chief Scientist at the California Institute for Telecommunication and Information Technology known as Cal-(IT)^2.

He is married to Fan Chung, a mathematician, and they have four children. He was a close friend of Paul Erdös, who often stayed with him. Graham had looked after papers of Erdös and even his money.

Graham popularized the concept of the Erdös number, which is the minimum number of links away from Erdös. A co-author of Erdös gets the Erdös number 1.

He has done important work in scheduling theory, computational geometry, Ramsey theory and guasi-randomness.

He received the AMS Steele Prize in 2003 and the first Pólya Prize in 1971 by the Society for Industrial and Applied Mathematics (SIAM). He served as president of the American Mathematical Society in 1993-94.

Graham is a highly skilled trampolinist and juggler serving as president of the International Juggler's Association.[101] He and a colleague mass-produced 100,000 Penrose tiles at Bell Labs.[94]

Heisuke Hironaka (廣中平祐)

Heisuke Hironaka was born on 9 April 1931 in Yamaguchi Prefecture, Japan. His parents had fifteen children. Hironaka's two older brothers were killed in combat during

World War II. He learned to play the piano while he was in high school to become a concert pianist.[107]

He graduated from Kyoto University in 1953 and continued his graduate study there under Akizuki. Hironaka went to Harvard University in 1957 and received his Ph.D. in 1960 for a thesis titled *On the theory* of *birational blowing-up* under Zariski's supervision.[10]

David Mumford and Michael Artin were his classmates at Harvard. Hironaka was also greatly influenced by Alexandre Grothendieck, who invited him in 1959 to the *Institute des Hautes Études Scientifiques* (IHÉS) in Paris.

After several years at Brandeis University and Columbia University, he became a professor at Harvard University in 1968. From 1975 to 1988, he had joint appointment at Harvard and Kyoto University.

He had supervised 13 Ph.D. students, 8 at Columbia, 4 at Harvard and 1 at Paris VII.[10]

He published *Resolution of singularities of an algebraic variety over a field of characteristic zero* in the Annals of Mathematics in 1964 in two parts. It was called "Hironaka's telephone directory", because of its tremendous length—217 pages.

In 1970, Hironaka was awarded a Fields Medal at the International Congress of Mathematicians at Nice for his generalizing Zariski's work for dimension <=3, the theorem concerning the resolution of singularities on an algebraic variety over a field of characteristic zero.

Mikhail Gromov of IHES said that Hironaka's resolution of singularities "is unique in the history of mathematics. It is one of the most difficult in the world that has not, to this day, been surpassed or simplified".[102]

Figure 369. Heisuke Hironaka

Hironake received the Asahi Prize in 1967 and the Japanese Academy Prize in 1970. He received the Order of Culture from Japan in 1975, which provides lifetime pension.

He is married to Wakako Hironaka and they have two children. She has been a member of the Japanese House of Councillors (Senate) since 1986 and was State Minister, Director-General of the Environment Agency in the Hosokawa Cabinet between 1993-94.

From 1996 to 2002 he was president of the Yamaguchi University at his hometown.

Figure 370.   David B. Mumford

Figure 371.   Daniel Gorenstein

David Bryant Mumford

David Mumford was born on 11 June 1937 in Worth, Sussex, England and came to the United States in 1940. His mother was born in the United States and his father was English.

He graduated from Harvard in 1957 and received a Ph.D. for his thesis entitled *Existence of the moduli scheme for curves of any genus* under Zariski's supervision.[10]

He was classmates with Hironaka, Michael Artin and Steve Kleiman at Harvard Graduate School. He was a Junior Fellow at Harvard from 1958 to 1961, an associate professor from 1962 to 1966, a professor from 1966 to 1977, and finally Higgins professor from 1977 to 1997. Since 1996, he has been a professor at Brown University.

Mumford was president of the International Mathematical Union from 1995 to 1998. Munford is a member of the National Academy of Sciences. Mumford was awarded a Fields Medal at the International Congress of Mathematicians in Vancouver in 1974 for his contribution to problems of the existence and structure of varieties of moduli, varieties whose points parametrize isomorphism class of some type of geometric objects. He also received the Wolf Prize in 2008.

He married Erika Jentsch on 27 June 1959 and they had four children, but Erika died on 30 July 1988. He married Jennifer Moore on 29 December 1989.

Mumford wrote many books including *Geometric Invariant Theory* (1965), *Abelian Varieties* (1970), *Introductions to Algebraic Geometry* (1976) and *2 and 3 Dimensional Patterns of the Face* (1999).[103]

Daniel Gorenstein

Daniel Gorenstein was born on 1 January 1923 in Boston, Massachusetts.

He graduated from Harvard University in 1943 and taught mathematics to Army personnel during World War II.

He received a Ph.D. at Harvard in 1950 for his thesis titled *An arithmetic theory of adjoint plane curves* supervised by Zariski.

Gorenstein taught at Clark University from 1951 to 1964, Northwestern University from 1964 to 1969, and Rutgers University from 1970 to 1992. He became Jacqueline B Lewis Professor of Mathematics at Rutgers in 1984.

He was elected to the National Academy of Sciences in 1978 and he received the AMS Steele Prize in 1989. His name is honored by the Gorenstein rings.[104]

Gorenstein died on 26 August 1992.

Claude Chevalley

Claude Chevalley was born on 11 February 1909 in Johannesburg, South Africa. He studied at the *École Normale Supérieur* in Paris, graduating in 1929. He also studied in Germany under Emil Artin and Helmut Hasse. Cheralley received a doctorate at the University of Paris in 1933 for his thesis entitled *Sur la theorie du corps de classes dans les corps finis et les corps locaux.*[3]

In 1934, he became the youngest member of the Bourbaki mathematicians in Paris. Four years later, Chevalley went to the Institute for Advanced Study at Princeton. From 1949 to 1957, he taught at Columbia University becoming an American citizen during this period.

After leaving Columbia, he was appointed to the Université de Paris VII.[3]

He wrote *Theory of Lie Groups* in three volumes (1946, 1956, 1955), *Theory of Distributions* (1951), *Introduction to the Theory of Algebraic Functions of One Variable* (1951), *The Algebraic Theory of Spinors* (1954), *Class Field Theory* (1954), *Fundamental Concepts of Algebra* (1956) and *Foundation of Algebraic Geometry* (1958).

He received the AMS Cole Prize in 1941. His name can be found in mathematical terms such as Chevally decompositions and Chevalley type of semi-simple algebraic group.

Chevalley died on 28 June 1984 in Paris, France.

Michael Artin

Michael Artin was born in 1934 in Germany as a son of Emil Artin. Michael was brought up in Indiana, where his father was a professor of mathematics.

He received a Ph.D. at Harvard University in 1960 for a thesis titled *On enriques surfaces* under Zariski's supervision.[10]

At Harvard Graduate School, Peter Falb, Heisuke Hironaka and David Mumford were his classmates.

In the 1960s, he was influenced by Alexandre Grothendieck at the Institut des Hautes Étude Scientifiques (IHES).

Artin supervised 33 Ph.D. students (31 at MIT). He served as president of the American Mathematical Society in 1991-1992 and is a member of the National Academy of Sciences.

Figure 372. Claude Chevalley

Figure 373. Michael Artin

(Courtesy of the American Mathematical Society)

He received the AMS Steele Prize in 2002. He is currently working on non-commutative rings and his name is included in mathematical terms such as Artin-Mazur zeta function and Artin approximation theorem.

Persi Warren Diaconis

Persi Diaconis was born on 31 January 1945 in New York City. At fourteen, he left his home to wander the world as a professional magician.

A friend recommended a probability book by Feller as the best and most interesting on the subject. Diaconis bought the famous two-volume book *An Introduction to Probability and Its Applications*, but could not read it. So, he decided to study at a college.

At twenty-four, he enrolled as a freshman at the City College of New York, graduating in 1971. He received a Ph.D. at Harvard University in 1974 for a thesis titled *Weak and strong averages in probability and the theory of numbers* supervised by Frederick Mosteller.

In 1974, he joined the faculty of the Statistics Department at Stanford University from where he takes a sabbatical every fourth year. In 1982, he won the MacArthur Prize which provided $40,000 a year, tax free for five years. He considers statistics as the physics of numbers.

His background in magic and statistics has proven useful in exposing several psychics, including Uri Geller.[94]

Figure 374. Einar C. Hille
(Courtesy of the American Mathematical Society)

Einar Carl Hille

Einar Carl Hille was born on 28 June 1894 in New York City. His parents separated before his birth. His original name, Carl Einar Heuman, was mistakenly altered to Einar Carl Hille. His mother brought him up. His parents were Swedish and Einar Hille moved to Sweden when he was two years old.[3]

He entered the University of Stockholm in 1911 to study chemistry, but changed to mathematics later. He received a doctorate from Stockholm in 1918 for his dissertation titled *Some problems concerning spherical harmonics* under the supervision of Marcel Riesz.[3]

Hille returned to the United States in 1920 and worked at Harvard and Princeton for thirteen years. In 1933, he was appointed a full professor at Yale University from where he retired in 1962.

He married Kirsti Ore in 1937, the sister of Oystein Ore, his colleague at Yale University. They had two sons.

Hille served as president of the American Mathematical Society in 1937–1938. He was elected to the National Academy of Sciences in 1953 and the Royal Academy of Sciences of Stockholm. The Government of Sweden awarded him the Order of the North Star.

He wrote 12 books including *Function Analysis and Semigroups* (1948), *Analytic Function Theory* Vol. 1 (1959), Vol. 2 (1964), *Lectures on Ordinary Differential Equations* (1969) and *Methods in Classical and Functional analysis* (1972).

Hille died on 12 February 1980 in LaJolla, California.

Figure 375.   Irving E. Segal

Irving Ezra Segal

Irving Segal was born on 13 September 1918 in the Bronx, New York. He graduated with highest honors with a bachelor's degree in 1937 from Princeton University winning the George B. Covington Prize in Mathematics. He received a Ph.D. at Yale University in 1940 for his thesis, *Ring properties of certain classes of functions*, supervised by Einar Hille.

Segal worked at the Institute for Advanced Study at Princeton from 1941 until 1943 and at the Aberdeen Proving Ground in Maryland conducting research in ballistics during World War II.

He taught at the University of Chicago from 1960 until 1989. He supervised 40 Ph.D. students at Chicago (15) and MIT (25). Isadore Singer, Henry Dye, Jr., Bertram Constant, Edward Nelson and Leonard Gross are some of his students.

Segal was elected to the National Academy of Sciences in 1973.

He became known for his work in quantum field theory, and functional and harmonic analysis, particularly his innovation of the algebraic axioms known as C*-algebra.

He died on 30 August 1998 while taking an evening walk.

Isadore Manuel Singer

Isadore Singer was born on 3 May 1924 in Detroit, Michigan. He graduated from the University of Michigan in 1944 and served three years in the US Army. He received a Ph.D. at the University of Chicago in 1950 for his thesis titled *Lie algebras of unbounded operators* supervised by Irving Segal.

Singer spent most of his professional career at the Massachusetts Institute of Technology beginning as an instructor in 1950 and eventually becoming Institute Professor in 1987. He taught at the University of California, Berkeley from 1979 to 1983. He has supervised 34 Ph.D. students at M.I.T.

He received the AMS Bôcher Prize in 1969, the AMS Steele Prize in 2001, the National Medal of Science in 1983 and the Abel Prize in 2004 jointly with Michael Atiyah.

Singer was vice president of the American Mathematical Society during 1970–72.

He is best known for the Atiyah–Singer Index Theorem.

Figure 376. Isadore M. Singer

Frederick Mosteller

Frederick Mosteller was born on 24 December 1916 in Clarksburg, West Virginia. When his parents divorced in 1928, his mother was his main source of support.

He received his B.S. in 1938 and Sc.M. in 1939 at Carnegie Institute of Technology (now Carnegie Mellon University). He received a Ph.D. at Princeton University in 1946 for a thesis, *On some useful inefficient statistics*, under the supervision of Samuel Wilks (1906-1964) and John W. Tukey (1915-2000).[10]

Mosteller's initial appointment at Harvard University in 1946 was in the Department of Social Relations. Nine years later, he became the founding chair of Harvard's statistics department.

In the early 1960s, Mosteller taught statistics to millions via early morning television for continental classroom. He used the book *Probability with Statistical Applications* jointly written by George B. Thomas, Robert E.K. Rourke.[89]

He wrote 65 books and nearly 350 papers, many of these were in collaboration with others.

He was president of the American Statistical Association in 1967 and received the Founders Award in 1992. Mosteller was a member of the National Academy of Sciences and received five honorary degrees.

He died on 23 July 2006 in Falls Church, Virginia.

Abraham Robinson

Abraham Robinson was born into a Jewish family on 6 October 1918 in Waldenburg, Germany (now Walbrzych, Poland). He graduated from the Hebrew University of Jerusalem in 1939.

During World War II, he was a Scientific Officer at the Royal Aircraft Establishment at Farnborough becoming an expert in aerodynamics.

In 1946, he received a Master's degree from the Hebrew University of Jerusalem and, in 1949, a Ph.D. from the University of London for a thesis titled *The mathematics of algebraic systems* supervised by Paul Dienes.[10]

He taught at the University of Toronto from 1951 until 1957, the Hebrew University of Jerusalem from 1957 to 1962, the University of California Los Angeles from 1962 to 1967 and Yale University from 1967 until 1974. Robinson was appointed to the prestigious Sterling Professor of Mathematics at Yale in 1971.[3]

His main work was on mathematical logic. He wrote *Introduction to Model Theory and to the Metamathematics of Algebra* in 1963 and *Non-Standard Analysis* in 1966.

He died on 11 April 1974 in New Haven, Connecticut.

Figure 377. Richard D. Brauer

(Courtesy of the American Mathematical Society)

Richard Dagobert Brauer

Richard Brauer was born into a Jewish family on 10 February 1901 in Berlin-Charlottenberg, Germany. His brother, Alfred T. Brauer (1894-1985), was also a famous mathematician too. Both brothers received their Ph.D. at the University of Berlin under Issai Schur, but Alfred had to serve in the German Army during World War I, so he graduated two years later than Richard.

Richard Brauer received a Ph.D. at the University of Berlin for a thesis, *Über die Darstellung der Drehungsgruppe durch Gruppen linearer Substitutionen*[10]

He married Ilse Karger, a fellow mathematics student, in September 1925. He went to the University of Königsberg as Konrad Knopp's assistant in the autumn of 1925 and received his habilitation there in 1927 to become a privatdozent.

At Königsberg in 1925, Kurt Reidemeister (1893-1971) and Konrad Knopp (1882-1957) were full professors and Gabor Szegö was an extraordinary professor.

Because of the Nazis takeover in Germany, Brauer left Germany and took an appointment at the University of Kentucky in November 1933. His wife and two sons arrived three months later and Alfred Brauer left Germany in 1939. But Brauer's sister, Alice was murdered in a concentration camp.

After one year at Kentucky, he worked at the Institute for Advanced Study at Princeton, the University of Toronto from 1935 to 1948, the University of Michigan from 1948 to 1952 and Harvard University from 1952 to 1971.

He received the AMS Cole Prize in 1949 and the National Medal for Scientific Merit in 1971. He was editor of many journals and elected to the National Academy of Sciences in 1955 and many academies. He also served as president of the American Mathematical Society in 1959-60.[3]

His main field was group theory and he introduced the theory of blocks. Almost half of his entire papers were written when he was over fifty years old.

Brauer died on 17 April 1977 in Belmont, Massachusetts.

Robert Phelan Langlands

Robert Langlands was born on 6 October 1936 in New Westminster, British Columbia, Canada. He married Charlotte Lorraine Cheverie on 13 August 1956. He received his B.A. in 1957 at the University of British Columbia.

He received a Ph.D. at Yale University in 1960 for a thesis, *Semi-groups and representations of Lie groups*, supervised by C.T. Ionescu Tulcea.

He taught at Princeton University from 1960 until 1967 and Yale University from 1967 until 1972. Since 1972, he has been at the Institute for Advanced Study at Princeton.

Langlands received the AMS Cole Prize in Number Theory in 1982, the National Academy of Sciences Award in Mathematics in 1988, and the Wolf Prize in 1995.

He is a member of the National Academy of Sciences and a Fellow of the Royal Society of London (1981). Langlands received several honorary degrees.[3]

The Deligne-Langlands Conjecture was finally proved by David Kazhdan and George Lusztig.

Figure 378. Robert P. Langlands

Figure 379. Alfred Tarski

Alfred Tarski

Alfred Tarski's original surname was Teitelbaum. He was born into a Jewish family on 14 January 1902 in Warsaw, Russian Empire (now Poland). At the University of Warsaw, he attended courses by Lesniewski (1886-1939), Siepinski (1882-1969), Mazurkiewicz (1888-1945) and Lukasiewicz (1878-1956).

In 1924, he was awarded a doctorate for a thesis entitled [10] *O Wyrazie pierwotnym logistyki* supervised by Lesniewski. Tarski taught logic at the Polish Pedagogical Institute in Warsaw from 1922 until 1925 as well as at the University of Warsaw and Zeromski's Lyceé in Warsaw from 1925 until 1939.

He married Maria Witkowski in June 1929 and they had one son and one daughter.[3]

When Germany invaded Poland in August 1939, Tarski was visiting Harvard University.

The family joined him in 1946. However, his parents, brother and sister-in-law all died at the hands of the Nazis during World War II.[3]

After many temporary jobs, he joined the mathematics faculty at the University of California, Berkeley in 1942. He was prompted to full professor in 1949 and became professor emeritus in 1968. From 1968 to his death in 1983, he supervised six Ph.D. students.[10]

Tarski wrote 19 monographs and his collected papers were produced in four volumes. He was elected to the National Academy of Sciences and many other academies.

In 1943, before his family came from Poland, he was invited to a department picnic, in Berkeley. Everyone was supposed to bring something so Tarski was asked to bring napkins. Tarski showed up at the picnic with a giant-sized box of Kotex.[50]

When Tarski joined Berkeley, Jerzy Neyman was already there as director of the Statistics Laboratory. Both of them were refugees from Poland, but they were both ambitious empire-builders. People used to say Tarski and Neyman were "Poles apart".[50]

Tarski died on 26 October 1983 in Berkeley, California.

Figure 380.   Donald C. Spencer

Donald Clayton Spencer

Donald Spencer was born on 25 April 1912 in Boulder, Colorado. He received a B.A. from the University of Colorado in 1934 and also the Massachusetts Institute of Technology in 1936 with a bachelor's degree in aeronautical engineering. He received a Ph.D. at the University of Cambridge in 1939 for his thesis entitled *On a Hardy-Littlewood problem of diophantine approximation* supervised by Littlewood.[3]

After returning to the United States, he taught three years at the Massachusetts Institute of Technology from 1939 until 1942. Then he worked at Princeton University (23 years) and Stanford University (13 years). At Princeton University he collaborated with Kunihiko Kodaira on the deformation of complex manifolds.

Spencer married Mary J. Halley in July 1936 and they had two children. After this marriage ended in divorce, he married Natalie Robertson Sanborn in July 1951 and they had one son.[3]

When Spencer was at Princeton University in 1951, the faculty members included S. Lefschetz, A. Church, W. Feller, A.W. Tucker, E. Artin, J.T. Tate, J.W. Tukey, D.C. Spencer. R.C. Lyndon, V. Bargmann, S. Bochner, R.H. Fox, N.E. Steenrod, E. Wigner and S.S. Wilks.

Spencer received the AMS Bôcher Prize in 1948 and the National Medal of Science from President Bush in 1989. He was elected to the National Academy of Sciences in 1961.

Spencer died on 23 December 2001 in Durango, Colorado.

Figure 381. Phillip A. Griffiths

Phillip Augustus Griffiths

Phillip Griffiths was born on 18 October 1938 in Raleigh, North Carolina. He graduated from Wake Forest University at the Winston-Salem campus in 1959. He married Ann Lane Crittenden in 1958 and they had two children. Griffiths received his Ph.D. at Princeton University in 1962 for a thesis, *On certain homogeneous complex manifolds*, supervised by Donald Spencer.

Griffiths taught at the University of California Berkeley (6 years), Princeton University (5 years), Harvard University (9 years) and Duke University (9 years). He was appointed as director of the Institute for Advanced Study at Princeton in 1992, and stayed for twelve years.

In 2004, he became a professor of mathematics at the Institute for Advanced Study. In 1967, his first marriage ended in divorce. The next year he married Marian Folsom Jones and they have two children. Griffiths wrote eleven books and papers of 2600 pages.[3] He supervised 26 Ph.D. students.[10]

Griffiths received the AMS Steele Prize in 1971 and he was elected to many academies, including the National Academy of Sciences in 1979. He also received the Wolf Prize in 2008.

Edward Witten

Edward Witten was born on 26 August 1951 in Baltimore, Maryland and graduated from Brandeis University in 1971. He received a Ph.D. in physics at Princeton University in 1976 for a thesis titled *Some problems in the short distance analysis of gauge theories* under the supervision of David Gross.[10] Gross received a Nobel Prize in Physics in 2004.

Witten was a Junior Fellow at Harvard University from 1977 to 1980. In September 1980, he was appointed professor of physics at Princeton. In 1987, he became a Professor in the School of Natural Sciences at the Institute for Advanced Study at Princeton.

He received a Fields Medal at the International Congress of Mathematicians at Kyoto in 1990 for proving the classic Morse inequalities and giving a proof of positivity of energy in Einstein's theory of gravitation.

Julia Hall Bowman Robinson

Julia Robinson was born on 8 December 1919 in St. Louis, Missouri. Her mother died when Julia was two years old. Her father remarried Edenia Kridelbaugh in 1922. In 1937, her father committed suicide because his savings had been wiped out by the Great Depression.

Julia graduated from the University of California, Berkeley in 1940 with a B.S. in mathematics and M.A. in 1941. Jerzy Neyman hired Julia as an assistant for $35 a month while she was working for her master's degree.[12]

She married Raphael Robinson who was a professor of mathematics at Berkeley on 22 December 1941.

Julia Robinson received her Ph.D. at Berkeley for a thesis, *Definability and decision problems in aristhmetic*, supervised by Alfred Tarski.

She started work on Hilbert's Tenth Problem to determine whether a Diophantine equation with integral rational numbers is solvable in such numbers.[6] Along with Martin Davis (1928-) and Hilary Putnam (1926-), Julia Robinson gave a fundamental result that contributed to the solution to Hilbert's Tenth Problem.

In 1970, Yuri Matiyasevich (1947-) solved the problem by showing that no such method exists. In 1971, Julia with her husband Raphael visited Leningrad to become acquainted with Matiyasevich and his wife, Nina, a physicist. Robinson and Matiyasovich wrote two papers together.[89]

In 1975, she was elected to the National Academy of Sciences as the first woman mathematician. In 1982, she received the MacArthur Prize Fellowship. In 1976, she was appointed a professor at the University of California, Berkeley.

She served as the first woman president of the American Mathematical Society in 1983-84.

Robinson died of leukemia on 30 July 1985 in Oakland, California.

Constance Reid, author of many books on mathematicians, is Julia's elder sister.

Figure 382. Julia H. B. Robinson

(Courtesy of the American Mathematical Society)

Figure 383. Edward Witten

Alexander Victor Oppenheim

Alexander Oppenheim was born on 4 February 1903 in Salford, Lancashire, England and graduated from Oxford University in 1927. He received a Ph.D. at the University of Chicago with his thesis titled *Minima of indefinite quadratic quaternary forms* supervised by L.E. Dickson in 1930. He also received the Doctor of Science degree at Oxford University.

After teaching at Edinburgh University in 1930-31, he went to Singapore and taught at Raffles College from 1931 to 1942. He was captured by the Japanese Army when they invaded Singapore during World War II. He and thousands of Allied prisoners of war were engaged as slave labour for the infamous railroad construction work in Burma.

After the war ended, he became Deputy Principal of Raffles College (1947-1949) and Dean of the Faculty of Arts at the University of Malaya. Oppenheim also served as Vice-Chancellor of the Universities of Malaya in Singapore and Kuala Lumpur from 1957 to 1965. Later, he was a Visiting Professor of Mathematics at universities in England, Ghana and Nigeria.

In 1929, Oppenheim formulated the Oppenheim Conjecture.[90]

Let Q be an nondegenerate indefinite quadratic form in n variables. Let $L_Q = Q(Z^n)$ denote the set of values of Q at integral points. The Oppenheim Conjecture states that if $n \geq 3$, and Q is not proportional to a form with rational coefficients, then L_Q is dense.

The Oppenheim conjecture was proved affirmative by Gregori Margulis in 1987 using methods of ergodic theory.

Oppenheim was knighted in 1961 and published *The Prisoner's Walk: An Exercise in Number Theory* in 1984.

He died on 13 December 1997.

Austalian School

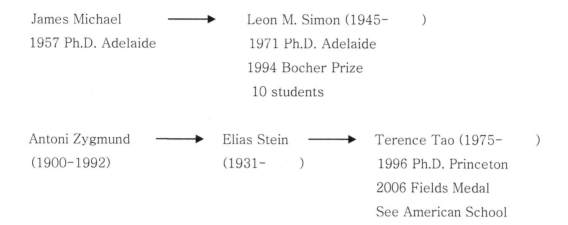

James Michael   ⟶   Leon M. Simon (1945–)
1957 Ph.D. Adelaide   1971 Ph.D. Adelaide
                         1994 Bocher Prize
                         10 students

Antoni Zygmund   ⟶   Elias Stein   ⟶   Terence Tao (1975–)
(1900–1992)   (1931–)   1996 Ph.D. Princeton
                                         2006 Fields Medal
                                         See American School

Austalian School

Leon Melvyn Simon

Leon Simon was born on 6 July 1945 in Adelaide, Australia. He received a B.Sc. in 1967 at the University of Adelaide and a Ph.D. in 1971 at Adelaide for his thesis, *Interior gradient bounds for non-uniformly elliptic equations*, supervised by James H. Michael.[3]

He was at Stanford University between 1973 and 1976. Subsequently, he worked at the University of Adelaide (1976-1977), the University of Minnesota (1977-1978), the University of Melbourne (1978-1981), the Australian National University (1981-1986), before returning to Stanford.

Simon received the Australian Mathematical Medal in 1983. He was awarded the Bôcher Memorial Prize by the American Mathematical Society in 1994 for his profound contributions towards understanding the structure of singular sets for solutions of variational problems.[3]

Terence Chi-Shen Tao (陶哲軒)

Terence Tao was born on 17 July 1975 in Adelaide, Australia. His father is a pediatrician and his mother was a secondary school teacher of mathematics in Hong Kong. Tao was a child prodigy to score 760 in SAT mathematics section at the age of eight. He won a bronze, silver and gold medal in the International Mathematical Olympiad in 1986, 1987 and 1988 respectively.

After receiving his bachelors and master's degrees in 1992 from Finders University in Adelaide, he went to Princeton University for graduate study. He received his Ph.D. in 1996 with his dissertation *Three regularity results in harmonic analysis* supervised by Elias Stein.[10]

In 2004, Tao along with Ben Joseph Green, solved a problem related to the Twin Prime Conjecture proving that it is possible to find, somewhere in the infinity of integers, a progression of prime number of equal spacing and any length.[117]

Tao was awarded a Fields Medal in 2006 "For his contributions to partial differential equations, combinatorics, harmonic analysis and additive number theory".

Tao is the first Australian, the first UCLA faculty member ever to be awarded a Fields Medal.[117]

Tao was elected a Fellow of the Royal Society in 2007.

Czech School

Edouard Cech ⟶ Ivo M. Babuska (1926-   )
(1893-1960)   1955 Ph.D. Czechoslorak Academy
1920 Ph.D. Prague   24 students (Maryland)
                     1994 Birkhoff Prize

Charles Loewner ⟶ Lipman Bers (1914-1993)
 (1893-1968)   1938 Ph.D. Prague
1917 Ph.D. Prague   1975-76 AMS President
23 students (Stanford)   53 students (Columbia, NYU)

Czech School

Ivo Babuska

Ivo Babuska was born in 1926. He graduated from the Technical University of Prague receiving his Diplom Ing. degree in 1949 and Doctor of Technical Science in 1951. From 1949 he studied at Mathematical Institute of Czechoslovakia Academy of Sciences where he received a Candidate of Science (Ph.D. equivalent) degree in 1955 and Doctor of Science degree (similar to habilitation) in 1960.

In 1968 he became a professor of mathematics at the University of Maryland, College Park where he was appointed Distinguished University Professor in 1995.

In 1995 he was appointed Robert Trull Chair in Engineering and Professor of Aerospace Engineering and Engineering Mechanics as well as Professor of Mathematics at the University of Texas at Austin.

He was awarded the Birkhoff Prize in 1994, J. von Neumann Medal in 1995, the Bolzano Medal of the Czech Academy of Sciences in 1996.

Babuska has published more than 230 papers and several books. He has supervised 35 doctoral dissertations, 24 at the University of Maryland.

Charles Loewner

Charles Loewner was born into a Jewish family on 29 May 1893 in Lany, Bohemia (now Czech Republic). His original name in Czech was Karel Löwner and German version was Karl Löwner. He received a Ph.D. at the Charles University of Prague (Univerzita Karlova) in 1917 under George Pick's supervision. He taught four and half years at the Technical University of Prague (Ĉeske Vysoke Uĉeni Technicke v Praze) until 1922.

Then he was appointed an Assistant at the University of Berlin in 1922, where Schur, Bieberbach, Schmidt and von Mises held chairs. Rademacher and Szégö were Privatdozent.[1]

Loewner received his habilitation at Berlin in 1923 and remained as privatdozent until 1928. Then he was appointed extraordinary professor at the University of Köln in 1928.[1]

The University of Köln (Cologne) was closed by Napoleon in 1798 and reopended in 1919.

Loewner moved to Charles University of Prague in 1930 and emigrated to the United States in 1939 after paying the emigration tax.

He married Elizabeth Alexander, a singer, in 1934 and they had one daughter born in 1936.

Lipman Bers was one of his students at Prague.

Von Neumann arranged a position for Loewner at Louisville University when he came to the United States. Von Neumann was a privatdozent at Berlin from 1927 to 1930.

During World War II Loewner worked on a program related to war work at Brown University in 1944. Then he taught five years at Syracuse University from 1946 until 1951.

In 1951 he moved to Stanford Unievrsity where he supervised 23 doctoral students. In the 1950s Stanford mathematics department had George Polya, Gabor Szegö and Stefan Bergman.

He proved a special case of the Bieberbach Conjecture in 1923 and he devoted to the properties of n-monotonic functions.[3]

Loewner died on 8 January 1968 in Stanford, California.

Figure 384.   Charles Loewner

Figure 385.   Lipman Bers

(Courtesy of the American Mathematical Society)

Lipman Bers

Lipman (Lipa) Bers was born into a Jewish family on 23 May 1914 in Riga, Russia (now Latvia). His parents were teachers.

He attended the elementary school of which his mother had been principal where he met his future wife, Mary Kagan. He also attended the gymnasium in Riga of which his father was director.[89]

Bers was involved in the socialist youth movements in Riga and he escaped to Prague through Estonia. He married Mary Kagan in May 1938 in Prague and they had one son and one daughter.

He received a Ph.D (Dr.rer. nat) in 1938 at Charles University of Prague supervised by Loewner.[89]

Since Germany annexed Czechoslovakia in 1939, Bers had to escape to France. Bers and his family arrived in the United States in 1940. There were many qualified refuge mathematicians from Europe, he was unemployed until 1942.

Then he worked at Brown University as a research instructor at $800 a year studying two-dimensional subsonic fluid flow which was important for aircraft wing designs.

When he taught at Syracuse University from 1945 until 1949, he looked on the problem of removability of singularities of non-linear elliptic equations. His major results in this problem were presented by him at the International Congress of Mathematicians held at Cambridge, Massachusetts.

Then he began to work on Teichmüller theory at the Institute for Advanced Study at Princeton.

Oswald Teichmüler (1913-1943) was a student Nazis leader boycotting Edmund Landau at Göttingen in 1933. Teichmüller received a Ph.D. at Göttingen and habilitation at Berlin University in 1938, but he was killed in the Russian front in 1943.

From 1951 to 1964 Bers was at the Courant Institute, New York University. Then he moved to Columba University where he retired in 1984 being appointed Davis Professor of Mathematics in 1972.

He served as president of the American Mathematical Society in 1975-76 and received the AMS Steele Prize in 1975.

Ber published *Theory of Pseudo-Analytic Functions* in 1953.[3]

He had a great passion for human rights.

Bers died on 29 October 1993 in New Rochelle, New York.

References

1. Mathematische Institute in Deutschland 1800–1945, Deutsche Mathematiker-Vereinigung 1989

2. Carl B. Boyer, A History of Mathematics, second edition, John Wiley & Sons 1991, p. 500

3. John O'Connor and E.F. Robertson, MacTutor History of Mathematics Archive

4. Ioan James, Remarkable Mathematicians, Cambridge University Press 2002

5. The Biographical Dictionary of Scientists, Oxford University Press 2000

6. Dirk J. Struik "A Concise History of Mathematics", Fourth revised edition, Dover Publications 1987

7. Jagdish Mebra and Helmut Rechenberg "The Historical Development of Quantum Theory" Vol. 6, Springer 2000

8. http://en.wikipedia.org/wiki/Heinz_Hopf

9. Constance Reid, Hilbert Courant, Springer Verlag 1986

10. Mathematics Genealogy Project, Department of Mathematics, North Dakota State University

11. Ioan James, Remarkable Physicists, Cambridge University Press 2004

12. Ben H. Yandell, The Honors Class, AK Peters 2002

13. George Bachman and Lawrence Narici, Functional Analysis, Academic Press 1966

14. http://en.wikipedia.org/wiki/Wilhelm_Cauer

15. Nancy T. Greenspan, The end of the Certain World The life and Science of Max Bonn, Basic Books 2005

16. E.T. Bell, Men of Mathematics, Simon & Schuster 1965

17. Erwin Kreyszig, Advanced Engineering Mathematics, John Wiley & Sons 1999

18. R.S. Millman and G.D. Parker, Elements of Differential Geometry, Prentice-Hall 1977

19. Annuaire du Collége de France 2001–02 Résumé des Cours et travaux

20. Andre Weil and Jennifer Gage, The Apprenticeship of a Mathematician, Birkhäuser 2002

21. Katswhiko Ogata, Modern Control Engineering, Prentice-Hall 1997

22. http://en.wikipedia.org/wiki/Grigori_Perelman

23. Obituary, Acta Arithmetica

24. http://en.wikipedia.org/wiki/Pavel_Alexandrov

25. Obituary George G. Lorentz, University of Texas

26. http://en.wikipedia.org/wiki/Yakov G._Sinai

27. Larry Riddle, Biography of Marina Ratner

28. http://en.wikipedia.org/wiki/Mikhail_Gromov

29. http://en.wikipedia.org/wiki/Andrei_Okounkov

30. Obituary Andrei A. Bolibrukh

31. http://en.wikipedia.org/wiki/Yuri_Matiyasevich

32. Walter Gautschi, Lecture at the Ostrowski Foundation, May 24–25, 2002

33. www.groups.dcs.st-and.ac.uk/~history/Davis/Indexes/yCambridge.html

34. http://en.wikipedia.org/wiki/Cambridge_Mathematical_Tripos

35. Steven G. Krantz, Mathematical Apocrypha, The Mathematical Association of America 2002

36. Robert Kanigel, The Man who knew infinity, Washington Square Press 1991

37. Encyclopedia Britannica/Barnes, Ernest William

38. http://en.wikipedia.org/wiki/John_Coates

39. Wolfram Mathworld/Swinnerton-Dyer Conjecture

40. Béla Bollobás (editor), Littlewood's Miscellany, Cambridge University Press 1986

41. Peter Harman and Simon Mitton (editors), Cambridge Scientific Minds, Cambridge University Press 2002

42. Who's who 2005, A&C Black, London

43. http://en.wikipedia.org/wiki/Simon_Donaldson

44. AMS Statement on 1998 Fields Medalist William Timothy Gowers

45. Encarta, Microsoft Corporation

46. Kazimierz Kuratowski, A Half Century of Polish Mathematics, Pergamon Press 1980

47. http://en.wikipedia.org/wiki/Stefan_Banach

48. Stan Ulam, Adventures of a Mathematician, University of California Press 1976

49. http://en.wikipedia.org/wiki/Dénes_König

50. Steven G. Krantz, Mathematical Apocrypha Redux, The Mathematical Association of America 2005

51. G.H. Hardy, A Mathematicians Apology, Cambridge University Press 1999

52. R. Ramachandran, Frontline, Vol. 23, issue 17, August 26, 2006

53. Arild Stubhaug, The Mathematician Sophus Lie, Springer 2002

54. David Abbott, Mathematicians, Peter Bedrick Books 1985

55. Britannica Online/Tullio Levi-Civita

56. Britannica Online/Giuseppe Peano

57. http://en.wikipedia.org/wiki/Francesco_Severi

58. http://en.wikipedia.org/wiki/Enrico_Bombieri

59. http://en.wikipedia.org/wiki/Laurea

60. Max Dresden, H.A. Kramers Between Tradition and Revolution, Springer-Verlag 1987

61. http://en.wikipedia.org/wiki/Masayoshi_Nagata

62. Chern and Hirzebruch, editor, Wolf Prize in Mathematics, World Scientific 2000

63. http://en.wikipedia.org/wiki/Mikio_Sato

64. http://en.wikipedia.org/wiki/Yutaka_Taniyama

65. http://en.wikipedia.org/wiki/Goro_Shimura

66. http://en.wikipedia.org/wiki/Shizuo_Kakutani

67. S. Iyanaga, On the life and works of Teiji Takagi, Teiji Takagi Collected Papers pp. 354-376

68. R.G.D. Richardson, The Ph.D. Degree and Mathematical Research, A Century of Mathematics in America Part II pp 361-378, American Mathematical Society 1989

69. John Parker, R.L. Moore Mathematician & Teacher, The Mathematical Association of America 2005

70. Paul R. Halmos, I Want to Be a Mathematician, Mathematical Association of America 1985

71. M.E. Van Valkenburg, Introduction to Modern Network Synthesis, John Wiley & Sons 1960

72. Faculty Portrait: Stephen Smale, 2006 Berkeley Science Review

73. http://en.wikipedia.org/wiki/Martin_Davis

74. http://en.wikipedia.org/wiki/Saharon_Shelah

75. http://en.wikipedia.org/wiki/Edward_Kasner

76. Richard Bellman, Eye of the Hurricane, World Scientific 1984

77. http://en.wikipedia.org/wiki/Albert_W._Tucker

78. J.M. Ball and R.D. James, The Scientific Life and Influence of C.A. Truesdell III, Arch Rational Mech Anal 161 (2002)

79. http://en.wikipedia.org/wiki/Ralph_Fox

80. http://en.wikipedia.org/wiki/William_Browder

81. http://en.wikipedia.org/wiki/Dennis_Sullivan

82. http://en.wikipedia.org/wiki/Curtis_T_McMullen

83. http://en.wikipedia.org/wiki/George_Lusztig

84. R.G.D. Richardson, The Ph.D. Degree and Mathematical Research A Century of Mathematics in America, Part II AMS 1989 pp 361-378

85. Constance Reid, The Search for E.T. Bell, The Mathematical Association of America 1993

86. Peter Duren, Editor, A Century of Mathematics in America Part III, American Mathematical Society 1989

87. R.G. Brown and P.Y.C. Hwang, Introduction to Random Signals and Applied Kalman Filtering, Third Edition, John Wiley 1997

88. Notices of the AMS, October 1996 p. 1111

89. D.J. Alberts, G.L. Alexanderson and Constance Reid, editors, More Mathematical People, Harcourt Brace Jovanovich, Publishers 1990

90. A. Eskin, G. Margulis, and S. Mozes, On a Quantitative Version of the Oppenheim Conjecture, Electronic Research Announcenents of AMS Vol. 1, Issue 3, 1995

91. Obitury www.math.rutgers.edu/kruskal/

92. http://en.wikipedia.org/wiki/Franz_Rellich

93. Constance Reid, Neyman from Life, Springer-Verlag 1982

94. D.J. Albert, G.L. Alexanderson, Mathematical People, Profiles and Interviews, Birkhauser Boston 1985

95. Howard Eves, Mathematical Reminiscences, The Mathematical Association of America, 2001

96. R.R. Coifman and R.S. Strichartz, The School of Antoni Zygmund, A Century of Mathematics in America Part III 1989, pp. 343-368

97. Obituary: Albert Calderon, The University of Chicago Chronicle, 30 April 1998 Vol. 17 No. 15

98. http://en.wikipedia.org/wiki/Elias_M_Stein

99. Obituary Nagoya Mathematical Journal

100. H.W. Leopolat, Obituary of Sigekatu Kuroda, Journal of Number Theory 7, 1-4 (1975)

101. http://en.wikipedia.org/wiki/Ronald_Graham

102. Notices of the AMS, Volume 52, Number 9 Interview with Heisuke Hironaka

103. Who's Who in Science and Engineering 2003-04, Marquis Who's Who

104. Daniel Gorenstein, The Classification of the Finite Simple Groups, A Personal Journey: The Early Years A Century of Mathematics in America Part I, p. 457

105. John Fauvel, et al, Oxford Figures 800 Years of the Mathematical Sciences, Oxford University Press 2000

106. Hölder's inequality in Wikipedia

107. Heisuke Hironaka, Joy of Learning, Kimyoungsa 1992

108. Encyclopedia Britannica

109. Arild Stubhaug, Niels Henrik Abel and his Times, Springer 1996

110. http://en.wikipedia.org/wiki/Shaun_Wylie

111. http://en.wikipedia.org/wiki/Jeong_Han_Kim

112. private communication with Jongil Park

113. http://en.wikipedia.org/wiki/Arzel-Ascoli_theorem

114. http://en.wikipedia.org/wiki/Lamberto_Cesari

115. private communication with Paul Butzer

116. private communication with Yongnam Lee

117. http://en.wikipedia.org/wiki/Terence_Tao

Appendix A Fields Medalists

2006	Madrid, Spain
	Andrei Okounkov, Russia
	Grigori Perelman, Russia (declined)
	Terence Tao, Australia
	Wendelin Werner, France
2002	Beijing, China
	Laurent Lafforgue, France
	Vladimir Voedvodsky, Russia
1998	Berlin, Germany
	Richard Borcherds, United Kingdom
	William Timothy Gowers, United Kingdom
	Maxim Kontsevich, Russia
	Cutis T. McMullen, U.S.A
	Andrew Wiles (Silver Plaque), United Kingdom
1994	Zurich, Switzerland
	Jean Bourgain, Belgium
	Pierre-Louis Lions, France
	Jean-Christophe Yoccoz, France
	Efim Zelmanov, Russia
1990	Kyoto, Japan
	Vladimir Drinfeld, U.S.S.R
	Vaughan F. R. Jones, New Zealand
	Shigefumi Mori, Japan
	Edward Witten, U.S.A
1986	Berkeley, United States
	Simon Donaldson, United Kingdom
	Gerd Faltings, Germany
	Michael Freedman, U.S.A
1982	Warsaw, Poland
	Alain Connes, France
	William Thurston, U.S.A
	Shing-Tung Yau, China/U.S.A

1978	Helsinki, Finland
	Pierre Deligne, Belgium
	Charles Fefferman, U.S.A
	Grigory Margulis, U.S.S.R
	Daniel Quillen, U.S.A
1974	Vancouver, Canada
	Enrico Bombieri, Italy
	David Mumford, U.S.A
1970	Nice, France
	Alan Baker, United Kingdom
	Heisuke Hironaka, Japan
	Sergei Novikov, U.S.S.R
	John Griggs Thompson, U.S.A
1966	Moscow, Soviet Union
	Michael Atiyah, United Kingdom
	Paul Joseph Cohen, U.S.A
	Alexander Grothendieck, France
	Stephen Smale, U.S.A
1962	Stockholm, Sweden
	Lars Hörmander, Sweden
	John Milnor, U.S.A
1958	Edinburgh, United Kingdom
	Klaus Roth, United Kingdom
	René Thom, France
1954	Amsterdem, The Netherlands
	Kunihiko Kodaira, Japan
	Jean-Pierre Serre, France
1950	Cambridge, United Kingdom
	Laurent Schwartz, France
	Atle Selberg, Norway
1936	Oslo, Norway
	Lars Ahlfors, Finland
	Jesse Douglas, U.S.A

Appendix B Wolf Foundation Prize in Mathematics

2008	Pierre R. Deligne, Institute for Advanced Study, Princeton, N.J. Phillip A. Griffiths, Institute for Advanced Study, Princeton, N.J. David B. Mumford, Brown University, Providence, R.I.
2006/7	Stephen Smale, University of California, Berkeley, CA. Harry Furstenberg, The Hebrew University of Jerusalem
2005	Gregory A. Margulis, Yale University, New Haven, Connecticut Sergei P. Novikov, University of Maryland, College Park, MD.
2002/3	Mikio Sato, Kyoto University, Kyoto Japan John T. Tate, University of Texas, Austin, Texas
2001	Vladimir I. Arnold, Steklov Mathematical Institute, Moscow Saharon Shelah, Hebrew University of Jerusalem
2000	Raoul Bott, Harvard University, Cambridge, MA. Jean-Pierre Serre, College de France, Paris France
1999	Laszlo Lovasz, Yale University, New Haven, Connecticut Elias M. Stein, Princeton University, Princeton, N.J.
1996/7	Joseph B. Keller, Stanford University, Stanford, CA. Yakov G. Sinai, Princeton University, Princeton, N.J.
1995/6	Robert P. Langlands, Institute for Advanced Study, Princeton, N.J. Andrew J. Wiles, Princeton University, Princeton, N.J.
1994/5	Jürgen K. Moser, Swiss Federal Institute of Technology, Zurich, Switzerland
1993	Mikhael Gromov, Institut des Hautes Etudes, Scientifiques, Bures-sur-Yuette, France Jacques Tits, College de France, Paris, France
1992	Lennart A.E. Carlson, University of Uppsala, Uppsala, Sweden John G. Thompson, University of Cambridge, Cambridge, U.K.
1990	Ennio De Giorgi, Scuola Normale Superiore, Pisa, Italy Ilya Piatetski-Shapiro, Tel-Aviv University, Tel Aviv, Israel
1989	Alberto P. Calderón, University of Chicago, Chicago, Illinois John W. Milnor, Institute for Advanced Study, Princeton, N.J.
1988	Friedrich Hirzebruch, Max-Planck-Institute and University of Bonn Lars Hormander, University of Lund, Lund, Sweden

1987	Kiyoshi Ito, Kyoto University, Kyoto, Japan
	Peter D. Lax, New York University, New York, N.Y.
1986	Samuel Eilenberg, Columbia University, New York, N.Y.
	Atle Selberg, Institute for Advanced Study, Princeton, N.J.
1984/5	Kunihiko Kodaira, The Japan-Academy, Tokyo, Japan
	Hans Lewy, University of California, Berkeley, CA
1983/4	Shiing S. Chern, University of California, Berkeley, CA
	Paul Erdös, Hungarian Academy of Sciences, Budapest, Hungary
1982	Hassler Whitney, Institute for Advanced Study, Princeton, N.J.
	Mark G. Krein, Ukrainian S.S.R. Academy of Sciences, Odessa, U.S.S.R.
1981	Lars V. Ahlfors, Harvard University, Cambridge, MA.
	Oscar Zariski, Harvard University, Cambridge, MA.
1980	Henri Cartan, University de Paris, Paris, France
	Andrei N. Kolmogorov, Moscow State University, Moscow, USSR
1979	Jean Leray, College de France, Paris, France
	Andre Weil, Institute for Advanced Study, Princeton, N.J.
1978	Izrail M. Gelfand, Moscow State University, Moscow, USSR
	Carl L. Siegel, University of Goettingen, Goettingen, Germany

Index